INDUSTRIAL SECURITY
FOR
STRIKES, RIOTS AND DISASTERS

Third Printing

INDUSTRIAL SECURITY

FOR

STRIKES, RIOTS AND

DISASTERS

By

RAYMOND M. MOMBOISSE

Riot Advisory Committee
President's Commission on Law Enforcement
Advisory Committee
California Peace Officers Association

CHARLES C THOMAS · PUBLISHER
Springfield · Illinois · U.S.A.

Published and Distributed Throughout the World by
CHARLES C THOMAS • PUBLISHER
Bannerstone House
301-327 East Lawrence Avenue, Springfield, Illinois, U.S.A.

© *1968, by* CHARLES C THOMAS • PUBLISHER
ISBN 0-398-01325-X
Library of Congress Catalog Card Number: 67-27932

First Printing, 1968
Second Printing, 1972
Third Printing, 1977

With THOMAS BOOKS *careful attention is given to all details of
manufacturing and design. It is the Publisher's desire to present books that
are satisfactory as to their physical qualities and artistic possibilities and
appropriate for their particular use.* THOMAS BOOKS *will be true to those
laws of quality that assure a good name and good will.*

Printed in the United States of America
R-1

To
Pearl Mitchell

An outstanding secretary
A marvelous person
A true friend

ACKNOWLEDGMENTS

IT WOULD BE IMPOSSIBLE to acknowledge herein all those who have aided in the preparation of this book. Over the last few years I have had the pleasure of working with literally hundreds of security officers, peace officers and members of the National Guard and Armed Forces. As a result of that close association, I have developed the utmost admiration for these men.

First and foremost, I owe a debt of gratitude to my wife, Mary Jane, and my three sons, Michael, Steven and Mark, for their patience and understanding during the long months of work on this book.

A special debt of gratitude is owed to the American Society for Industrial Security and the numerous members with whom I have had the pleasure of associating during the last few years. The outstanding work of this society has furnished much of the inspiration for this present undertaking.

There are certain individuals I must also single out for the assistance which they supplied. They are L. Rudy Weber, Von's Grocery Company, whose work on preventive measures for industry during riots was of substantial assistance in the preparation of this book; Raymond C. Farber, Security World Publishing Company, Incorporated, and Don D. Darling of Don D. Darling Associates, who have done so much to further research and education in industrial security; and Frank Montfort, William Bennett and Paul Hannigan for their advice and constructive criticism, which was always tempered with encouragement.

Others who have been of great assistance are as follows:

Dale O. Simpson, Dale Simpson and Associates Incorporated,
Leo H. Jones, Saber Laboratories, Inc.,
Robert H. Blundred, International Association of Amusement Parks,

Richard A. Walther, General Electric Company,
Charles K. Moody, Texas Instruments Incorporated,
William South, Southern California Edison Company,
Jack W. Farrar, Super Market Institute Incorporated,
John W. Davis, Los Angeles County and Cities Disaster and
Civil Defense Commission,
Dr. Robert B. Stegmaier, Department of Defense,
George E. Nielsen, U.C. Lawrence Radiation Laboratory,
N. Carter Hammond, Diebold, Incorporated,
Thomas Reddin, Los Angeles Police Department,
Frank E. Young, Bank of America NT&SA, San Francisco
Floyd E. Purvis, Texas Instruments Incorporated,
Robert L. Dennis, System Development Corporation,
James H. Davis, General Electric Company,
Robert Donovan, United Technology Corporation,
William B. Alderdice, LTV Electrosystems Incorporated, and
John R. Gabbert, Collins Radio.

RAYMOND M. MOMBOISSE

CONTENTS

SECTION II
DISASTER ORGANIZATION AND PROCEDURE

SECTION III

SPECIFIC PROBLEMS AND PROCEDURES

INDUSTRIAL SECURITY
FOR
STRIKES, RIOTS AND DISASTERS

SECTION I
INDUSTRIAL SECURITY

Chapter 1

PHYSICAL SECURITY

INTRODUCTION

THE MAGNITUDE of the threat to key industry because of violence, sabotage or espionage has grown with the increase in the capitalization and the importance of industry to the security of our society. The danger cannot be ignored. It must be recognized and appropriate remedial steps taken to prevent, or at least lessen, the danger. This is the function of commercial and industrial security.

Firms large and small, manufacturers, wholesalers, retailers, and servicers, must invest adequately in security just as they allocate sufficient funds to advertising, sales, production, planning, maintenance and research. Commercial or industrial security is not a business expense but a business investment. Security neglect not only permits but encourages losses which frequently spell the difference betwen profit and bankruptcy, between competitive advantage and disadvantage in pricing, between satisfied, dividend-rich stockholders and irritated dissidents hostile to management.

INDUSTRIAL SECURITY AND THE POLICE

Often, business management expects local police to provide protective services which are impossible for a police department to deliver. Even the best police departments cannot provide the constant surveillanc required to protect adquately major industries and business establishments. Big business must protect itself by providing capable and responsible security officers.

A suggestion that management make some provision for its own protection should not be interpreted as an unwillingness on the part of law enforcement to provide a tax-bought service. Nor should overtures by business to provide their own internal security be viewed as a vote of "no confidence" in the local police department.

It is therefore well to stop for a moment and consider how one can complement the other.

A good plant security officer aids police in numerous ways. For instance, crime prevention and investigation are the major concern of industrial security. The establishment of a fire protection system and the training of employee fire brigades is certainly another vital part of the plant protection workload. Of course, the assistance rendered to the local police in the removal of employees wanted for crimes with a minimum amount of notoriety falls into the lap of industrial security. The control of pedestrian and vehicular traffic in and around the property of the industry is also important. Enforcement of health, sanitation and safety rules, while not police related, may have a remote connection to the ultimate preservation of peace.

Industrial security's responsibility of establishing and maintaining an adequate system of locks and keys is an outstanding example of crime prevention at work. The system for employee and visitor identification, which is usually developed and controlled by industrial security, is another act of crime prevention. Of course, the patrol of buildings and premises and the inspection of special hazards and fences on a continuing basis affects the ultimate security of the plant and the community. Even an unarmed guard by his mere presence on the premises, is a deterrent to some types of crime.

Law enforcement has a responsibility to advise and assist private security constantly in the performance of its job. To accomplish this, it is good to review from time to time with the industrial community methods and ways in which its members can help.

Law enforcement and industrial security should develop a checklist of needs and contributions which dovetail into each other. Crime prevention techniques in the hands of the professional police can provide a great deal of assistance to plant security. The police can, for instance, patrol areas where employees park and develop programs to prevent stealing from autos.

The physical facilities and equipment of the department, the labs and special investigative equipment and so forth, should be made available to industrial security units. Many departments permit industrial security personnel to take police training in the regular police academy. This, of course, is probably the best way of insuring a good liaison between security people and local police. It also builds each group's confidence in the other.

When department policy permits, records and arrest registers should be offered without hesitation. They are invaluable to plant security people. Labor disputes, fraud cases and shoplifting are all incidents which require close work between industrial security and local police. Police specialists, particularly in the case of shoplifters, can be of great assistance to store security people by pointing out and identifying known thieves.

DEFINITION OF PHYSICAL SECURITY

Physical security means those physical measures designed to protect security interests against unauthorized access, espionage, compromise, theft, sabotage, loss or other intentional damage.

Physical security is concerned with the safeguarding of industrial and government facilities against espionage, sabotage, and other subversive activities. Among the measures which are taken to provide physical security are (1) the prevention of unauthorized entry by means of guard service, perimeter barriers (fencing or walls), protective lighting, and alarms; (2) control of authorized entry by means of employee and visitor screening and identification, truck and railroad car control, and incoming and outgoing package control; (3) fire prevention and control; (4) prevention of accidents, and (5) prevention of contamination of air, food and water.

Facility

The term *facility* encompasses a physical plant, bridge, tunnel, pipeline, pumping station, laboratory, office, college or university, manufacturing, producing or commercial structure, with its warehouses, storage areas, utilities and components, which, related by function and location, form an integral operating entity. These utilities may include communication systems, gas and oil pipelines, water systems, power systems, highways and bridges upon which the facility is dependent. One or more facilities make up an organization.

EVALUATION OF FACILITIES

Since it is not economically possible or necessary that facilities of every kind and character achieve the same degree of protection, the determination of the degree of protection warranted in any particular facility is predicted upon an analysis of two factors, criticality and vulnerability.

Criticality

The portion of the facility which is considered to be of high criticality is one whose partial or complete loss would have an immediate and serious impact on the ability of the facility to provide continuity of production or service for a considerable period of time. The criticality of a portion or part of a facility may bear no direct relationship to its size or whether it produces an end-product. Management is fully familiar with the widespread curtailment of production or service resulting from the loss of a relatively small part of a facility. Hence, criticality must be determined upon the basis of the importance of part of the facility to the whole facility as a unit. Criticality is modified by many factors. For example, an item of equipment or machinery readily available on the open market, or for which parts are readily available, is certainly less critical than custom made one-of-a-kind items.

Criticality is often confused with essentiality. An item may be essential, such as air for the operation of a steam generator in a thermal electric plant; but air is plentiful, available and cheap, and the thought of providing secondary supplies of air is ridiculous. Electric motors, fans, conduits, and other required items of equipment may also be essential to the proper processing of this plentiful ingredient, but as long as they are of common size and standard specification, the expenditure of funds to guard or even stockpile replacements for them would hardly be feasible. On the other hand, the enormous chimney structures required to carry off waste gases are equally essential and replacement time may be excessive.

The relative criticality of various portions of a facility must be determined by management. In determining the relative criticality of a part of a facility, management must consider whether or not it is a vital link in a chain of production or whether it makes vital or critical parts upon which the production of an important produce or series of products depends. This may be done by using reverse thinking. They should evaluate the facility as if they were an informed person striving to stop production. An uninformed troublemaker might damage transmission lines or easily replaceable small transformers or pumps, where as an informed professional might go for the steam drum at the top of the generator. Dropping this heavy drum down through the structure beneath it would put a generating plant out of operation for a longer period.

Vulnerability

By vulnerability is meant the susceptibility of each part of a facility to espionage, sabotage, thefts, slowdown, or work stoppage from any cause. In determining relative vulnerability, management must determine the susceptibility to damage of each part of its facility, and the means by which the part of the facility may be damaged.

One essential ingredient of any evaluation of vulnerability must be an awareness of the inherent danger within the item itself. This capability for self-destruction may take the form of rapidly moving heavy parts, such as a large turbine generator operating with supercritical steam pressures at a very high rate of speed. It may consist of safety equipment such as governors, automatic control devices and relief valves being inoperative. Many of these highly critical and hard-to-replace items can be caused not only to destroy themselves, but much of the equipment in the vicinity. This self-destructive capability is an important factor, both in determining how to interrupt production and how to prevent such interruption. Natural hazards such as combustibility of products, materials or structures must be considered. The wooden cooling tower structures in steam electric generating plants not on large bodies of water are good examples. Simple, homemade incendiary devices, properly placed on these structures, can stop production as effectively as more involved instruments such as explosives, abrasives, or mechanical means of sabotage.

HAZARDS

Introduction

Having determined the critical and vulnerable portions of the facility, management must determine the probability of damage to these portions or parts of a facility from the following:

1. Hazards inherent in the production or service operation.
2. Espionage.
3. Sabotage.
4. Mob action.
5. Direct enemy attack.
6. Theft.

Types of Hazards

Hazards to installations and property are acts or conditions which may result in the leakage of information, loss of life, damages, loss or destruction of property, or disruption of the installation. These acts or conditions may be constantly present or recur at infrequent intervals. The very possibility that they sometimes do exist is sufficient to justify their being considered risks to security. The degree of risk involved is dependent upon two factors: (1) the probability of adverse effects occurring as the direct result of the hazard, and (2) the degree to which the facility will be affected if the hazard becomes a reality. Hazards are of two types: natural and manmade.

NATURAL HAZARDS. Natural hazards are usually the consequence of natural phenomena, although some can be induced by human action. Examples of natural hazards are the following:

1. Earthquake.
2. Tidal Wave.
3. Flood (can also be caused by human action).
4. Fire (can also be caused by man).

5. Storm (cyclone, hurricane, tornado and typhoon).

6. Extremes in temperature.

MANMADE HAZARDS. Manmade hazards are of the following two categories:

1. Human weaknesses which result in damage, loss, or compromise.

2. Intentional acts of commission or omission.

EVALUATION

A part of a facility may be critical but not vulnerable. Conversely, a part of a facility may be vulnerable but not critical.

Criticality and vulnerability must be cross-examined to determine those items needing greater protection or consideration for replacement. For example, transmission lines are exceedingly vulnerable because of the difficulty of giving them real protection from molestation. Towers are easily damaged, lines can be shorted or grounded, but these hazards are always present and repair is usually accomplished in a short time. These structures are essential and highly vulnerable, but their criticality factor is low because of the relatively easy replacement or repair.

The degree of protection afforded to a facility or to a part of the facility is determined by the relationship of relative criticality and relative vulnerability. A facility which is both highly critical and highly vulnerable is the ideal target for enemy action. Hence, an extensive security protection program becomes necessary for those facilities or parts of a facility which are both highly critical and highly vulnerable.

In determining the physical protection to be afforded a facility, consideration shall be given to the following factors:

1. The relative importance of the facility to the overall program, including consideration of alternate facilities which could be used in an emergency and the availability of stockpiles of raw or finished materials.

2. The classification or strategic importance of information, material or activities located therein.

3. The vulnerability of vital equipment or materials to intentional damage or theft.

4. The location, size and arrangement of the facility, and numbers and character of personnel involved.
5. Character or kind of products (size, value, desirability and its explosive and inflammable qualities).
6. The need for integrating security measures with other operations.
7. The probable duration of operations.
8. The probability of expansion, curtailment, or other significant change in operations.
9. Alternate methods of providing effective protection.
10. Need for compartmentalization of information or activities.
11. The imminence of a hazard.
12. The probable risk.
13. Effect of physical security measures on efficiency.
14. Practical limitations imposed by the physical characteristics of the facility.
15. Availability of capital.
16. Alternate measures or techniques which will provide adequate physical security.
17. Evaluation of potential damage.
18. Evaluation and appraisal of physical security capabilities with the effective use of available resources.

Constant review and revision of criticality and vulnerability factors are necessary due to changes in technology, alternate means of production available, redundancy of means of accomplishing the mission and the relative importance of the plant or facility itself. The criticality of an entire facility may be reduced considerably by the construction of other productive means within the company system or the supplying of the facility through alternate transmission or supply means.

Chapter 2

SECURITY SURVEY

INTRODUCTION

T HE FIRST STEP in any security program is an evaluation of the present state of security. This means that a security survey must be initiated.

A security survey is a critical on-site examination and analysis of an industrial plant or business to ascertain the present security status, identify deficiencies or excesses, determine the protection needed, and make recommendations to improve the security where necessary.

The extent or degree of detail to be encompassed by the security survey will depend upon the need as indicated by the apparent lack of security measures, or it may depend upon the limits which have been placed on the study by plant management. The survey should include not only a study of the surrounding area, but also an outline of the present physical barriers, a description of the security measures in effect, key control, protective lighting, security of docks, safety for personnel, organization for emergencies, a study of the watchmen and guards organization, alarm systems, handling of classified information and materials, theft control, personnel control and organization for fire protection.

This survey may be conducted by the staff of the plant or by private independent specialists hired for this purpose. Surveys by the latter generally provide a more objective appraisal.

Whichever method is used, care should be exercised in choosing personnel to conduct a physical security survey. They must have not only technical ability but also be able to gain the cooperation of the plant personnel concerned.

Although it is possible for a single person to make the survey, it is recommended that the survey be made by a team.

The size of the team is dependent on the extent of the installation. The team system saves time and as a result of added experience and viewpoints assures a more complete job.

Also, the use of a survey team permits specialization by the members that develops expertness in surveying the various aspects of physical security. One member may examine the employment and training of the guard force while another surveys the perimeter barriers and the lighting system. Any division of duties that is expedient at a particular facility may be made.

PREPARATORY STEPS

Before conducting the survey, several preliminary steps should be taken to insure an adequate and practical estimate of the security situation. The first step is to enlist the whole-hearted support of top management to assure the cooperation that surveying personnel need at lower levels. Management must be convinced of the importance of the survey and be made to realize that without complete cooperation the survey and the resulting security program will suffer.

Survey personnel should be familiar with the operations involved at the installation to be surveyed. This basic knowledge permits the formulation of effective plans for making the survey and indicates the technical advise necessary to establish the most sensitive areas.

If previous surveys have been made, survey personnel should carefully review and analyze them and inspection reports to determine the potential and actual security hazards. They should note previously recommendations, descriptions of premises, security measures, and other relevant data that may contribute materially to the security survey.

MAKING THE SURVEY

A survey is made to verify current data and to obtain new facts. It should be conducted not only when the installation is in operation, but also at other times, including the hours of darkness. It should provide data for a true evaluation of existing hazards and the effectiveness of current protective measures.

As the survey is an evaluation of the present state of the

defenses and will be used to formulate future plans, it is essential that it be conducted in as thorough and systematic a manner as possible. This can be achieved if, in advance, a detailed check list is prepared to lessen the chance of omission. This list will vary with the facts of each case so that the following check list is intended merely as a guide.

SECURITY SURVEY CHECK LIST

Every plant and business is unique and different from all others, thus it is impossible to establish a universal checklist. The list that follows is therefore but a guide to aid in conducting any such study.

General

1. The purpose of the survey.
2. Name of plant.
3. Location (street address, city, county, state).
4. Jurisdiction (city, county, federal).
5. Security responsibility (private or some governmental agency).
6. Names of the principal officers and their duties. (Attach as an exhibit an official roster of the installation if readily available).
7. Type of business.
8. Product or service rendered.
9. Brief history of plant and products.
10. Size and physical characteristics of the plant and labor force.
11. Statement of hazards caused by type of industry, neighborhood and terrain.
12. Previous surveys, including the date and conducting agency, and a statement of the corrective actions taken.

National Importance of Plant

1. Is the product or service vital to national defense?
   a. Military importance.
   b. Civilian importance.
2. Do foreign nations desire information about the plant?

3. Is any part of its operations or product of a classified nature?
4. How critical is the product?
   a. Demand for product.
   b. Number of plants producing and/or personnel engaged in such production (ease of replacement or substitution).
5. Is the plant a probable target for espionage or sabotage?
   a. In peacetime?
   b. In wartime?

Physical Description of the Plant and Surroundings

1. Layout of the installation. (Attach as exhibit maps, sketches or photographs showing the area, types and location of buildings, roads, railroad sidings and piers.)
2. Roads or highways passing or leading to the plant. Description of railroads in the vicinity (spurs to plant), nearby airports serving the plant, dock facilities and steamship lines.
3. Physical description of surrounding area, topographical features, adjoining plants or buildings, etc.
4. Area of plant premises. Type of boundary (fenced?).
5. Description of buildings and other facilities on the premises, including docks, roads, canals, loading platforms, sheds, storage areas, repair shops, motor pools, fuel pumps, warehouses, power plants, communication systems, transformer stations, laboratories, fire stations and fire fighting equipment, etc.

Identification of Critical Areas

1. Has management determined and identified critical and vulnerable areas, both manufacturing and non-manufacturing?
2. Is management's evaluation of critical and vulnerable areas up-to-date, considering changes in products, production processes, vulnerability factors, etc.?
3. Is management aware of the physical vulnerability of

the facility in terms of location, type of construction and equipment, degree of combustibility, etc.?

Security of Buildings

1. Type of construction (wood, steel, concrete). Are the buildings fireproof, earthquake proof, protected for lightning?
2. Are the buildings locked at night and on holidays and weekends when not in use? If so, are the doors, windows, and other openings adequately protected?
3. How often are the buildings checked during closing hours for fire, burglary, damage by storm, etc.?
4. Are the buildings equipped with burglar alarms? If so, are they local alarms or the central station type?
5. Is valuable or critical machinery or equipment stored in the buildings?
6. Do the buildings house critical or secret products, plans, formulas, etc.? If so, are extra precautionary measures taken?
7. Are those buildings or rooms which contain safes or valuable or secret materials given extra attention by the guards?

Restricted Areas

1. Number, size and location of restricted areas.
2. Degree of restriction in each.
3. Are restricted areas fenced independently? If so, what type of fencing is used?
4. Type of pass system used for admittance to restricted areas.
5. Is an "access list" maintained of those authorized admittance?
6. Are electronic alarm or intrusion detection devices used?

Classified Information and Material

1. Have all personnel been properly cleared to handle the classified material with which they come in contact?

2. Is all government classified material of any kind properly secured at closing time according to government regulations and standards?
3. When in use, is all classified material kept under surveillance so that it will not be compromised by strangers, visitors, or uncleared personnel?
4. When not in use, is classified material returned promptly to safe storage places?
5. Are safe combinations known only by cleared personnel? Are combinations changed every six months, or whenever anyone knowing the combination leaves the organization or changes jobs to another activity?
6. Are there any specific recommendations for the tightening up of security controls over classified information and material?

Receiving and Shipping

1. Are employees' and visitors' packages checked at the entrance?
2. What examinations or spot checks are made of lunchboxes, toolboxes and packages?
3. Is such inspection manual or fluoroscopic?
4. If the installation is classed as sensitive,
   a. are incoming mail, express and freight packages inspected before distribution?
   b. is provision made for inspecting trucks and railroad cars from below road or track level?
5. Describe method of sealing railroad cars.
6. Are truck and rail shipments adequately braced, lashed, and delivered before movement?
7. Is the shipping and receiving area separated from the rest of the installation by a fence or other barrier? If so, is access through a controlled gate?
8. If there is storage area, is it open or covered?
9. Describe the physical security of each type of storage.
10. What supervision is given the docks, wharves and loading platforms? Is it adequate?
11. Is the area open to visitors or loitering?

12. Is there guard protection for the area? Is it adequate? How could it be improved?
13. Is incoming and outgoing material handled quickly and efficiently, or is there a backlog of crates and boxes piled up on the dock or platform area? What is the loss experience from pilferage here? Is it excessive?
14. Are freight receivers, shippers and handlers supervised for possible conspiracy with truck drivers or others for theft of merchandise?
15. What specific recommendations can be made to improve security on the docks, wharves and loading platforms?

Perimeter Barriers

1. Type of barrier, including height and presence or absence of a top guard.
2. Do buildings or bodies of water form any part of the perimeter, and if so, what protective measures are employed?
3. Are utility openings and outside windows and doors properly secured?
4. Frequency of inspections by guard and maintenance personnel.
5. Are clear zones maintained on inside and outside of perimeter fence? If not, are they feasible?
6. Are electric or electronic alarm or intrusion detection devices used on the perimeter? Describe.
7. Are warning signs used on the perimeter? If so, give size and location. (Attach scale drawing or photograph as an exhibit.)
8. Are there any unprotected areas?
9. Present condition of perimeter fence. Any holes, breaks, posts down, etc.?
10. Any holes or tunnels under fence?
11. Any lumber, boxes or refuse piled near fence?
12. Can vehicles drive up to fence? If so, at which points? Do vehicles park near fence? What is the relative ease of using vehicle near fence as a "stepladder" for climbing over?

13. Weak places in perimeter caused by railroad spur tracks, a stream or body of water, sewer line, coal chute or other opening through which unauthorized persons could enter?
14. Are there watch towers or guard huts for specific posts?
15. Are there roving guard patrols? If so, what means of transportation, if any, is used? What is the frequency of patrol route activity?
16. What means of communication is used?

Entrances

1. Number, type and location of active gates, vehicle or pedestrian.
2. Railroad siding gates.
3. Hours gates are open.
4. Inactive gates, and occasions on which they are used.
5. Type of locking devices on gates.
6. Location of keys for the locking devices.
7. Is there an adequate key control system in effect?
8. Are active gates properly illuminated for guard's inspection of passes and the occupants or contents of vehicles?
9. When gates are not in use, are they securely locked?
10. Are entrance and exit gates designed so as to facilitate the proper checking of vehicles and credentials during rush periods?
11. Are gates guarded?

Alarm System

1. Describe completely the protective burglar alarm system for the premises. Is it an off-premises central station reporting system or a local reporting system which sends the alarm to personnel standing by on the plant premises?
2. Does the perimeter fence have an alarm system to detect trespassers?
3. Do the safes, vaults, or rooms containing valuable or vital information have adequate protective devices?
4. What specific protection is provided to prevent burglary, fire, robbery? Is it adequate?

5. Following an alarm, are the communication and transportation facilities sufficient to insure a rapid follow-up by plant guards? By local police?

Lighting

1. Description, method and location of control, and the power facilities available.
2. Are the protective lighting system and the working lighting system separate?
3. Is there an auxiliary power source for protective lighting? If so, what type and what capacity?
4. Are the lights controlled by automatic timer or manually operated?
5. Are switchboxes and automatic timer boxes secured?
6. What emergency lighting is available?
7. Are the perimeter areas adequately floodlighted during hours of darkness?
8. Are the light sources directed to aid the guard and hinder the trespasser, i.e., into the face of the trespasser and leaving the guard in softer light or semishadow?
9. Are the entrance and exist gates sufficiently lighted?
10. Are lights at the gate arranged to light up the interior of vehicles entering and leaving?
11. Are buildings or areas which contain classified material or valuable material well illuminated?
12. Are power stations, transformers and other critical machinery and equipment adequately lighted at night?

Guard Force

1. What are the authorized and actual strengths? How many are on duty at any one time? Is this adequate for conditions?
2. How are watchmen and guards selected? What are the qualifications for the job (age, physical, mental, education, background)?
3. Are the requirements such as to bring good men into the job? If not, how should the qualification standards be revised?

4. What are the clearance requirements?
5. What are the duty assignments of guards pending clearances?
6. Training. (Attach training schedule in effect as an exhibit.) Is it sufficient for them to carry out their duties and responsibilities properly?
7. Is there a set of written rules and regulations for the night watchman and guard force? If not, should there be? If so, is it adequate?
8. What specific instructions are given them pertaining to: their daily duties, identity checks or personnel and visitors, vehicle checks, box car and train checks, package checks, personal searches, making arrests, handling drunks, handling mental cases, first aid, money escorts, etc.
9. Organization of the guard force.
10. What are the working hours for watchmen and guards? Are the hours too long?
11. What is the salary for watchmen and guards? Is it high enough to attract qualified men? What salary is recommended?
12. Are the guards alert and efficient in carrying out their duties?
13. How is the morale of the guards and watchmen? If inadequate, what recommendations can be made for improvements?
14. Uniform and weapons. Attach photograph of uniformed guard as exhibit.)
15. What type of firearms and other equipment do they carry? Are reserve firearms and emergency disaster equipment available?
16. If firearms and ammunition are issued, how are they accounted for?
17. What instructions and target practice are given guards and watchmen before firearms are issued to them?
18. Are the guards and watchmen thoroughly schooled as to when and under what circumstances they are to use their firearms? Do they understand the criminal and

civil consequences of misuse of their firearms on another person?

19. System of supervision.
20. Communication system with civil and military agencies.
21. Number and location of guard posts, punch clock stations. (Attach schematic diagram of installation, showing guard posts.) Frequency of guard post changes and reliefs.
22. Number of patrols (foot and motor) and their frequency.
23. Guard communications. Give details, radio, telephone, signal.
24. Is the guard communication system separate from other systems?
25. Location and layout of the guard headquarters.

Control of Entry

1. Identification media for assigned and employed personnel. (Attach sample of card or badge, or both.) Are they of tamper-proof construction? Do they contain the photo of the bearer? Are they serially numbered?
2. Control and issue of identification (describe briefly).
3. Loss of identification (describe control system).
4. How carefully are employees screened by the personnel department at the time of submitting their application? Is there close cooperation between the personnel department and the security department in the examination and clearance of questionable applicants?
5. Are all applicants' character references and previous employment history completely checked? If there are gaps or blanks in an applicant's previous employment record, are these carefully investigated?
6. Does the application questionnaire provide adequate information regarding the applicant's character, reliability, stability, honesty, etc.?
7. Visitor control (describe system, registration, badges, escorts).
8. Privately owned vehicle registration and control procedure.

9. Parking lots: location? fenced? lighted? guarded?
10. Are contractors' employees permitted free access to all plant areas? What controls are exercised over them?
11. What controls are maintained over salesmen, vendors and solicitors?
12. Are there restricted areas within the plant premises? If so, what additional controls are exercised over personnel entering these areas? Are these controls adequate?
13. How are visitors screened before admittance to restricted areas? Is a daily log of visitors "in and out" maintained by the gate guard? Are visitors controls adequate for the conditions?

Key Control

1. Are building keys and gate keys maintained in a secure place when not actually in use?
2. Are all keys accounted for? Are they handled only by authorized personnel?
3. Is a record kept of all persons holding building or gate keys? Is this record reviewed periodically and kept up to date?
4. Is provision made for the replacement of locks when a key is lost?
5. Are master keys closely controlled?
6. Are the locks of a type which offer adequate protection for the purpose for which they are used?

Security Education

1. Are all newly assigned or employed personnel given a security orientation?
2. What follow-up security instruction is given?
3. Is the instruction applicable to various types of employment and degrees of responsibility?
4. What personnel give the instruction and what are their qualifications?
5. Are security signs and placards on display? Is any other media used?
6. Does the installation have a security operating plan? (Attach a copy.)

7. Are periodic tests on security given? How? With what results?

Theft Control

1. Is an effort made to interest foremen and supervisors in reducing the amount of theft in the plant?
2. How prompt and cooperative are employees in reporting thefts or suspicious incidents?
3. Are trash collectors properly supervised by security personnel?
4. Is the disposal or sale of scrap and salvage material frequently checked for honesty and accuracy?
5. Are janitors, maintenance men, and clean-up crews properly supervised for control of theft activity?
6. If there is a general lack of theft control, what is the reason? Inadequate accounting methods? Lack of supervision? Improper attitude of employees? Lack of interest on the part of management? Poor package pass control?
7. What type of theft is most prevalent? What is being done about it? Is it actually known how great the theft problem really is?
8. Is the tool checking system adequate? Are tool inventories made regularly? Are maintenance men checked by security personnel when they leave the premises for jobs in other areas?

Safety for Personnel

1. To what extent is management interested in the safety of plant personnel?
2. What organizations have been established by management for the specific purpose of reducing the number of employee accidents and improving safety conditions in the plant? How do they operate? What is their program? How effective have they been?
3. Are plant safety rules vigorously enforced? By whom?
4. What is the general attitude of the workers and supervisors toward plant safety? What creates this attitude?
5. Are safety posters displayed in the plant? Are there any safety notices on plant bulletin boards? Are safety films

ever shown? Are safety contests ever held? What is the general attitude toward such activities?

Fire Protection

1. Is there an employee fire brigade organized? Does it have the full support of management? Is the roster kept up to date? How often does it hold meetings and drills? Who heads the organization?
2. Are there an adequate number of fire alarm boxes located throughout the premises? Are they conspicuously located?
3. Is the type and number of fire extinguishers adequate? Are they frequently inspected to ascertain that they are in good working order, filled, sealed, etc.?
4. How far is it to the nearest public fire department? Have the firemen ever visited the plant to study the layout of buildings and equipment, the locations of fire doors, stairways, and the special problems involved in the plant?
5. Are the buildings equipped with automatic sprinkler systems? Are they equipped with automatic alarm systems for fire control?
6. Are the physical fire barriers in the buildings adequate?
7. Are the fire doors installed where needed? Are they in proper working condition, with fusable fire links or automatic closing devices working properly and free of obstructions of any kind?
8. Is the houskeeping of the plant well maintained? Are oily rags kept in covered metal containers? Are paper, lint and other inflammable deposits frequently cleaned from floors and from the insides of exhausts, ventilators, air-conditioning ducts, ovens and from around heaters, conveyors and pipes?
9. Are the "No Smoking" rules enforced?
10. Are electric cords roped over nails or otherwise improperly placed? Are worn or defective electric cords immediately replaced?
11. Are electric fuses replaced by fuses of too high amper-

age, or by coins, wires, or nails?

12. Are proper measures taken to safeguard the storage and handling of gasoline, kerosene, alcohols, fuel oils and other explosive or toxic liquids and gases?

13. Are there adequate precautions taken for fire protection whenever the welding operations take place?

Emergency Plans

1. Are there separate emergency plans for fire, explosion, riots, mass civil disobedience, demonstrations, major accidents and air alert?

2. Is there a sabotage alert plan?

3. Is there a natural disaster emergency plan, if applicable? (Attach a copy.)

4. When was each of the above plans last reviewed or revised?

5. Do the above plans provide for evacuation of the installation?

6. Have the plans for installation disaster been coordinated with the area civil defense plans?

7. Do any of the above plans have classified annexes?

8. Have the plans been tested? Give dates and results.

9. Who is in charge of the operation of these plans during an emergency?

10. Are the emergency plans kept up to date with changing conditions and personnel?

11. Has an organization for handling emergencies actually been established? Does it hold drills? How often? Are these drills successful? Why? How are the emergency squads notified of a drill for emergency?

12. What are the plans for carrying on intraplant communications during emergencies?

13. What part do the employees play in an emergency situation? How are they notified of an emergency?

14. How are outside agencies such as the fire department, police department, Red Cross, doctors, ambulances, etc., notified in the event they are needed? Who is responsible for seeing that this is done?

15. Are there an adequate number of exits (doorways, fire escapes, etc.) in the buildings to meet an emergency situation? Are they kept clear of obstructions?
16. What instructions have been given the guards and watchmen regarding their duties during emergencies in the plant?

Liaison with Police and Military

1. Have contacts with local police been established?
2. Are the police advised of the plant's security plans?
3. Have arrangements been made for police aid in emergencies?
4. Have contact men been designated?
5. Is there an exchange of intelligence relating to danger to the plant or a disturbance in the area?

Minimizing Effect of Damage

1. Have emergency shutdown procedures been developed?
2. Have the key departments of the facility been relocated or rearranged to increase the independence of individual departments?
3. Has management investigated the use of protective construction, such as underground construction and blast protection?
4. Does management maintain a supply of spare parts, tools, standby equipment, raw materials and components, safely stored or otherwise protected?
5. Has management arranged for secondary supplies of water, gas, electric power and other utility needs for continued production.
6. Has management arranged for an emergency supply of electric power to meet minimum needs while the facility is inoperative?
7. Does the facility participate in an area program to pool machinery, equipment, supplies and the like to resume production following a disaster?
8. Has advanced reconstruction planning been accomplished?
9. Has facility dispersion been considered?

Restoring Production

1. Have tests of production restoration measures been made to check their effectiveness against hypothetical damage?
2. Does the facility have a well-equipped and trained emergency repair crew?
3. Have alternate sources of supply for critical materials and components used for production in this facility been identified?
4. Are any sources of supply for materials and components used for production in this facility known to be the sole producer of such items?
5. Is there a current, safely stored listing of special purpose or long lead-time tools and equipment?
6. Are manufacturing process and engineering data, plans and specifications vital to production duplicated and safely stored in an alternate location?
7. Has management estimated the time for replacement or repair to restore full production in the event of complete loss of the most critical equipment or utilities?
8. Have alternate production means been developed?
9. Could all or part of this facility's production be transferred to other facilities either within or outside the company organization?
10. Has advance production planning been accomplished with regard to flow charts, operations analysis, plant layout and internal arrangements for postattack production?
11. Have there been any major incidents or damage not previously reported?

Measures for Personnel Continuity

1. Have alternates or successors for management positions for this facility been designated?
2. Have alternate or successors for key production, maintenance and administrative personnel been designated?
4. Have vital administrative, personnel and accounting
3. Has an emergency facility headquarters been selected?

records been duplicated and safely stored in an alternate location?

5. Has management developed a procedure to reassemble its employees after a disaster?

6. Has an inventory been made of the job skills which are essential to maintain production after an emergency?

7. Has an estimation been made of the extent to which essential skills may be available among persons other than those employees currently performing the job?

8. Is there an active on-the-job training program which will provide for a rapid increase in availability of skilled personnel in event of mobilization?

9. Has any provision been made for the emergency housing, feeding and financing of employees in the event of emergency?

10. Have employees at all levels been made sufficiently aware of the continuous need for industrial defense and been encouraged to participate in these activities?

11. Is literature made freely and easily available to educate employees concerning all aspects of industrial defense and production continuity, including defenses against blast, radiation and chemical warfare?

REPORT AND RECOMMENDATIONS

The concluding step in making the physical security survey is the preparation of a report. The report itemizes the conditions which are conducive to breaches of security, records the preventive measures currently in effect and, when required, makes, specific, practical and reasonable recommendations to bring the physical security to the desired standard. The report may be in any form desired although normally the narrative form is used.

EVALUATION AND PLANNING

With the survey as the basis, evaluation and planning now can begin. The plan will cover such matters as classification of areas, guard force, barriers, lighting, alarm systems, communication and liaison. The first question is how much security is needed and what can be afforded, for the hard facts of life are that there are limitations on the funds, materials and personnel available. You must therefore utilize available resources to achieve maximum results. This demands thought and planning.

Even after such a plan is developed you cannot rest. You must continually examine and reevaluate that plan and the security measures being employed to maintain security at the desired level.

Resurvey

One should not stop with one survey. Rather, surveys should be repeated at regular intervals and whenever any major improvement in the existing facility is planned.

INSPECTIONS AND REPORTS

Management should perform frequent inspections and receive reports to assure itself of the sufficiency of the security program. This procedure will be of assistance in maintaining the efficiency and continuity of the security program and providing revisions as required by changing conditions, as well as creating and maintaining interest.

When governmental agencies make a security inspection, they should first arrange with top management to review the overall program and decide which facilities should be inspected. The inspector should be accompanied by a representative of top management. Any reports prepared by the inspector should be in cooperation with management, which should be furnished with a copy.

Chapter 3

ORGANIZATION FOR PHYSICAL SECURITY
INTRODUCTION

Eᴀᴄʜ ꜰᴀᴄɪʟɪᴛʏ sʜᴏᴜʟᴅ have an alert and well-trained plant protection organization. This organization should be responsible not only for planning a comprehensive plant protection program but also for implementing it so far as required, or preparing for its implementation upon short notice. Safety must be considered one of security's responsibilities.

Size of Organization

The scope and exact pattern of the protection organization will vary with local conditions and with special requirements of the plant. Even a small plant should have an embryonic plant protection organization capable of being further developed with increases in the operations of the plant or changes in the security requirements.

However, successful facility protection depends upon the interest and skill of those who devise and administer the program. It requires complete coordination between management and workers, and between the facility and the community and government.

Cost of Security

Obviously, it is not economical to spend thousands of dollars on plant security when the return on such investment is only a fraction of that amount. The theoretical or ultimate point at which greater expenditure for security should cease is that point beyond which diminishing returns is a factor. In other words, that point beyond which each dollar spent on security will not bring back a dollar's worth of added protection. This point is indeed theoretical, for there is no known method for discovering that exact place at which any given plant should stop spending more money for security protection. There are several practical considerations, however, which will serve as criteria for determining when that point is being approached.

One such criterion is that provided by insurance companies. If the plant is not adequately protected for fire hazards, the insurance rates will be higher. A greater expenditure by management for fire protection will mean lower insurance rates on the property.

Another criterion is that offered by evidence of thefts of company property. If reports show that thousands of dollars worth of tools and other company property are being stolen each month, it will pay management to spend more money for plant guards and better check-out systems in order to control the situation.

Still additional evidence of inadequate plant security is that offered by the statistics on plant injuries and accidents. If the volume of injuries and the expenditure for personal disabilities is high, an investigation may disclose that money spent on ordinary safeguards and plant safety education will be more than returned through increased production and lower injury costs.

FUNCTIONS

A protective organization should provide the following functions:

1. Analyzing the criticality and vulnerability of the facility to determine how great are the hazards and where the hazards are most likely to develop.
2. Analyzing the existing physical security measures to determine their adequacy in the light of hazards involved and to establish such additional measures as may be necessary.
3. Training employees to observe physical security regulations and to understand the reasons and purposes of the various regulations.
4. Training key personnel in physical security functions so that they may organize and direct workers under them

in an emergency.

5. Establishing and maintaining cooperative procedures between management and workers to secure suggestions and information from workers, assign physical security responsibilities, and aid in the harmonious and efficient operation of the program.

6. Establishing coordinated working relationships with the local government official who will direct emergency operations in the event of attack.

The proper discharge of the foregoing functions, as well as the vital importance of the matter to management and workers, requires that the responsibility for physical security be placed with a person in a position of top-level authority so that physical security problems will receive equal consideration with production, sales and other management functions.

SECURITY DIRECTOR

Because of the nature and importance of the problem, the responsibility for the preparation of physical security plans, administration of physical security activities, and execution of appropriate measures must be delegated to a specific individual.

He should be thoroughly trained in all security matters. He should be a ranking management official of sufficient position to command respect and obedience, and with sufficient authority to make the quick decisions independently that will be forced upon him in an emergency. He should report directly to the plant manager or other key management official. A suggested title for this individual is Security Director or Physical Security Director. The suggested duties of the security director are the following:

1. To analyze the need for and to devise physical protection measures.

2. To administer and supervise the operation of the physical protection measures.

3. To organize and train employee service groups.

4. To educate employees in physical protection and to develop employee cooperation and assistance in all phases of the physical protection program.

5. To maintain liaison with the local government official who will direct emergency operations in the event of attack, and to integrate the emergency plan of the facility with that of local government.

6. To maintain relations with other facility security directors to provide mutual assistance in the solution of problems and the sharing in an emergency of supplies, equipment, productive capacity and personnel.

Emergency and disaster programs should be based by the Security Director, to the extent feasible, upon existing programs maintained as part of the normal business practices of the facility involved.

An emergency or disaster function for which no counterpart is found in existing facility programs must be developed to provide the necessary emergency or disaster services for the protection of employees and property. In the establishment of the emergency or disaster protection program, full use should be made of the existing service departments such as fire, medical, repair and guard organizations. The departments should continue to function as usual, coordinated and guided by the Security Director, except in emergency or disaster, at which time they should operate as a unit of the emergency and disaster program under the full administration of the Security Director. The personnel of these units should be augmented as necessary, and trained to cope with all types of emergencies and disasters.

Chapter 4

GUARD FORCE

INTRODUCTION

T HE GUARD FORCE at a facility is the most important single element in the physical security of the facility. It should consist of personnel who are specifically organized, trained and equipped to protect the security interests of the facility.

It is for management to decide upon an extensive use of mechanical means of protection as opposed to a larger guard and watch service, or to use a combination of the two as the best method of providing the degree of protection desired. The need for night watchmen or for a guard force at an industrial plant or business establishment must be determined after a careful study of all factors bearing on the security interest at the plant.

DEFINITIONS

A distinction is often drawn between "watchmen" and "guards." "Watchmen" are considered more passive. As the term implies, they are to watch something and make periodic reports or sound an alarm if some unusual condition or incident develops. More often than not they are to perform their duties after the plant has closed operations for the day. They usually conduct an inspectional tour to make sure that all doors and windows are properly closed, office fans, lights and heaters turned off, that no burning cigarette stubs are left lying around, etc.

"Guards," on the other hand, deal more directly with people. A "guard" is an armed employee who is specifically and primarily assigned to prevent unauthorized access to security areas and to protect security interests against espionage, theft, sabotage, or other intentional damage. A guard takes an aggressive role in enforcing the plant rules and regulations, checking

36

employees' badges and credentials, escorting visitors, checking on thefts of company property.

Unfortunately, the title "watchman" or "plant guard" is not an appealing one to the prospective applicant. Generally it has become a descriptive title representing a menial job for the lazy and nonqualifying type of individual. Thus it is now common to designate the position with a title such as "security officer," or "plant protection officer." This title will appeal to the applicant as well as establish prestige for the department.

FUNCTIONS

In accordance with established plans and procedures at the facility or facilities to which they are assigned and within the scope of their authority, guard forces prevent unauthorized entry by a combination of actions, consisting principally of the following:

1. Implement and enforce the system of personnel identification and control.
2. Observe and patrol designated perimeters, areas, structures and activities of security interest.
3. Apprehend persons or vehicles gaining unauthorized access to security areas.
4. Check designated depositories, rooms, or buildings of security interest during other than normal working hours to determine that they are properly locked or are otherwise in order.
5. Report to the supervisor, as a matter of prescribed routine under normal conditions, and as necessary in the event of unusual circumstances.
6. Perform essential escort duties.
7. Implement and enforce the established system of control over the removal of documents or material of security interest from security areas.
8. Respond to protective alarm signals or other indications.
9. Act as necessary in the event of situations affecting the security of the facility including fire, industrial accidents, internal disorders and attempts to commit es-

pionage, sabotage, or other criminal acts.
10. Otherwise generally safeguard data, materials, or equipment against unauthorized access, loss, theft, or damage.

NEED EVALUATION

In determining the need for a guard force in the security program for a facility, all factors bearing on the security interest at such facility must be considered, as well as conditions relating to the nature, location and layout of the facility and the incorporation and effectiveness of other elements in such program. As these factors may differ widely in individual cases, it is not possible to prescribe criteria which will specifically fit each situation.

However, the need for guards is roughly in proportion to the scope of activities, the number of personnel involved, and the size of the facility concerned. As these factors increase, the safeguards provided by individual employees tend to decrease, resulting in the need for a guard force to control access to the facility or restricted area involved. Guard forces are seldom required at facilities of individual consultants and small laboratories involving only a few individuals. The security interest at such facilities usually consists of documents or materials of such type and amount as to be adequately safeguarded by appropriate depositories and by personal custody of cleared personnel using such matter in the performance of their work.

When guard forces are desirable but prohibitive because of expense, e.g., in small isolated vital facilities or installations such as communications transmitters, receivers, etc., the operating personnel should be responsible for providing, to the greatest possible extent, physical security to these properties.

Standby facilities may require guards to control access and protect against malicious or accidental damage when the facilities constitute essential insurance for the continuity of production or service.

Where the nature and scope of the industrial security program of a facility are such as to include a guard force, it constitutes one of the most important single elements in the program. At the same time, the continuing cost of guarding a facility in most cases represents the largest item of security expense.

The use and deployment of a guard force should, therefore, be carefully planned and continuously reviewed to the end that the most effective and economical utilization of manpower, commensurate with security needs, may be obtained.

Due to the widely differing types of situations and extreme range of security interest at facilities, the guard forces requirements, utilization and deployment at each facility must be locally determined after evaluation of all factors, including the importance, classification and vulnerability to damage of the data, materials, equipment, or activities involved; the arrangement, size and location of the facility; and the effectiveness of other security measures in effect.

USE OF OTHER SECURITY ELEMENTS

The efficiency and overall effectiveness of the guard force will be increased by the application of such measures as the most effective location of barriers, adequate protective lighting, properly designed guard shelters, use of protective alarm devices where their use is practical, adequate communication facilities, appropriate emergency procedures, etc. The proper application of these measures will help to keep the size of the guard force to a minimum.

TYPE OF GUARD FORCE

A company can set up its own guard force, or can contract for a guard service from world-famous firms, or from any of a number of regional and local companies. These firms will supply one man or an entire guard force, uniformed, trained and ready to go. They also offer mobile guards, who will check buildings periodically, looking for signs of break in or fire. The advantages of contracted guard services may include lower cost because of the elimination of administrative, scheduling, bargaining and emergency substitution headaches.

SELECTION

General

The effectiveness of the guard force depends on the quality of its members. The job of a plant guard is such that he handles more of the day-by-day problems concerning the interest of the plant as they affect employees and visitors—particularly in a regulatory sense. Guards enforce rules and regulations as well as many of the laws of the community and state on the company premises. They should therefore possess the quality of being able to handle people effectively.

The standards developed must meet the complex, technical demands made on the protection department. Prior to the establishment of any recruitment standards, a satisfactory wage must be allocated. This will attract the type of individual who is properly qualified. Otherwise, it would be a wasted effort to set up a satisfactory standard and expect to receive applicants who would be willing to receive a meager salary. The first step in selecting guards is the establishment of minimum physical and mental standards. A system must be established that will assure that all guards are carefully screened to assure that they meet those minimum qualifications.

General Qualifications

In order to perform necessary, continuing and emergency functions adequately, guards should be limited to individuals who meet the qaulifications set forth below:
1. Loyalty.
2. Intelligence.
3. Physically qualified to perform the required duties.
5. Cooperativeness.
6. Have the ability to exercise good judgment; possess courage, alertness, tact, self-reliance; be of even temper; and be able to maintain good performance.

Physical Qualifications

Guards must be fully capable of performing duties requiring moderate to arduous physical exertion under either normal or emergency conditions. They must possess good distant vision in each eye, normal fields of vision and good depth perception, and be able to read typewritten material without strain, glasses permitted, to distinguish basic colors, and to hear, without the use of a hearing aid, the conversational voice at a distance of fifteen feet in both ears. Physical examinations of guards shall be conducted at time of employment and annually thereafter to assure their physical fitness. During these examinations, close attention should be given to the mental condition of the applicant as well as his physical fitness. Needless to say, any hint of mental instability should disqualify the applicant. He should be old enough to have sound judgment and young enough for the physical stamina required on the job.

It is possible that there may be assignments that less vigorous and even the physically handicapped can do. Thus, waivers may be effected, but it should be remembered that such personnel cannot be included in active emergency plans.

Mental Qualifications

Guards should be mentally alert, capable of understanding and implementing all normal instructions and receptive to specialized training which may be necessary in an emergency. Emotional and mental stability is essential since duties normally require contact with the public and the carrying of firearms and, under emergency situations, may involve long periods of duty without relief. Guards should have good judgment, courage, tact, resourcefulness and should be composed at all times.

In addition to judgment, tact and logic, the general mental attitude towards life and the job is very important. Uncompromising interest in and loyalty to the job are particularly applicable to security guards.

Personal qualities include, but are not limited to, poise, tact and the ability to work harmoniously with others. They should be developed thoroughly in pre-employment interviews.

Character

The importance of acceptable character in a security guard increases with the sensitivity of the position he is to fill. His integrity and moral code must be above reproach.

All guards should be investigated and cleared. Guards occupy a position of high trust. By the very nature of their duties and responsibilities in the security program, they may, through necessity or the inadvertence of others, come into knowledge and/or custody of vital information, other classified matter, or material of high strategic and/or momentary value. Unquestioned loyalty and integrity is, therefore, an essential requisite for all guards.

Such an investigation should start with the submitting of the individual's fingerprints to the local, state and federal identification bureaus to check for any previous criminal record. His loyalty to the country must be determined next. This applies not only to those plants that are performing classified work, but to all others. The plant, even though it is not performing classified work, may be in times of national emergencies a very critical source of materials.

The next step is to check the applicant's personal habits and past associations. This can be accomplished by contacting neighbors, employers, teachers and clubs who know the applicant. Any evidence of poor credit, excessive drinking, narcotics, bellicosity or the like should disqualify the applicant.

If the guard is bonded, the company issuing the bond will make an investigation. The bonding company report may be accepted, or a further check may be made by a private investigating agency or by the employer. Police records are usually open to legitimate inquiry and much other useful information can be gleaned from a neighborhood check. Such an investigation normally includes inquiry of former employers and checking of credit ratings. When special circumstances require a pre-employment investigation which is beyond the normal scope of the plant's own investigative personnel, the services of a reputable personal investigative agency should be employed.

Experience

Previous military, industrial, or civil police or investigative experience is desirable. Termination from such previous employment should have been on an honorable basis and not because of age or physical handicaps.

GUARD UNIFORMS

Guard personnel should be distinctively uniformed while on duty and identified with their function by appropriate emblems or badges. This should include a minimum of cap, shirt or blouse, trousers and badge. Deviations from the prescribed uniform requirements should not be made except for such additional items as are necessary to protect the health, comfort and safety of the individual. A uniform serves a twofold purpose: it readily identifies the wearer as the individual who is charged with protecting lives and property in the plant and gives him a distinctive symbol as a guardian of the public welfare. The wearer himself is given a feeling of prestige and pride in his organization.

A word of caution is needed in this regard. The uniform should not be similar to that of the regular police in the jurisdiction in which the plant is located. If it is, there is the possibility that embarrassing situations may develop when the plant guard is mistakenly identified as a member of that police force. Such copying ads nothing to identity with the company. On the contrary, they lead outsiders to believe that members of the plant guard are trying to masquerade as police.

The duty uniform should be worn during all tours of duty. Normally, it may be worn during off-duty hours only between the place of residence and place of duty. Each member of the guard force should maintain high standards of personal and uniform appearance. Each guard should wear a neat, clean and well pressed uniform, shoes should be polished, and the badge and leather well shined. Further, the uniform should be well tailored and designed to impress not only those who see it, but even more important, those who wear it. All guards should be equipped with a flashlight, police whistle and notebook.

GUARD WEAPONS

Whether or not guards should carry firearms is for management to decide. Various factors will influence this decision, such as the product and type of security interest involved, the number and kind of persons employed at the plant, and the character of the community or area in which the plant is located. In most cases, the weapons of issue should be .38 caliber revolvers, as its use would lend ease in procurement of ammunition and interchangeability of weapons and parts. However, at key posts of that weapon is in wide usage for police and guard service, and critical facilities, semiautomatic weapons, riot guns and other emergency weapons may be provided. Such weapons and ammunition supply should be available at strategic points, properly controlled and maintained in operating condition for emergency use. All members of the guard force should be legally authorized to carry firearms in the performance of official duties and in the areas where such duties are performed. If guards are armed, the question of bonding should be thoroughly explored by the employer. Guards may be deputized by local police departments to make arrests, if necessary, in the vicinity of the facility.

The following is suggested as a guide for armament of a guard force employed at a facility requiring plans for its emergency deployment:

1. One handgun per guard.
2. Small supply of riot shotguns.
3. Supply of tear gas grenades or shells.
4. Limited supply of special weapons for unusual activities and posts, such as the following:
   a. Short-barrelled handguns for shipment escorts.
   b. Submachine guns for protective work where maximum firepower is needed at short ranges.
   c. Rifles for lookouts, isolated posts or other positions where long-range accuracy is a consideration.
5. Small reserve in excess of actual requirements to provide for repairs, variations in strength or other contingency.

Regulations

There should be strict regulations regarding the issuance, accountability, use and general care of weapons. All guards should be required to undertake a series of familiarization courses with the weapon. These courses should cover the general use and care of the gun as well as actual firing tests on the target range. Even though many guards are former police officers and are familiar with this type of gun, the familiarization courses will serve as a refresher for these men and will also initiate those who are wholly uneducated in the proper use of a gun.

Generally speaking, a gun should be used only in the actual defense of the guard's or another person's life, or when there is strong belief that someone's life is in danger. Guards wearing side arms should not display or draw their revolver except for inspection, cleaning, or actual use in the line of duty. Guards must report in writing any incident involving the accidental or intentional discharge of their revolvers while on plant property.

Safety First with a Gun

Following are several simple safety rules in the handling of a gun:

1. Never point a revolver at any one you do not intend to shoot or disable, nor in any direction where an accidental discharge may do harm.
2. When at a shooting range, do not snap the hammer of the gun for practice while standing behind or back of the firing line.
3. Never place a finger within the trigger guard except to fire a shot.
4. Never place a revolver in the holster with the hammer cocked.
5. While running, keep all fingers off the trigger and the thumb off the hammer.
6. Never discharge a revolver while running, especially in pursuit of a criminal. Always stop, cock the hammer, take careful aim squeeze steadily on the trigger until the shot is fired. Bullets upon striking a hard surface are very likely to ricochet and cause serious injury or death to innocent persons.
7. Always take extreme care to avoid injuring any person who may be within range of fire. If necessary, the shooting arm may be steadied by using the left or right arm and hand as a brace.
8. If the revolver is fired as a call for assistance, it should point straight up in the air.
9. If the revolver is carried in one's hand, the cylinder should be dropped and open, with the fingers thrust through the frame.
10. The barrel of the revolver should be clean and free from any foreign substance such as rags, dirt, etc., for when

it is so obstructed and a round is fired, the barrel is very likely to bulge or burst.

11. The revolver should always be unloaded when it is handled for any purpose other than to fire it. To make certain of this, release the latch, push out the cylinder and examine it to see that there are no shells or cartridges in the chambers.

12. Never turn around at the firing point while holding a loaded revolver, nor point it in any direction other than at the target.

Care of a Gun

A gun should never be left uncleaned over night. After firing the weapon, a good cleaning solvent should be used with a brush and a rag so that all dirt, powder and grease is removed from the bore, cylinder and other surfaces. The gun should then be wiped with a clean rag so that all of the solvent is removed. Finally, the bore and cyclinder should be lubricated with a light gun oil and the outside polished with a clean dry rag to remove excess oil and fingerprints before replacing the gun in the holster. This will not only keep the gun free of rust but will prevent it from gathering dust and from functioning improperly due to the presence of old, gummy oil.

A guard should inspect his gun every day before beginning his tour of duty to see that it is functioning properly. During this inspection it should be wiped clean. At least onc a month, regardless of whether or not a gun has been fired, it should be thoroughly cleaned and lubricated. This will insure the gun being kept in top condition.

Before cleaning the revolver all cartridges should be removed. This is not only a safety factor, but also permits more thorough cleaning of the gun and prevents oil from running over the cartridges where it might work its way inside and cause a misfire.

A gun should never be placed on the ground where sand or dirt can enter the mechanism. Neither should a gun be abused by throwing or dropping it or by using the butt as a hammer.

Although guns are sturdily built, they must be properly maintained. When a gun is needed, a person's life may depend on it.

TRAINING

General

The objective of training is to enable an organization to perform routine duties competently and to meet emergencies speedily and efficiently. To maintain interest, the training must be applicable to the job. When this application is not evident, the connection should be explained before the training begins.

Need

Efficient and continuing training is the most effective means of obtaining and maintaining maximum proficiency of guard force personnel. Regardless of how carefully a supervisor picks his guard force, it is seldom that they will initially have all the qualifications and experience necessary to do the job well. In addition, new or revised job requirements frequently mean that guards must be retained for different jobs and skills. The gulf between ability and job requirements can be bridged by training.

It is also well for supervisors to remember that all guard personnel do not have the same training needs. It is a waste of valuable time to train an individual in subject matter which he has already mastered, and it is a source of dissatisfaction to the man when he is subjected to instructions which he knows are not appropriate to his skill level. Past experience, training, acquired skills and duty assignments should be evaluated for each man as an aid in planning an effective training program.

Extent and Type

The extent and type of training required for guard forces will vary according to the importance, vulnerability, size and other factors affecting a particular installation. The objective of the training program is to insure that the guard is able to perform routine duties competently and to meet emergencies speedily and efficiently. However, each member of a guard force should be required to complete a course of basic training and thereafter periodic courses of in-service or advanced training. All such courses should include necessary phases of on-the-job training prior to initial or new assignments and appropriate supervision and follow-through thereafter.

Benefits

A good training program has benefits for both the facility and the guard force. Some of these benefits are the following:

FOR SUPERVISORS

The task of supervising the guard force is made easier. There is much less wasted time. There are fewer mistakes made. The resulting economies of motion or action are of benefit to the facility. There is also less friction with other agencies. A good program also helps to instill confidence which is most valuable to a guard force.

FOR GUARDS

Training benefits the guards to the extent that their skills are increased; it provides increased opportunities for promotion; and it provides for better understanding of their relationship to management.

FOR THE GUARD ORGANIZATION

Good training helps to provide for more flexibility and better physical protection; fewer persons may be required; and less time may be required for men to learn guard requirements. Training also helps to establish systematic and uniform work habits. An effective program will help to create better attitudes and morale.

Basic Training

As a minimum, guard personnel should rceeive training in the subjects listed below. Other subjects may be added to fit the needs of the particular installation.

1. General orientation, to include organization of the guard force, physical geography of the installation and the location of all security units and restricted areas on the installation.

2. Authority of the individual guard.
3. Guard orders, general and special.
4. Discipline, including conduct and appearance.
5. Weapons instruction, to include safety, maintenance, qualification and firing on practice range.
6. Self-defense.
7. Traffic control, when appropriate.
8. Communications facilities and procedures.
9. Elementary first aid and fire protection.
10. All security measures and procedures, with emphasis on the pass and identification system.
11. Package and vehicle searches.
12. Employee and public relations.
13. Purposes and principles of the system of security.
14. Security as applied to the local installation.
15. Organization of the guard force.
16. Functions of the guard force.
17. Specific duties of the individual, including sufficient "breaking-in" training.
18. Operation and care of motor vehicles.
19. Report writing.
20. Riot control.
21. Operation and use of special equipment.

This training should make the guards understand their relationship to employees. They have certain duties to carry out in respect to employees, but bad employee relationship can result if guards become officious and assume powers which are not rightfully theirs.

All guards should receive training in procedures necessary for the implementation of emergency and disaster plans formulated for such facilities. Training should include periodic practice alerts and rehearsals of such procedures. Such training should include coordination with outside agencies which may be called in the event of emergencies beyond the capabilities of local security forces. In most instances, this will involve civil law enforcement agencies. All personnel of guard forces should be required to undergo periodic in-service training to include necessary review of basic material and such other subjects as may be applicable to the specific installation.

Special and Advanced Training

Completion of the basic training should fit the guard to take his place in the organization. Special and advanced training should be required at all installations where the duties of the guard force are more varied and complex. In such cases, it will be necessary to conduct additional specialized training for selected individuals and supervisory personnel of the guard force. Advanced or special training may be given either in solid blocks following the basic training period, or in short periods of instruction fitted into the duty schedule. Appropriate subjects for inclusion in advanced training are the following:

1. Weapons instruction to include range firing of the submachine gun, rifle, riot type shotgun, tear gas grenades and other appropriate weapons.
2. The prevention and detection of espionage and sabotage.
3. Operation and use of special equipment.
4. Riot control.
5. Disaster procedures, to include rapid evacuation of the installation.

In-service Training

When a new guard is assigned, he must be given instruction in conditions peculiar to his post. Whenever possible the first assignments should be with an experienced guard. Additional in-service training is a continuing process by supervisors.

Testing

Testing serves a dual purpose in that the results are indicative of both the quality of the instruction and the alertness and retentiveness of the student. Short, unannounced tests at frequent intervals during the training are usually preferred to longer end-of-course tests.

Plant Geography — Knowledge of the Plant

In order to perform successfully his routine duties as well as any emergency task, the guard should be very familiar with all parts of the premises for which he is responsible. A check list of the various items should be carefully studied to insure that he understands everything he may have occasion to use:

1. Location of all stairways and where they lead to.
2. Location of all emergency exits.
3. Location of fire alarm boxes.
4. Location of telephones.
5. Location of all emergency fire fighting equipment, such as hand extinguishers, water pails, sand boxes, shovels, etc.
6. Location of light switches, so that in an emergency lights may be turned on in any part of the premises without delay.
7. Location of water standpipes, fire hydrants and hose reels, so that firemen can be directed to this equipment without delay.
8. If the building has a water sprinkler system, the location of all valves controlling the supply of water to the sprinkler system and the location of extra sprinkler heads.
9. Location of electrical control switches for turning on the elevator motors.
10. Location of fuse boxes, power control switches, steam valves and any other additional devices for the machinery or other operations within the premises. In cases of fire or other emgergency, it is often important for the firemen or other emergency crews to use such devices, and the information should be available without delay.
11. Operation of any equipment which might be needed in any emergency situation. If this is not known, the supervisor should arrange for the plant engineer or other competent person to give the guard any instruction needed.
12. Location of control rooms and shutoff switches for air conditioning or ventilation systems. In case of fire, it may be necessary to use the air conditioning system to clear out smoke; or in the event it is in operation during a fire, it should be turned off so not to spread the

flame.

13. A knowledge of the various stocks of materials stored on the different floors and in the various sections of the building. This knowledge may be very important in the event of fire.

14. If the guard's duties include attending to the heating system, he should understand its operation thoroughly, so that if anything goes wrong he can detect it and take proper action. He should always call for assistance if he is unable to handle any situation.

15. If the guard's duties are always during hours of darkness, it should be arranged for him to make a complete tour of the premises during the daytime with his supervisor, so that he may be supplied with information concerning machinery, operations, or any other fact that may assist him in his work.

16. Location of emergency information such as

    a. telephone numbers of the police department, fire department, doctor, ambulances, supervisor, etc.

    b. addresses and telephone numbers of any other employees who should be called under special conditions or in emergencies, such as the plant superintendent or engineer. This information should be supplied by the guard's supervisor and personally checked to make sure it is up to date.

    c. location of the nearest street fire alarm box if the plant is not equipped with fire alarm boxes.

GUARDS' DUTIES

Although much can and must be said concerning specific tasks of the guard force, there are many simple tasks that are strictly speaking not guard duties which cannot be overlooked. These countless tasks are acts of courtesy and consideration to fellow employees. These tasks must be performed, for today people look upon the uniformed protection officer as an individual who will help him in his immediate difficulties. Time and time again the employee will turn to the security officer. The guards must help when they can. Indeed, they should be admonished to look for ways constantly in which they may render help, even though their regular assignments are not within this scope.

The duties of a guard are as follows:

1. Preserve the peace, protect life and property, prevent crime, apprehend violators, recover stolen property and enforce all plant rules and regulations.
2. Observe and control designated perimeter areas, structures and activities of security interest.
3. Implement and enforce systems for personnel identification, visitor control, vehicle and package control, and at entrances and exit gates and throughout the premises, as necessary.
4. Check persons and vehicles entering and leaving the premises or other established security areas, according to instructions.
5. Apprehend persons or vehicles gaining unauthorized access to security areas.
6. Conduct security inspections of designated depositories, rooms, or buildings of security interest during other than normal working hours to determine that they are properly locked or are otherwise in good order.
7. Regulate and direct vehicular and foot traffic, as necessary.
8. Perform essential escort duties.
9. Enforce the established system of control over the removal of documents or materials of security interest from the premises.
10. Upon discovering a fire, immediately turn in an alarm by the quickest method possible.
11. Respond to protective alarm signals (fire alarm, burglar alarm) or other indications of suspicious activity.
12. Act as necessary in protecting the best interests of management in the event of disaster or disorders such as fires, accidents, riots, or the commission of criminal acts, including espionage and sabotage.
13. Otherwise generally safeguard data, materials and equipment against unauthorized access, loss, theft, or damage.

14. Maintain a control system for locks and keys used for company property.
15. Report all unusual incidents that occur in the plant or on company property.
16. Report all information received concerning the violation of any plant policy or rules and regulations. Under no circumstances should the guards conceal, repress, ignore, or alter the facts of any such violations.
17. Report all employees found loitering, sleeping and gambling, or who are visiting outside their respective departments during working hours.
18. Report promptly all cases of sickness among employees.
19. Make routine or special reports to supervisors, as prescribed, concerning matters of unusual circumstances.
20. Tag and turn in to plant protection headquarters all lost, stolen, or unclaimed property coming into their possession. This property must be recorded and identified by the guard in charge.

Guard Surveillance

Guards at each facility should be trained, adequately instructed and encouraged to be constantly on the alert to detect acts of sabotage, espionage, and suspicious acts of all kinds.

Detection of subversive acts may be accomplished in line of normal guard duty by the following:

1. Checking ingoing and outgoing personnel and packages.
2. Careful surveillance of sensitive points which are susceptible to sabotage.
3. Patrolling plant area, particularly during hours when plant operations are closed down.
4. Watching carefully the actions of personnel rumored or alleged to be engaged in acts detrimental to the company or the government.

Guards should immediately report any suspicious acts or circumstances to the plant protection officer.

Attitude

No discussion of a guard's duties would be complete without some consideration being given to the conduct and attitude of the individual officer. The officer must never lose sight of the fact that he is a representative of the company. His appearance, speech, demeanor and deportment are a reflection of company policy just as much as the product manufactured by the company.

This company representative (for that is what he is), dressed in a distinctive uniform, is an animate emblem of their product. He is constantly in the eye of employees, management, visitors and the public. His daily duties afford him many opportunities to evoke a feeling of genuine affection. He will be called upon to demonstrate his courage in emergencies, his calmness during disorder or confusion, his genuine interest in the safety and welfare of employees.

Occasions will inevitably arise when an officer must reprimand someone for a misdemeanor or an error in judgment. Such situations should be handled in private with tact and diplomacy. If the offender's dignity is destroyed, many times through an explosive "bawling out," the officer has not only failed to correct the fault, but has also made an enemy of the force. Tact is the essential requirement.

The officer will be constantly approached by employees with complaints and problems even though a great percentage of them may not apply to the basic functions of his department. These requests for information or help should always be given the greatest consideration and interest. If they pertain to matters not within the jurisdiction or knowledge of the guard, he should direct the complainant to the proper person for satisfaction. Always the motto should be to serve.

The guard who greets visitors, salesmen and the public has a great opportunity to improve public relations. The officer should be advised to remember that nothing pleases a visitor so much as unexpected courtesies.

SUPERVISION

A security supervisor has the task of overseeing and directing the work and behavior of other members of the guard force. The effective supervisor needs a complete understanding of the principles of leadership and how to apply them to obtain maximum performance from members of the guard force.

The supervisor is called upon to think and act in terms of many different jobs. He is often responsible for the selection, induction, training, productivity, safety, morale and advancement of the guard force. He must understand these and all other employment aspects of his force.

Supervision of a guard force is necessary to assure the effectiveness of the mechanics of guarding and detecting and taking appropriate action in case of emergency or unusual circumstances. The morale and general efficiency of the individual guard is largely dependent upon the quality of his supervision.

At facilities where guard forces of several men per shift are engaged, full-time personal supervision is needed. The ratio of supervisory personnel to guards at larger facilities and elsewhere, where practicable, should not exceed one to twelve. Personal supervision should include the following:

1. Each guard shall be inspected by his supervisor upon reporting for duty to determine his apparent physical fitness therefor and the condition and adequacy of his weapon, uniform or other equipment. At such time he shall be given special instructions or orders which may be necessary.

2. Each guard post, patrol and other activity should be personally inspected by supervision at irregular intervals throughout each shift to determine that personnel and the system in general are functioning properly.

Various means and devices may be successfully used as supplements to personal supervision or, in the case of small facilities or remote areas, to supplant personal supervision as a means of assuring that necessary areas are patrolled and that other functions are performed. These include the following:

RECORD TOUR SYSTEMS, under which guards record their patrols or presence at strategic points throughout the installation by use of portable watch clocks, central watch clock stations, or other similar devices. A watch clock is merely a portable clock which the watchman carries with him as he makes his rounds. At various locations distributed about the premises are a series of stations containing keys which the watchman inserts into the clock and thereby records upon a paper roll or chart inside the clock the time each station is visited. These clocks are of tamper-proof construction so there can be no manipulations of the record. Upon completing his tour of duty, the watchman turns in his clock which is then inspected by his supervisor the following day to make certain that the watchman was at his various stations at the proper time and in the order prescribed.

If for any reason the watchman is unable to visit his stations on time, he should make sure that he leaves a written explanation, describing in detail the reason for his delay. By so doing, providing he has a legitimate reason, he may avoid the censure of his supervisor. If the watchman was injured during his rounds or had to handle an emergency situation, this would be considered a valid excuse for failing to "punch a station" on time.

CENTRAL STATION REPORTING SERVICE. With this type of service the watchman is in contact with a central agency which records the time of his rounds. Should the watchman fail to reach a station on time or otherwise fail to notify the central station of his activities, the central station immediately takes steps to investigate the cause or failure of the watchman to report.

The value of such a system is readily seen in that the watchman cannot be out of service for any length of time before an emergency check is made to ascertain whether or not he needs help. Should he have encountered unforeseen danger during his rounds, the central station service is of great value in preventing prolonged danger to the watchman or plant.

These have application at a limited number of the most vital facilities to supplement personal supervision or at facilities with small guard forces, to supplant personal supervision. These systems provide instantaneous supervision of the guard, plus a means of detecting interferences with his normal activities and a means of initiating an investigation and/or other appropriate action.

All guards should be required to report regularly to headquarters by usual means of communication. The frequency of such reports will vary, depending on a number of factors including the importance of the installation.

Records of tours and reports to headquarters should be carefully checked. Failure on the part of a guard to record a visit at a designated station, to report to headquarters as required, or other deviation from established reporting procedures, should be immediately investigated.

RULES AND REGULATIONS

Every plant with its own guard force should establish definite rules and regulations prescribing standards of conduct and deportment for its guard personnel, both while on duty and off. By so doing, the guards know what is expected of them. Such preventive action will go far towards eliminating certain types of conduct which are considered undesirable in the general duties of an efficient plant guard.

Some of the points which should be covered are the following:

1. OBEDIENCE — ORDERS AND RULES. The guard must know and personally obey all company rules and regulations. The guard should promptly obey all lawful and legitimate commands of his supervisor. He should never adopt a haughty or insolent attitude toward such orders. If he doubts the wisdom or lawfulness of any order, he should point out his reasons for this belief to his supervisor, if appropriate. He should then obey the order, providing it is not an obvious violation of the law, in which case he should consult his supervisors' superior or the plant manager for instructions.

2. PERSONAL DEPORTMENT. Guards should be attentive to their duty at all times. They should act with dignity and maintain a soldierly bearing, avoiding a slouchy appearance of the body. They should never act in any manner which will bring discredit upon the guard force. Guards must be civil, courteous and orderly in their conduct and deportment. They must refrain from violent, harsh, coarse, profane, or insolent language.

While on duty a guard must gave undivided attention to his assignment. Prolonged or unnecessary conversations with visitors or fellow workers distract his attention from his responsibilities and should be avoided. Should any person attempt to engage the guard in unnecessary conversation during his tour of duty, he should courteously excuse himself and point out that he must attend to his assignment.

A plant guard should never read or smoke when on duty in the public view. During break periods a guard may be allowed to smoke or read a newspaper if he wishes, in the guard locker room, guard office, or other especially designated places. The use of chewing tobacco or snuff by the plant guards should be discouraged.

Guards should not sell papers, books, tickets, or any form of merchandise to employees or fellow officers while on duty or on plant property. Nor should a plant guard accept gifts from visitors, employees, or other members of the guard force. The acceptance of gifts leaves one obligated to the giver and therefore places the guard in a compromising position from which it is difficult for him to have complete freedom of thought and action in performing his duties properly. Certain unscrupulous employees are quick to sense that certain guards are "easy marks" who will look the other way instead of reporting a theft or other violation. The security director must be constantly alert in order to detect such guards among the security personnel. This type of guard is not suited for security work and should be released or transferred to other duties.

In order to prevent guards from becoming too friendly with employees, their duties should be rotated, both in terms of time and place of operation. Short in-service training sessions should be given periodically in order to reemphasize the most important aspects of their duties and to remind them of the correct way of performing those tasks in which they may have become negligent.

3. APPEARANCE. Guards should be neat and clean in appearance while on duty. Boots and shoes should have a high polish. The uniform must be kept kept clean and pressed.

4. LIQUOR AND NARCOTICS. A guard should never use intoxicating liquors while on duty nor should he report for duty with the smell of alcohol on his breath. Narcotics should at no time be used except upon the advice of a physician. If a guard must use narcotics for medicinal purposes, this fact should be reported to his supervisor so that management is fully aware of the circumstances surrounding its use. Under such circumstances the guard should never be given an assignment which he would

be incapable of performing because of his mental or physical condition.

GUARD INSTRUCTIONS

General and special orders should be issued in writing covering the duties of each post and assignment. Such orders should be carefully and clearly worded to include all necessary phases of each assignment. They should be reviewed not less than monthly to be certain that they are currently applicable. Periodic inspections and examinations should be conducted to determine the degree of understanding of and compliance with all orders.

A guard manual or handbook, setting forth policies, organization, authority, functions, procedures and miscellaneous operating information, should be prepared and distributed to all members of larger guard forces. Each guard should be held responsible for full knowledge and understanding of its contents.

These instructions will normally cover such duties as enforcement of the pass and identification system, observation of designated perimeter barriers to prevent unauthorized entry, and other duties related to the protection program in general.

All guards should be fully and frequently briefed on the exact nature and scope of their duties. Guards should be assigned definite duty posts, patrols, or patrol areas.

Instructions in Case of Fire

When a plant is equipped with fire alarm boxes, the guard should transmit an alarm immediately in an emergency. If there is no fire alarm boxes, the telephone or other available means should be used to call the fire department. In calling by telephone, if the guard does not remember the number of the fire department, he should dial the operator and say "I want to report a fire." He should then give slowly and distinctly the name of the plant and the address, repeating it if necessary, and should not hang up the receiver until he is certain that the fire department has the location correctly.

After giving the alarm, the guard should do whatever he can to put out the fire himself. He should make sure, however, that the firemen are properly directed to the blaze and are admitted through any locked doors or gates as soon as they arrive.

Fire statistics have shown that men occasionally commit grave mistakes in handling fires that break out in buildings with automatic sprinkler systems. By turning the shutoff valve at the wrong time, a guard may unwittingly cause the fire to spread. Following are rules for the guard in the event fire breaks out in a building protected by wet-pipe or dry-pipe sprinkler systems:

1. If the fire is discovered before the sprinkler system operates, he should turn in th alarm immediately and then proceed to take whatever measures are necessary in controlling the fire.

2. Even if a fire is discovered and the sprinkler system is already in operation, he should first call the fire department before attempting to control the blaze or take other action. Some sprinkler systems operate with an automatic central station alarm system, so that upon a flow of water the alarm is automatically transmitted. If this is the case, he should then meet the fire trucks at the entrance of the plant or building and direct them to the blaze.

3. If the sprinkler system has extinguished the fire before the fire department arrives, the guard should shut off the sprinklers, but only when he is absolutely certain that the fire is completely extinguished. If there is any doubt about the fire being out, the guard should wait until the firemen arrive. Experience has shown that large losses have occurred because overzealous guards have shut off the sprinklers before the fire was completely extinguished. It is better that a plant suffers some water damage than that it is completely burned out by a disastrous fire.

4. When the fire is out and the sprinklers have been shut off, the next step is to restore the sprinkler protection immediately. In many communities the firemen will help in restoring new sprinkler heads. However, if the blaze has been extensive and numerous sprinkler heads have been fused, the watchman should call for help in restoring the system.

5. If a leak is discovered in the sprinkler system, the guard

should first make certain that it is a leak and not a reoccurrence of the fire. He should then close the valve controlling the sprinkling section where the leak occurs and notify the proper person or authority.

6. After a fire or leak in the system, everything possible should be done to stop water damage. Water should be swept or directed towards floor drains, when possible. The use of large amounts of sawdust has been found valuable in soaking up pools of water or in making small dikes on the floor around where merchandise is stored.

Instructions for the Tour of Duty — Making the Rounds

It is to be expected that the guard's rounds will be completed on schedule unless circumstances prevent it, however time should be taken to observe each area carefully so that any unusual condition is thoroughly investigated. Hazardous areas should be completely covered, and no potentially dangerous situation should be left unchallenged merely because it would delay the scheduled completion of the patrol.

The guard should remain alert at all times. He may be required to meet an unexpected situation at any moment, even during his rest period. Alertness on the first round, or patrol, is of primary importance. Hazardous conditions or dangerous situations should be discovered as soon as possible so that they may be immediately corrected. The following is a check list of some of the things to watch for during the first round:

1. Test the telephone to make sure that it is in working order. This is especially important if the telephone is connected through a switchboard, since the operator may forget to plug in the night line on the guard's extension. If this is not done, the telephone will only operate for calls within the building. The guard should familiarize himself sufficiently with the switchboard so that he can plug in a line and receive a call.

2. Turn off all fans or other motors or machinery of any kind that have been left running which should have been turned off at quitting time. The guard can quickly and easily turn off the swtiches of such items as he comes to them on his first round; however, if in doubt, he should consult his supervisor as to what he should do.

3. Water taps left open so that water is running, fires left burning in gas heaters or under coffee makers — such items should be turned off, unless instructions have been given to the contrary.

4. Unnecessary lights which are left burning should be turned off.

5. Check all doors, windows, skylights, etc. to make sure that they are closed and locked. This not only serves as a protection against intruders, but prevents damage from wind, rain and snow.

6. If a window has been broken, it should be covered temporarily with boards or cardboard, and the condition should be reported in writing to the supervisor the following morning.

7. Fire extinguishers, fire pails, sand pails, etc. should be inspected to see that they are in their proper places and in operating condition.

8. Check all aisles and passageways leading to exits and fire doors to make sure that they are not obstructed by merchandise or machinery which would hamper firemen in case of fire. The condition should be corrected if possible, and if not, it should be reported.

9. Check fire alarm boxes, fire extinguishers, fire hydrants, fire hose reels, or other emergency equipment stations for obstructions which may make it difficult to get to them in an emergency. Correct the condition, if possible, or report it in the morning.

10. Any cigar butts or cigarette stubs or matches found in areas where smoking is not allowed should be reported in writing. Management depends on the guard to protect the plant from the carelessness of others. Unless the guard reports such incidents promptly, the detrimental carelessness of fellow employes who deliberately disobey fire regulations could result in a serious fire which would destroy the plant sufficiently to throw all the employees out of work.

11. Check all fire doors and shutters to see that they are closed properly.

12. Investigate immediately any unusual odors, smoke, or gas. If the cause cannot be determined within a reasonable length of time, it should be reported to the supervisor.

13. Inspect employees' locker rooms for any sign of smoke caused by lighted pipes, cigarettes, etc., carelessly left by employees. Also check for pieces of oily rags or clothing in lockers or which have been tossed in the corners of the locker rooms.

14. All heating devices which are operating on the premises, whether using gas, electricity, coal, or kerosene, should be checked regularly on every round to make sure they are operating properly.

15. See that all gas or electric appliances such as soldering irons or pressing irons have been turned off.

16. Remove all boxes, paper, rubbish, or other inflammables from the vicinity of the boilers, stoves, heating vents, or steam pipes and deposit all oily rags in metal containers provided for such use. Report any hazardous conditions thus found.

17. Check to see that all containers holding gasoline, alcohol, paints, varnish, etc., are properly covered and in their proper storage place. Fireproof containers should be available for the storage of all such inflammable materials.

18. During cold weather make sure that the heat is turned on in all areas where sprinkler pipes or other water pipes might freeze.

19. Remove all papers, books, or other combustible materials which are stored on or against radiators.

20. Report any other hazardous condition found during the tour of duty and correct it, if possible.

GUARD SHIFTS

Guard forces are normally organized into three or four shifts, usually on duty for eight-hour periods. Changes of shifts should occur before or after peak periods of activity in the normal operation of the installation. Guard shifts should be so scheduled that they will not coincide with employee shifts. The minimum requirement of guard personnel for each shift should be established by dividing the total number of man-hours needed by the hours in the shift. To this number must be added sufficient manpower to provide relief, which is usually based on one-half hour per man needed for each shift. If there is a post or patrol requiring less than eight hours duty during a shift, this service must be provided by drawing a guard or guards from a less essential mission. If this is impossible, a man must be added to the shift and his services utilized in some other post, patrol, or guard service, or as relief during extra time.

ORGANIZATION

One individual should be placed in charge of each shift of the guards. In small facilities, the chief of guards may assume this function on one shift with subordinates on the other shifts. Clear and definite understanding should exist as to seniority and who is in charge of the guard force. Guards may be organized by the following:

1. Fixed post development.
2. Patrol deployment.
3. Response to calls for assistance.
4. Any combination of the above three.

GUARD POSTS

Local conditions will dictate the number and location of guard posts at each facility. During the hours of darkness all guard posts, foot patrols and guard shelters should be located in the darkened areas behind the protective lighting screen. Consideration may be given to providing where possible at least one guard post located at some high point within each vital area. This post should be provided with a high caliber rifle, or similar weapon, a manually operated searchlight, and protection from small-arms fire.

Guards who are assigned to fixed posts should have some designated method of securing relief if such is required. Where fixed posts do not permit the guard to move at all, such as guards in watch towers, arrangements should be made so they may leave their posts not less than every two hours.

Guards should remain at their posts until relieved by another guard. The failure to receive relief shall be reported to the officer in charge by telephone or other available means.

If on duty at a location where a chair is provided, a guard should not remain seated when approached by someone who wishes recognition or who seeks information. Under such conditions it is considered discourteous and slovenly for a guard to remain seated when talking to another person. More than that, while in a seated position, he will be at a disadvantage should the person attack him.

GUARD SHELTERS

Guard shelters should be basically designed to provide occasional temporary protection from severe weather. The design should include space for one guard only; facilities such as heat, ventilation, storage space for essential guard accessories and lighting which will not expose the guard; good visibility in all directions; windows which can be opened and used as gun ports; provision for adding barricades such as concrete to make the shelter bullet-resistant when necessary, and providing maximum height barricades. Guard shelters should be painted to render them inconspicuous.

PATROLS

In addition to fixed posts assignments, guards should be available for specific or roving patrol duty.

Critical Area Patrol

Guard patrols should cover critical or sensitive areas of the plant frequently but should not follow a fixed pattern or time schedule.

Perimeter Barrier Patrol

The complete barrier should be checked at least once each shift by the guard patrol. When patrol cars are used, they should be equipped with two-way radio or telephones connected with the police or control stations. Perimeter guards should never leave their post to check a commotion in areas other than their own. They may thus be drawn away from an attempted penetration of their area.

Preparation

At the beginning of the patrol the officer must check his equipment to determine that it is available and in working order. He must know the entire plant and the area to which he has been assigned in detail. He should know the location of buildings, roads, alleys, tanks, water system, electrical system, fencing and the lighting system. He should know every means of exit and the quickest possible way of going from one point of the plant or grounds to another in the event of an emergency; he should be able to recognize an emergency when he sees one, and react to it effectively.

Mapping the Facility

A complete plan of patrol should be laid out prior to the actual patrol assignment. A survey should be made of the facility. The area should be divided into sections roughly twenty by forty feet. Sections running north and south, say, may be given numbers 1, 2, 3 and so on. Sections running east and west designated by letters A, B, C and so on. Thus any section of the plant can be located quickly.

How you zone depends on the particular area. A vertical building might be best zoned by floors; a horizontal layout could be zoned by rooms, buildings, or areas.

Timing

Most firms and insurance carriers require that hourly tours be made of vacated premises. This is the minimum and should be considered as such. The time of the routes should vary; this can be arranged by constant reversal of the route, back-tracking and varying the time of commencing the patrols.

Procedure

Particular attention must be given to any condition or hazard that can play a part in endangering life or property. Many of these conditions are fires; strange lights; frozen or plugged fire hydrants, hoses, extinguishers, alarms, etc.; jammed machinery or conveyors; unattended running machinery, heaters, etc.; gas leaks; plugged sewers; open manholes; defective platforms; plugged elevators; broken steam or oil lines; loose or poor signs; debris; loose wires; broken fences; obstacles and obstructions; improper or poor lighting conditions; ditches; gullies around railroad sidings; lockers and rest rooms; unoccupied or abandoned buildings and ruins; strange heaps or bundles; tampered phones, skylights, fire escapes, exits, or windows.

All of the above conditions should be investigated and reported. If they are very dangerous they should be guarded while immediate word is sent to a superior.

Should the officer on duty observe any action or condition outside of plant jurisdiction that would call for investigation, he should report such events to his superior. In the event the condition is serious and calls for immediate action, the officer, if he has no "'special officer" rights, should act as a "citizen."

All cars, trucks or persons strange to the area or acting suspiciously should be stopped and looked over. After stopping a car or a truck, the make, ownership, occupants, license numbers and reasons for being there should be noted in the notebook.

If the officer suspects the presence of an unauthorized person on the premises, he should treat him as such. The guard should be careful not to make any noise or in any way frighten or startle the intruder. Help should be requested, and preparations made to use any force which may be required. After help is called, the guard should try to cover the possible escape routes of the criminal.

The search of the area should be made with great care for the safety of the guards. Weapons should be at the quick position. The search should be from one side of the area to the other. All objects that could be used as a cover should be noted. Overhead exposed beams offer a convenient and excellent hideaway. Behind large cartons, under desks and behind doors also make convenient spots to hide.

Use light in conducting the search. It will search out the intruder and destroy his greatest asset, the concealment of darkness. Flashlights should be held in front at arms length.

If the suspect is located, attempt to blind him by placing the beam directly in his eyes. While informing the suspect that he is under arrest, make certain that his hands are free and that he can in no way cause harm or attack. Order him to raise his hands above his head and proceed toward the nearest wall.

Have the subject face a wall, spread his feet far apart, and place his hands far apart against the wall. At this position direct him to step back a step with his head down. This will place him in an off balance position.

The searching officer with gun drawn, but not in cocked position, keeps it pointed at the subject. With the hand not holding the revolver, he frisks the subject. The subject may be dropped to the ground if he should offer any resistance. By placing one of the guard's feet in front of the subject, only a slight kick against his foot will drop him.

With the free hand (the other holding the revolver), remove the subject's hat, feeling through the sweat band and lining for any weapon that may be brought into use. Many experienced police officers employ just the sense of touch in frisking the suspect. They first feel the half of the body nearest them; then they cover the other half. This permits the searcher to keep the weapon hand free and his eyes at all time on the movements of the subject.

In order not to overlook any possible "hides" the frisking is done in a systematic procedure. Proceed with the head and work down to the feet of one side, with the same order being followed with the other side of the body. Never "pat" for a concealed object; knives are often ground flat and thin with this thought in mind. Commence with the shoulder blades, working down the back to the waist, the waist band, secret pockets that may be sewn in the band, upwards over the belly, armpit down the arm to the tips of the fingers. Do not overlook the possibility of a concealed object in the palm of the hand pressed against the wall. If the subject is wearing a jacket or coat, do not expose yourself by offering him a degree of movement in having him remove it. It is then necessary to make a careful search of these garments. In addition to the normal pockets that such outer garments contain, make certain that your hand covers every inch of the material, the lapels, the sleeves, tie, pockets that may have been sewn in other unsuspected places such as back through the linings. After the upper portion has been thoroughly covered, proceed to the lower extremities. Next feel all the material of the trousers with enough pressure of the hand so that the sense of feeling penetrates to the flesh. Never overlook the usual pockets and any additional ones. The groin, right up to the privates, will in many instances uncover a weapon. Feel the leg thoroughly right to the inside of the sock and shoe. Cuffs of pants are an unsuspected place of concealment. When the one side of the extremities is searched, proceed in like order to the other. If another officer is assisting, make certain to keep clear of his firing line. The covering officer always takes a stance from the opposite side.

RESERVES

The individual guard on post has little effectiveness unless he knows that he can call for and receive support. Whether the reserve consists of a single supervisor, a radio patrol, or a platoon or armed men, it should be ready for immediate response when the need arises.

Reserve strengths will vary with installations. Emergency plans should include provision for increased reserve forces in event of emergency conditions.

GUARD HEADQUARTERS

The location of the guard headquarters varies with the size and layout of the installation. The objective is efficient control of the guard force, and adequate and prompt security of vital activities.

The guard headquarters should be the control point for all matters pertaining to physical security of the installation. The headquarters should contain the control equipment and instruments of all alarm, warning and guard communications systems and should have direct communication with municipal fire and police headquarters.

A list of telephone numbers for use in emergency should be in guard headquarters. Written records covering all orders and assignments should be maintained.

Control Station

When the perimeter enclosure is very large, additional guard control stations should be maintained.

VULNERABILITY TEST

A vulnerability test is a means of pointing out weaknesses in the security system. Additionally, it gives security an idea of the vulnerability of the installation, tests the effectiveness of the guard force and other personnel, alerts people to the techniques that could be applied by an enemy, and provides material for corrective instruction. It consists of penetration or attempted penetration of a restricted area by means of deception.

The person or persons to make the test should be unknown to the guard personnel of the installation. A simple disguise, such as a guard uniform or a set of electrician's tools and appropriate clothing, false or altered credentials, a plausible story, and a confident attitude are sufficient for the purpose. The tester should also carry an article, such as a simulated sabotage bomb, that can be left within the protected area to prove a successful penetration.

The tester must be thoroughly trained and the test carefully planned. He must play the part he assumes, but must stop short of physical violence. He must not resist apprehension because it is the gullibility of the guards and not their physical powers that are being tested.

WRITTEN REPORT

Written reports should be required concerning guard activities. These should be prepared by each guard and delivered to a chief of guards for necessary action.

Duty Log

A continuous written record of all guard force activity, including details of any matters or occurrences having a bearing on the security of the facility, shall be maintained for each shift or day and retained for reference for an appropriate priod.

LIMITATIONS OF GUARD FUNCTIONS

Guards should have no firefighting or other duties. Such emergencies offer an excellent diversion to cover the entrance of a saboteur. Consequently, during such times guards should be more than normally alert in the performance of their guard duties.

It cannot be too strongly emphasized that guards are for guard duties and should not be given other functions except in small plants of insufficient size to warrant even one full-time guard.

EMERGENCY PLANS

A simple but effective plan of operation should be worked out for the general force to meet every foreseeable emergency. Practice alarms (like fire drills) should be run from time to time to test the effectiveness of this plan and the understanding of it by the guard force. Such plans should be particularly designed to prevent a ruse at one point in the facility drawing off the guards and distracting their attention from another section of the facility where unauthorized entry may be made.

At all facilities employing guard forces, plans shall be established and maintained to provide protection in the event of a strike or walkout by guard personnel. These plans shall include, as may be required and to the extent available, assistance of city, county or state police organizations.

DOGS

Dogs may be used to supplement the guards' perception. Indeed, sentry dogs are valuable assets to the security force since their extreme sensitivity to odor and sound alerts handlers to intruders at ranges of 150 to 200 yards during hours of darkness and under adverse weather conditions. While the dogs may be used at any hour of the day, nighttime is the dog's natural hunting time, and he will usually be more inquisitive and alert when working at night than during the day.

There are other reasons why dogs are of great value in security work. First, dogs, unlike human beings, are considered to be incorruptible and completely loyal. Second, guards who work with dogs are given a sense of security because of the animal's ability to sense danger and warn the guard long before he would know of it. Dogs also produce a psychological effect on the potential criminal, for he must not only fear their ability to sense his presence, but their ability to outdistance him and thus make his capture more likely. In addition, there are many who have a greater fear of the slashing teeth of a growling dog than of the fists or club of a guard. All of this gives the security guards aided by a dog a distinct psychological advantage. Thus dogs will generally offset any numerical advantage enjoyed by intruders and will enable the security guard to apprehend violators who might have otherwise escaped detection.

One point that must be stressed is that guard dogs should not be used unless they have been specifically trained. They have been trained to do many things in military and police work, but the types that have been most useful in physical security operations are "attack dogs" and "sentry dogs." Each dog is trained to work with one person and to dislike all others. An "attack dog" will run down and attack any person other than its handler. These dogs are easy to train, but due to their viciousness should be used where extreme security is considered essential. The "sentry dog" is essentially a detector. He becomes agitated at the presence of anyone except his handler and is trained to apprehend and hold an intruder without otherwise attaching him unless the intruder resists or attacks the handler. Sentry dogs should be trained to disarm violators on command, to climb ladders and other obstacles, and perform other aggressive feats to protect his handler.

Chapter 5

SECURITY AREAS

INTRODUCTION

SECURITY AREAS SHALL be established when the nature, size, revealing characteristics, sensitivity or importance of the security interests are such that access to them cannot otherwise be effectively controlled.

TYPES OF SECURITY AREAS

The first step in the development of the plan is to classify the areas with regard to their degree of security importance.

RESTRICTED AREAS. Restricted areas are considered to be those areas which must be subjected to special restrictions and controls for reasons of security. The reasons for security may include the safeguarding of property and material. Access to restricted areas is characteristically rigidly controlled.

LIMITED AREAS. A limited area is a restricted area which contains a security interest or other matter, in which uncontrolled movement will permit access to such security interest or matter, but within which access may be prevented by escort and other internal restrictions and controls.

EXCLUSION AREAS. Exclusion areas are defined as restricted areas which contain an item of security interest which is of such a nature that access to the area constitutes access to the item. They may also be defined as areas containing items of security interest which are of such vital importance that the proximity resulting from access to the area is equivalent to access to the item.

When possible, consideration must be given to the possibility of locating operations of like sensitivity in the same restricted area. Only those measures should be established which, because of the revealing characteristics, sensitivity, or importance of the area, are required to protect the installation. Care must be exercised that the restrictive measures do not impede the operation of the installation.

ESTABLISHMENT OF RESTRICTED AREAS

Areas within a facility which are vital to the continued operation of the plant or to the production of an essential product, or areas which are especially vulnerable to damage, should be provided special protection and movement of personnel therein restricted. Management should establish such areas as restricted areas and access to them should be limited to authorized personnel.

Normally, areas established as restricted areas should be physically separated from adjacent nonrestricted areas. Warning signs should be posted on such enclosures or boundaries forbidding entry of unauthorized personnel.

Entrances to restricted areas should be limited to the minimum number necessary for the safe and efficient operation of the area. Entry thereto should be controlled by guards or other supervisory personnel. Only persons who are recognized as authorized to enter the restricted area or persons bearing an identification badge valid for that area should be allowed to enter. Persons who are required to enter the restricted area due to an emergency may do so if escorted or if they carry the written authorization of the security officer. Such authorization should state reason for visit and the time limit permitted in the area.

Lighting

Restricted areas should be adequately lighted and so devoid of deep shadows that a person in the area may be easily recognized.

Electric Power Supply

Protection of on-the-premises electric power generating stations and substations should be provided commensurate with their importance to the continued production of a plant's critical items and their susceptibility to sabotage.

Protection of Power Generating Units

The plant's own power generating units, including steam boiler and hydroelectric installations, should be established as restricted areas and only authorized personnel should be permitted access thereto. In addition, special protection should be afforded, such as the following:

1. Use of submerged torpedo nets and floating booms to protect the forebays of hydroelectric plants or the condenser water intakes of steam-electric plants against torpedo type bombing and floating explosives.

2. Provision of an auxiliary diesel engine driven set or other suitable means for use in starting up a cold steam-electric plant without benefit of transmission interconnections to outside sources of energy.

3. Increase of fuel supply storage, increase of protection for fuel in storage and provision for the use of alternate types of fuels.

Protection of Power Substations

Those substations on the plant premises, whether owned by the facility or by a public utility company, which supply all of the electric energy used at the plant should be restricted areas and only accessible to authorized personnel. When a facility's production is dependent on electric energy, the substations are more critical than individual transformers and should receive protection equivalent to their importance. If the substation is an off-site one, it should be given protection equal to that of sensitive points within the facility's perimeter. If the off-site substation is owned by a public utility company, officials of that company should be encouraged to protect it properly.

Transformer Installations

Transformer installations should be well protected. Transformer installations located on a facility's premises should be restricted areas, should be fenced or screened and openings therein locked, and should be included in the itinerary of the guard patrol. Buildings in which transformers are located should be locked and only authorized personnel permitted entry.

Transformer enclosures should be lighted at night unless complete darkness is required to prevent detection of location. Transformers are vulnerable to rifle fire and should be adequately shielded by sand bagging or other protective measures when required. When enclosed in buildings or shielded adequately, ventilation should be provided.

Transformer enclosures should be kept free of debris, weeds and grass. Large transformers should be equipped with electric alarm pressure and temperature gauges to give warning of an injurious condition within the transformer.

Oil-filled transformers within buildings should be in safe locations, should be well drained and provided with curbed pits for the collection of oil. Such transformer sites should be provided with a foam, dry powder or carbon dioxide type of fire extinguisher of sufficient capacity to control an oil fire originating thereat. Oil-filled transformers outside should be at sufficient distances from buildings to minimize damage to the facility in event of fire. Patrols should be on the alert for oil losses resulting from rifle fire and other causes.

The security of transformers located on poles near the property lines or within the plant area should be frequently checked by guard patrols.

Electric Power Transmission

Power lines, terminals and switches and controls located on the premises of an industrial plant should be provided adequate protection to assure the continuous and uninterrupted flow of electric energy to the plant. Exposed overhead electric lines over 600 feet long entering or leaving power house substations or transformer banks should be protected with lightning arrestors. When justified economically, underground power lines should be used. When electric power cables are in close proximity, they should be protectively separated by nonconducting and nonflammable materials.

The main switches, power, terminals and power controls

should be located in restricted areas and included in the same protection plans as transformers and electric power substations since they are usually at the same location. Secondary switches located in operating departments should be readily accessible so that power can be shut off in an emergency.

Communication Centers and Equipment

The communication center and allied communication equipment which is essential to the operation of the plant should be adequately protected to prevent sabotage and tampering. The telephone exchange, the teletype and/or the short wave radio room and such other control centers as the guard headquarters should be restricted areas. When not individually manned, each such installation should be securely locked or guarded. Dispersion within the plant of the several means of communication, such as teletype, telephone switchboard, public address system and short wave radio may be advisable to afford adequate protection.

Frequently, communication installation areas are used individually or jointly as an emergency control center; as such, they are of the most vital importance and should be adequately protected. When an emergency control station is established in a shelter area for use in case of air attack, its communication equipment should be a restricted area and protected against tampering or unauthorized entry.

Valves and Regulators

Main control valves and regulators should be protected to prevent tampering and unauthorized manipulation, but should be accessible to authorized personnel for emergency use. Locations should be restricted areas and enclosed. If exposed, such valves should be locked in the position required for normal operations. Manholes and pits containing control valves should be secured by covers locked in place. Equipment of control valves with electric signalling devices should be considered as an additional means of protection against tampering.

Gas valves and regulators should be within a noncombustible locked enclosure, adequately ventilated to prevent accumulation of gas and provided with vapor-proof electric equipment. Gas valves used infrequently should be locked or sealed. Guards should check the enclosures and the valves and regulators periodically.

Water Tanks and Equipment

Water tanks, water pumps and allied equipment essential to the operation and production of the plant should be afforded adequate protection to prevent curtailment, contamination or damage of the water supply.

Water tanks, pumping stations, pumps and equipment should be within restricted areas and checked frequently by guards. Water tanks, pumps and equipment preferably should have an electrical supervision system to check water supply and indicate water failure. Elevated water tanks should be fenced and the gates locked. Roof tanks should be secured by screening or locked roof doors.

Pumping equipment located in a building or room should be protected by having all doors and windows of the enclosure adequately barred or locked. Pumps should have more than a single source of power for their operation.

PROTECTION OF CRITICAL PRODUCTION EQUIPMENT AND AREAS

Critical production equipment, industrial processes and critical production areas should be protected from sabotage, theft or other irreparable damage which could cause a production slowdown or work stoppage. The saboteur will know the location of the vital portions of the targets which he wishes to attack. Therefore, if at all possible, shift the vital targets around and camouflage and disguise vital points of installations so that a sabotage agent, briefed according to an old plan, will be unfamiliar with the new.

Production equipment, such as dies, jibs and patterns whose loss or damage would seriously affect production should be stored in restricted areas. Control of these items during working hours should be the responsibility of selected personnel, and such equipment should be locked up when unattended. Combustible equipment should be stored in a sprinkler protected area. High temperature process equipment, such as chemical baths, and melting furnaces require additional protection including suitable barriers and overriding thermostatic safety controls and trained operators on constant duty during the operating period.

Areas containing special equipment or machinery, the loss of which would cripple or stop production of one or more critical products, should be carefully studied and receive priority protective treatment. Areas and equipment should be limited to those which are in themselves relatively critical, such as special presses, stills, furnaces, control equipment, assembly lines and similar items of special construction. Systems of industrial process supervision should be considered for their effectiveness in preventing interruptions in processing or manufacturing.

Warehouses, tool cribs and stockrooms should be established as restricted areas and rigid control established on the issuance and use of stocks therefrom.

Frequent inspection of machinery is also a valuable deterrent. The following will prove of value:

1. Check lubrication, pipes and valves.
2. Check bearing surfaces of machinery for indications of wear due to use of abrasives.
3. Check to see that low pressure oil is not substituted for high pressure oil in the compression cylinders of pumps.
4. Check for liquids in cylinder heads of air compression pumps.
5. Check cooling systems on all water-cooled engines.
6. Check filters and strainers of fuel and lubrication systems.
7. Check oil and oil sumps for presence of abrasives.

Certain special precautions for protection of machinery are the following:

1. Make certain that no tools are left on machines. Tool rooms, under supervision, should be provided for tools not in use.
2. Keep machine and engine cover plates firmly bolted or provide some locking device for them.
3. Make certain that no cotton waste, oil cans, tool boxes, lunch boxes, or receptacles of any kind are left on machines when not in use.

4. Examine regularly and carefully cast-iron surfaces of machinery.
5. Fix locks on valves, strainers, or filters, making them difficult to remove or turn.

PROTECTION OF MANUFACTURING DATA

Plant management should take the required steps to protect essential engineering data, plans and specifications from loss or damage.

Engineering Data

Plans, specifications and experimental data should be used and stored within restricted areas and protected from loss by fire, theft, espionage and sabotage. Data which are vital to production should be copied on microfilm and stored at some removed or off-site secured area.

PROTECTION OF DOCKS, DRYDOCKS AND LOCKS

The following protective steps should be taken:
1. Floodlight locks.
2. Protect pumps, hydraulic equipment and rams.
3. Guard high water side of locks; heelpost can be effectively attacked from this side by underwater explosive charges.
4. Spot weld nuts on all manhole covers if gates are double skinned or of the caisson type.
5. Extend walks on lock gates beyond heelpost to prevent access to heelpost and heelpost strap.
6. Prevent barges or boats from tying up for the night near the locks.
7. Screen bypass openings. An explosive charge placed into bypass openings will blow away or badly damage underpinning framework.

DISPERSION OF PRODUCTION

Alternate locations, either within or without the plant, should be considered for critical production.

The principle of dispersion of production should be adapted to the individual characteristics of the plant. When production is dependent on a limited number of specialized machines, management should consider the possibility of segregating or isolating individual units so that all production would not cease if a single or limited area of the plant were damaged.

The same principle may be accomplished even more effectively by splitting the production of a critical item among two or more facilities operated by the same company but located at a considerable distance from each other.

STORAGE

Open Storage

Whenever possible, open storage is limited to those items of supply not subject to damage by weather conditions. Sound physical security practice requires that stacks and lines of equipment be a minimum of fifty feet from the perimeter barrier that they be as symmetrical as possible, and that the aisles be wide and straight. All cases should be tightly and neatly stacked so that any attempt to introduce incendiaries will be noticeable. These factors contribute to visibility by the guards, as variations from an even pattern attract attention. Additional security may be provided by wide fire lanes between subareas, and by stocking each subarea with a proper proportion of the different items stored. These are simple methods of giving maximum protection against the complete destruction of any one item.

Fixed-position lights in a storage area should be of a diffused type that does not cast deep shadows. Each guard patrol should be furnished with a flashlight.

Covered Storage

The same principles of even stacking and adequate aisle space used in open storage are applicable to covered storage. In case of stockpile storage where stored items are infrequently moved, the stacks may be placed to conform with existing lighting, or the lighting may be arranged after the stacks are in place. The objective is to reduce deeply shadowed areas to a minimum. In warehouses where the stored items are moved frequently, more emphasis should be placed on security guards than upon structural or mechanical protection especially during working hours.

Classified and sensitive items in storage should be kept separate from other material. Sensitive items are those that can be pilfered and disposed of easily. The most satisfactory method is to store such items in a separate building with a higher degree of physical protection. Where a separate building is not available, or where its use is not warranted by the quantity of classified or sensitive storage, a room, cage, or crib may be constructed within a warehouse building. The floors and roof of the enclosure must be of comparable strength to the walls. The walls must run from the floor to the ceiling, rather than only part of the distance. All windows should be checked periodically to insure that they have not been opened to provide a draft. Likewise fire-fighting equipment should be checked, as any indication of tampering should serve as an alert for a possible sabotage attempt. At an issue warehouse, back-to-back storage bins and a counter separating the material from persons receiving it will minimize pilferage.

Improve Inventory and Auditing Procedures

In many companies, the losses due to theft may be covered up by inadequate inventory and auditing procedures. If the installation of a tight company-wide inventory control system costs an excessive amount of money, it may be possible to establish controls on a test basis in certain departments or areas in order to ascertain whether or not the resulting saving pays its own way. Unless this is done, management will continue to remain in the dark concerning its actual theft expense.

TOOLS AND SUPPLIES

Supplies of all tools and equipment should be kept under lock and key in a tool crib or supply shop with one person designated as having the authority and responsibility for control of such property. These areas should be designed and constructed to prevent the unauthorized entry of personnel or the removal of any item contained therein without proper approval. Company tools should be distinctly marked by the use of special dies or stampings and should be checked out individually to company personnel in exchange for tool checks or other means of identification. It is the practice in some plants that all tools and equipment used during the shift be checked back into the tool crib at the end of the shift. This is a desirable practice unless there are existing circumstances which prevent it. Periodic inventories of all tools and equipment charged out to departments or remaining on hand in supply should be a regular practice. In addition to regular inventories, frequent checks should be made to see that each employee still retains those tools charged out to him which have not been returned. Supplies of other materials should likewise be carefully supervised in order to ascertain that there is no excessive damage or waste and that some of the items do not find their way outside the company premises.

The control over tools used by employees on company jobs should be especially accurate in cases where the workmen supply many of their own small tools. A complete written inventory of a workman's personally owned tools should be made on the first day he reports for work. This list should be checked against his tools when he leaves the company.

Outside contractors may also intentionally or unintentionally remove company owned property or actually exchange some of their own used equipment in better condition. As in the case of company personnel, outside contractors should be checked off the premises by security personnel or some other dependable person who has sufficient knowledge to identify company property.

TRASH, SCRAP AND SALVAGE CONTROL

Unless properly controlled, valuable company property may be lost through the disposal of salvage merchandise to junk dealers or outlet buyers, trash that is to be burned, or scrap metals or other material which is to be sold by weight or returned to the smelter for reprocessing. Unless the personnel who handle salvage materials and trash are properly supervised, new merchandise may intentionally or unintentionally find its way out of the plant in the trucks or tubs conveying this type of material. Supervisors and security guards should check constantly on the disposal of this material and should make sure that all new or useable merchandise recovered in this manner is promptly returned to stock.

The disposal of scrap should also be closely controlled. Losses to the company may occur through the following: (1) the substitution of one material for another on the load; (2) deliberately shorting the weight of the load by lightening the truck between the time of weighing empty and weighing loaded, to benefit the scrap buyer, and (3) fraudulently disposing of part of the scrap between the time of loading and the time of weighing.

Protective Procedure

1. Supervise the loading, weighing and disposition of valuable salvage or scrap.
2. Check scrap dumps at unannounced times during the period when company trucks containing scrap or refuse are unloading.
3. Fence in the company dump.
4. Make provision for selling or giving certain kinds of scrap or salvage to employees for their own use.
5. Make sure that unrepairable or defective merchandise or parts are effectively destroyed so that they cannot be returned for credit.
6. Make sure that broken or scrapped tools are effectively destroyed so that employees cannot turn them in for good tool replacements.
7. Check the trash-burning operation regularly to make certain that no valuable property is hidden in the rubbish trucks when removing trash from the premises.

Chapter 6

BARRIERS

PURPOSE

F ENCES AND OTHER antipersonnel barriers are the physical media by which the boundaries are physically defined for protection and control. The basic purpose of a physical barrier is to deny or impede access to security areas by unauthorized persons. Additionally it facilitates effective and economical utilization of guards and directs the flow of personnel and vehicles through designated portals.

In establishing physical barriers, special consideration must be given so that neither operating efficiency will be sacrificed through lack of planning, nor will essential barriers be sacrificed for operating expedience. In some instances, the temporary nature of the security interest makes the construction of costly physical barriers impracticable and unjustifiable. In such cases, the security interest may be protected by other means such as additional guard forces, patrols and other compensating protective measures.

TYPES OF BARRIERS

There are two general types of physical barriers — natural and structural. Natural barriers include rivers, seas, cliffs, canyons, or other terrain difficult to traverse. Structural barriers are manmade devices such as fences, walls, floors, roofs, grills, bars, roadblocks, or other structures which deter penetration. Physical barriers will delay but will not stop a determined intruder. Therefore, such barriers to be fully effective should be augmented by guard personnel. Although natural barriers aid in maintaining security, few, if any, present a sufficient barrier in themselves but need the suport of structural and human aids to realize their maximum potential as deterrents to unauthorized entry. Thus, a body of water, regardless of its depth, is not an adequate perimeter barrier. Additional security measures must be provided that portion of the perimeter bordering the water. This can take the form of fencing, lighting, or guard patrols. If the water is navigable and shipping and receiving docks are a part of the installation, the entire area should be fenced off from the rest of the installation with access only through controlled gates. To be fully effective, barriers must be under the surveillance of guards.

PERIMETER BARRIERS

A perimeter barrier defines the physical limits of a protected area and restricts or impedes access to the area by unauthorized persons. It performs the following functions:

1. Creates a physical and psychological deterrent to entry into the area.
2. Delays intrusion into an area and aids in the detection and apprehension of intruders by guards.
3. Facilitates the effective utilization of guard forces.
4. Channels the flow of persons and vehicles through designated entrances.

All buildings containing essential manufacturing, production and storage areas, all allied open storage and all work shops and utilities serving as appendages to the operation of the facility should be completely enclosed within perimeter barriers of man proof design. Administration buildings and vehicle parking areas may be located outside such enclosures; and when so located, entry to the enclosed area should be by means of controlled pedestrian gates.

Selection of Proper Type of Barrier

The type of barrier used should be determined after a study of local conditions. In cases of extreme criticality and vulnerability of a facility, it may be desirable to establish two lines of physical barriers at the facility or restricted area perimeter. Such barriers should be separated by not less than fifteen feet and not more than 150 feet for optimum protection and control.

FENCES

The three types of fencing normally used to protect a security area are chain link, barbed wire and concertina. Choice of the type or combination of types is dependent upon the individual situation and upon the funds and time available for construction. Specific information on the use of the three types of fencing follows:

Chain Link

Wire fences should be of chain link design, mesh openings not larger than two inches square, and made of No. 9 gauge or heavier wire with twisted and barbed selvage top and bottom. The minimum height of the chain link portion of the fence should be seven feet. Fences should extend to within two inches of firm ground or below the surface if the soil is sandy and easily windblown or shifted. Fence mesh should be drawn taut and securely fastened to rigid metal posts set in concrete, with additional bracing as necessary at corners and gate openings.

Wire fences should be topped with a 45° outward and upward extending arm bearing three strands of barbed wire stretch taut and so spaced as to increase the vertical height of the fence by approximately one foot.

Fences should be provided with culverts, troughs, or other openings where necessary to prevent washouts in the perimeter barrier and to permit carry-off of excessiv drainage and small streams. Such openings, when larger than ninety-six square inches in area, should be provided with physical barriers equivalent in protective capabilities to those of the perimeter barrier and so designed as to minimize impedance to water runoff.

Barbed Wire

Standard barbed wire is twisted double-strand, No. 12 gauge wire, with four-point barbs spaced four inches apart. Barbed-wire fencing including gates intended to prevent human trespassing should be not less than seven feet in height plus top guard and should be firmly affixed to posts not more than six feet apart. Distance between strands should not exceed six inches. Where the fencing is also intended to exclude small animals, the bottom strand should be at ground level to impede tunneling, and the distance between strands should be two inches at the bottom gradually increasing toward the top.

Concertina

The standard concertina barbed wire is a commercially manufactured wire coil of high strength steel barbed wire clipped together at intervals to form a cylinder weighing fifty-five pounds. Opened, it is fifty feet long and three feet in diameter. It can be laid quickly and used repeatedly because of its elasticity. It is more difficult to cut than standard barbed wire. When used as the perimeter barrier for a security area, concertina should be laid with one roll on top of another, making a total height of approximately six feet, or in a "pyramid" with two rolls on the bottom and one on the top. In pyramid arrangement, it is the most difficult of all fencing to penetrate. Ends should be staggered or fastened together and the base wires picketed to the ground. Because it is unsightly and hampers ground maintenance, it is seldom used at permanent installations except as a temporary expedient when the regular fencing has been breeched, or as a temporary barrier.

TOP GUARD

A vertical fence does not present any particular difficulty to a trespasser with intent to cross it. For this reason a top guard should be constructed on all vertical perimeter fences, and on interior enclosures when added protection is desirable. A top guard is an overhang of barbed wire along the top of a fence, facing outward and upward at an angle of 45°. It may be on both sides, the Y type, or only on the outside. The supporting arms are affixed to the top of the fence posts and are of sufficient length to increase the overall height of the fence at least one foot. Three or four strands of standard barbed wire, spaced six inches apart, are usual, but the length of the supporting arm and the number of strands of wire can be increased when required. Where a building of less than three stories forms a part of the perimeter, a top guard should be used along the outside coping to deny access to the roof.

During the war in Europe, captured German sabotage agents stated that the most difficult type of fence to scale was that with corrugated sheet iron at the top. This type of fence was difficult because it was noisy.

WALLS

When masonry walls form a part of the perimeter or where they are used to prevent observation from the outside, they should be at least seven feet high and should be surmounted with a barbed-wire top guard in the same manner as a chain link fence. An alternate, but less satisfactory, method for preventing scaling is to set broken glass on edge in concrete along the top of the wall. Where this method is used, the minimum height of the wall should be eight feet. Where a fence joins a masonry wall as part of a perimeter barrier, the height of the fence should be gradually increased to the overall height of the wall.

BUILDINGS

Building walls which form a part of the perimeter barrier should provide protection equivalent to that provided by the rest of the barrier. Where a fence joins a building wall, it should extend to within two inches of the building wall. Under some circumstances it may be desirable to double the height of the fence gradually to the point where it joins the building.

Openings in Perimeter Barriers

Openings in the perimeter barrier should be limited to the number necessary for the efficient and safe operation of the facility. Due consideration should be given to providing emergency exits and entrances in case of fire. All openings should be constantly guarded, locked, or otherwise secured.

SECURITY OF WINDOWS

Windows and other openings which penetrate the perimeter barrier and have an area more than ninety-six square inches in area and over six inches in smallest dimension should be protected by securely fastened bars, grills, or other equivalent means, when located less than eighteen feet above the level of the ground outside the perimeter barrier or less than fourteen feet from structures outside the perimeter barrier.

SECURITY OF SIDEWALK ELEVATOR OPENINGS

Sidewalk elevators which provide access to areas within the perimeter barrier should be locked, guarded, or otherwise provided with security equivalent to that of the perimeter barrier.

SECURITY OF UTILITIES OPENINGS

Sewers, air and water intakes, exhaust tunnels and other utilities openings which penetrate the perimeter barrier and have a cross-section area of ninety-six square inches or greater should be protected by bars, grills, water-filled traps, or other structural means providing security equivalent to that of the perimeter barrier. Interior manhole covers should be secured to prevent unauthorized opening.

AUTHORIZED ENTRANCES

Active entrances to an installation must be so designed that the guard force maintains full control without unnecessary delay to traffic. This is largely a matter of sufficient entrances to accommodate the peak flow of both pedestrian and vehicular traffic, and adequate lighting for rapid but efficient inspection. When gates are not manned during nonduty hours, they should be securely locked, illuminated during hours of darkness, and periodically inspected by a roving guard patrol.

Semiactive entrances, such as extra gates for use during peak traffic flow and railroad siding gates, should be locked at all times when not guarded. The keys to such gates should be in the custody of the security officer or the chief of the guard force and should be strictly controlled.

Inactive gates, those used only occasionally, should be kept locked.

LOCKS, KEYS AND COMBINATIONS

A key or combination lock and padlock should be considered delay devices and not bars to entry. Although some types of locks require considerable time and expert manipulation for covert opening, all will succumb to force and the proper tools.

In selecting locks to be used for the various gates, building and storage areas, consideration should be given to the amount of security desired. A good tumbler-type lock is money well spent in comparison to cheaper disc-type locks or others so constructed that they may be opened with almost any small piece of metal or wire. A reliable locksmith should be consulted regarding the lock requirements of the plan prior to their purchase and installation.

In addition to padlocks being susceptible to manipulation and "picking," there is also the danger of an identical lock with a known key or combination being surreptitiously substituted for a lock in use. For this reason, padlocks should always be snapped shut on one of the locking eyes of a container while it is open.

It is essential that combinations or keys be accessible only to those persons whose official duties so require. Likewise, records containing a combination are placed in the same classification as the highest classification of material in the container, or are protected in a manner consistent with the value or sensitivity of the material in the container. A combination or lock and key should be changed when any person having knowledge thereof leaves an organization, at any time when there is reason to believe that such knowledge has been compromised, or at least every six months. The clearance of the person changing a combination or a lock and key must be as high as the highest classified material protected by the lock. The safest procedure is to assure that at least one person having authorized access has been instructed in the method of changing the combination.

Lock and Key Records

A written record should be provided for all company padlocks, to whom they are issued or where they are installed. Records should show the name of the person and department who carries or controls the keys for the various locks. The records should show not only the name of the person to whom the key is assigned, but also his department number and the date the key was issued. It is duty of the security department to inspect this record periodically to insure that it is kept up to date.

Key Depository

A key depository should be provided at installations where keys are secured during non-working hours. Keys that are issued daily and turned in at quitting time each night should be suitably marked with a metal tag denoting the department or location of use and stored in a metal cabinet or locker under the close supervision of security personnel. This key control cabinet should be of fireproof construction and should be kept locked at all times when not actually in use. Supervisors should be required to sign a register for the keys at the beginning of each working day and to turn in keys at end of working day. Guards should check the keyboard and register to insure that all keys are accounted for. Duplicate or reserve locks and keys should be stored in a safe place under the control of the security department.

Lost Keys

If keys are lost, an immediate investigation should be made in an attempt to locate them. A written record of this investigation should be kept by the security department, setting forth the circumstances surrounding the loss. If the lost key is for an outside door or gate or if it permits access to an important building area, steps should be taken to have the lock changed or replaced without delay and a notation to that effect made in the key record.

One method of helping an employee to be aware of the importance of a key is to notify him at the time it is assigned to him that he will be charged a specified amount, such as $5.00, if it is lost, and it must be replaced.

Master Keys

Master keys should have no markings that will identify them as such. They should be issued only to personnel especially designated by management, and the list of holders of these keys should be frequently reviewed to determine the continuing necessity for the individuals having them. When possible, sub-master keys should be used instead of master keys.

Loaned Keys

The security department should have a form to provide for the recording of all temporary loans to keys. This form should include the date, department and/or badge number of the person to whom the key was loaned and the date of return. Failure to return a loaned key at the specified time should be promptly investigated.

WARNING SIGNS

"No Trespassing" and warning signs should be posted on or adjacent to the perimeter at such intervals that at least one sign will be visible and legible at any approach to the perimeter.

Except at public entrances of an installation where special signs are required, it is advisable to limit the wording on perimeter boundary signs to "No Trespassing." Signs containing any more detailed prohibitions merely confuse the public and serve to focus attention on the area being protected. Such notice definitely defines the perimeter of the area, facilities control, and makes accidental intrusion inexcusable, putting the whole burden of explanation on the trespasser.

CLEAR ZONES

An unobstructed area should be maintained on both sides of the perimeter barrier. A clear zone of twenty feet or greater should exist between the perimeter barrier and exterior structures, parking areas and natural or cultural features which may provide concealment or assistance to a person seeking unauthorized entry. When a twenty-foot clear zone is not possible due to property lines, location of the facility, or adjacent structures, the perimeter barrier should be increased in height or otherwise designed to compensate for the proximity of such aids to concealment or access. When possible, a clear zone of fifty feet or greater should exist between the perimeter barrier and structures within the protected area, except when a building wall constitutes the perimeter barrier.

Clear zones on both sides of the perimeter barrier should be such as to provide an unobstructed view of the barrier and the ground adjacent thereto, and should be kept clear of weeds, rubbish, or other material capable of offering cover or assistance to an intruder attempting to climb, cut through or tunnel under the perimeter barrier.

When broken terrain or other unusual features provide cover for the approach of an intruder or a potential saboteur, guard patrols should frequently inspect the area of additional security should be provided.

PATROL

Any barrier to be effective must be under visual observation and periodic inspection. When the perimeter barrier encloses a large area, an interior, all-weather perimeter road should be provided for the use of the guard.

Reliance on guard towers for observation of a perimeter is usually considered unsatisfactory. The height of a tower increases the range of observation during daylight hours and at night with artificial illumination. During inclement weather and during blackout, towers lose their advantage and must be supplemented by on-the-ground observation. Psychologically, on one hand, the mere elevation of the observer has an unnerving effect on a potential intruder, while on the other hand, the isolation of the tower tends to reduce the alertness of the occupant. Mobile towers are useful in some temporary situations. All towers, however, must have a support force available for emergencies. Towers have a place under certain conditions but the expense of their construction and operation necessarily limits their use to only the most critical installations.

INTRUSION DETECTION DEVICES

Intrusion detection devices may be used in perimeter protection either alone or in conjunction with conventional fencing. They give only a warning and must be supported by an alert force or guards in radio equipped vehicles. Electrical or electronic systems normally operate on 110 volt source with an automatic self-contained battery power source for temporary use in the event of power failure. The installation of any electrical or electronic type system should be supervised by technical personnel to prelude interference with existing radio communications or other electronic equipment.

PROTECTION IN DEPTH

On a very large installation, it is obviously impracticable to construct an expensive perimeter fence and to keep it under the necessary observation. Such an installation is usually established in a scantily inhabited area. Its comparative isolation and the depth of the installation itself give reasonable perimeter protection. Under these circumstances the posting of warning notices, the reduction of access roads to a minimum, and periodic patrols in the area between the outer perimeter and the conventionally protected vital area of the installation may be sufficient.

MAINTENANCE

In addition to daily observation by perimeter guard patrols, perimeter barriers should be checked for damage or deterioration by maintenance crews at least every thirty days. Necessary repairs must be accomplished promptly.

Chapter 7

PROTECTIVE LIGHTING
INTRODUCTION

G ENERALLY, protective lighting is inexpensive to maintain, and when properly employed may reduce the necessity for additional guards and may provide present guards with personal protection by reducing the advantages of cover and surprise by a determined saboteur. Protective lighting systems provide a means of continuing, during hours of darkness, a degree of protection of a facility approaching that which is maintained during daylight hours. They serve this purpose (1) by enabling the guards and other employees to observe activities in the area and to take appropriate defensive action when necessary, and (2) by acting as a psychological deterrent to potential intruders.

Requirements for protective lighting at facilities will depend upon the situation and areas to be protected. Some of the factors to consider are the nature of the tools and products and their vulnerability to sabotage or theft. The nature of the area and its vulnerability because of shadows cast by building angles, stacks of material, or parked vehicles, and the location of the plant will affect the problem. Thus industrial buildings and yards located in built up areas within the city will usually have fewer perimeter lighting problems. Street lights on surrounding intersections and along the perimeter sidewalks may furnish adequate illumination for ordinary protection of the plant boundaries, leaving only the plant gates and interior areas to be provided with additional illumination. On the other hand, if plants are in isolated locations, complete illumination must be provided by management for access roads, as well as for adequate perimeter protection.

Further consideration must be given to the intensity or quantity of light and the location of luminaires that is best for

97

any particular activity. Each situation requires careful study to provide the best visibility that is practical for such guard duties as identification of badges and people at gates, inspection of vehicles, prevention of illegal entry, detection of intruders both outside and inside buildings and other structures, and inspection of unusual or suspicious circumstances.

TYPES OF PROTECTIVE LIGHTING

Continuous Lighting

This consists of a series of fixed luminaires arranged to flood continuously a given area with overlapping cones of light during the hours of darkness.

Standby Lighting

This system is similar to continuous lighting but is turned on manually or by intrusion detection devices or other automatic means.

Movable Lighting

This consists of stationary or portable manually operated searchlights. They may be lighted continuously during the hours of darkness or only as needed, and are usually a supplement to either of the types described above.

Emergency Lighting

This type of system may duplicate any or all of the above systems. Its use is limited to time of power failure or other emergencies which render the normal system inoperative.

TYPES OF LIGHT SOURCES

The two principal types of lamps used in protective lighting are the following:

Incandescent Lamps

These are common glass light bulbs of high wattage in which the light is produced by the resistance of a filament to an electric current. Special purpose bulbs are manufactured with interior coatings to reflect the light, with built-in lens to direct or diffuse the light, or the naked bulb can be enclosed in shade or fixture to give similar results.

Gaseous Discharge Lamps

Mercury vapor lamps emit a blue-green light caused by an electric current passing through a tube of conducting and luminous gas. It is more efficient than an incandescent lamp of comparable wattage and is in widespread use for interior and exterior lighting, especially where people are working.

Sodium vapor lamps are constructed in the same general principle as mercury vapor lamps but use metallic sodium instead of mercury and emit a golden yellow glow. They are more efficient than the other two types and where the color is acceptable, such as on streets, roads and bridges, they are widely used.

The use of gaseous discharge lamps in protective lighting is somewhat limited because they require, when cold, a period of two to five minutes to snap on and, when hot, a slightly longer period to relight after a power interruption.

TYPES OF LIGHTING UNITS

Several types of lighting units are adaptable to protective lighting systems. They are the following:

1. The reflection type in which a parabolic mirror directs the light required. They are available in narrow, focused beam, spotlight types, and in wide angle types up to 180°.
2. Pendant lighting units with a refractory lens designed to direct most of the light in the direction needed.
3. Combinations of types 1 and 2 above.

Generator or battery powered portable and/or stationary lights should be available at key control points for use of guards in case of a complete failure which renders even the secondary power supply ineffective. When vapor type lights are used, provision should be made for emergency lighting of the entire perimeter since the return to full illumination of vapor type lights is slow when the light circuit is broken. Switches and controls for emergency lights should be located in locked enclosures and should come under the supervision of the chief guard on duty.

POWER SOURCES

Usually the primary power source at a facility is a local public utility. As installation control seldom extends beyond the perimeter, the interest of the installation guard force begins at the points which power feeder lines enter the installation. To compensate for this gap in security, an alternate source of power should be provided to afford automatic illumination during failure of normal power source.

All facilities or restricted areas provided with protective lighting should also have an emergency lighting system and a secondary source of power provided by generator equipment or batteries located within the restricted area. The standby source should be adequate to sustain the protective lighting of all vital areas and structures, and should be arranged to go into operation automatically in the event of failure of the primary power. There should also be generator or battery-powered portable and/or stationary lights available at key control points for use of guards in case of a complete failure which renders even the secondary power supply ineffective.

WIRING SYSTEMS

Both multiple and series circuits may be used to advantage in protective lighting systems, depending upon the type of luminary used and other design features of the system. The circuit should be so arranged that failure of any one lamp will not darken a long section of a restricted area perimeter line or a major segment of a critical or vulnerable position. Connections should be such that normal interruptions caused by overloads, industrial accidents and building or brush fires will not interrupt the protective system. In addition, feeder lines should be located underground or sufficiently inside the perimeter, in the case of overhead wiring, to minimize the possibility of sabotage or vandalism of feeder lines from outside the perimeter barrier. Light fixtures should not be easily accessible and normally should only be accessible from within the perimeter barrier.

The design should provide for simplicity and economy in system maintenance and should require a minimum of shutdowns for routine repairs, cleaning and lamp replacement.

MAINTENANCE

Periodic inspections should be made of all electrical circuits to replace or repair worn parts, tighten connections and check insulation. Luminaires should be kept clean and properly aimed. Lamps of the protective system should be scheduled for replacement at about 80 per cent of their rated life. Lamps so replaced can be used in less sensitive locations. This will reduce the loss of lighting protection by reducing the number of burned out lamps.

GENERAL CONSIDERATIONS

Protective lighting is not used merely as a psychological deterrent. It serves the important function of aiding the guard force in detecting intrusion and in apprehending the invader. To best achieve this, the following general considerations should receive attention at fixed luminary installations:

1. Insofar as possible, the cone of illumination from lighting units should be directed downward and away from the structure or area protected and away from the guard personnel assigned to such protection. The lighting should be arranged to create a minimum of shadows and a minimum glare in the eyes of guard personnel.

2. Lighting units for perimeter restricted area fence lighting should be located a sufficient distance within the protected area and above the fence so that the light pattern on the ground will include an area on both inside and outside of the fence. Generally, the light band should illuminate the restricted area barrier and extend as deep as possible into the approach area. Adjacent waterways, highways, railroads, residences, etc., may limit the depth of the light band.

Shadowed areas caused by structures within, or adjacent to, security areas should, when deemed necessary, be illuminated to the extent required to preclude easy concealment of an intruder. Lighting equipment, poles and electrical auxiliaries should be located inside the property fence, when practicable, or where they are not readily accessible to intentional damage. Protective lighting systems should be designed to provide overlapping light distribution, and equipment selected should offer optimum resistance to weather, corrosion or mechanical damage.

Consideration should be given to color contrast and its effect on visibility. There should be contrasting backgrounds in areas of the plant where guards on night duty are charged with the responsibility of detecting the movement of any intruder. This contrast may be achieved by painting stripes on the lower part of the buildings, by contrasting colors on sidewalk areas, and by the proper direction and intensity of illumination. This color contrast will enable the guards to distinguish detail better and thus will improve their ability to discover any intruder.

IMPLEMENTATION

The objectives to be achieved in providing protective lighting for boundaries, entrances, structures and areas are prescribed below. Unless otherwise stated, references to footcandles pertain to the minimum horizontal illumination at ground level. To attain these objectives, types of luminaires, height of mounting and spacing of units which will produce the recommended footcandle values are listed in the Protective Lighting Guides.

Fenced Boundaries

For effective planning of perimeter lighting systems, fence lines are categorized according to the depth of the approach area and the nature of the normal activities within this approach area.

1. Isolated fenced boundaries are those having a wide approach area, clear of obstruction for one hundred or more feet outside the fence. Acceptable light for these boundaries may be provided by conventional street lighting and floodlighting luminaires or glare projection systems. The application of a glare projection system is suitable only where there are no adjacent roads, railroads, navigable water or other properties which may suffer a glare hazard. If a glare projection system is selected, Fresnel units mounted on individual poles are desirable

when the fence line is some distance from adjacent structures; however, if suitable structures are available, floodlights can be mounted on top of the structures. With the proper luminaires a glare projection system should illuminate a strip from ten feet inside to 200 feet (where feasible) outside the fence line, with the portion of the strip from ten feet wide inside to twenty-five feet outside the fence line receiving enough spill light to assure ready detection of intruders. Since the use of a glare projection system is predicated on the desire to provide glare in the eyes of an intruder and make him readily visible to a guard through the use of high intensity horizontal beams, the footcandle value in the strip twenty-five feet outside the fence line to the 200 foot limit must be high and should be 0.07 minimum at any point, measured on a vertical plane three feet above ground and parallel to the fence. If conventional street lighting or floodlighting luminaires are used, a strip eighty feet wide, extending from ten feet inside to seventy feet outside the fence line, should receive 0.02 footcandle minimum at any point. For the more critical security areas 0.04 footcandle minimum is recommended.

2. Semi-isolated fenced boundaries are those having an approach area clear of obstruction for sixty to one hundred feet outside the fence. Such boundaries should be lighted by a system providing only a nominal amount of glare. If the boundary is located some distance from adjacent structures, street lighting luminaires mounted on individual poles are most desirable. If suitable structures are available, floodlights mounted on these structures can be used. With either method, a strip eighty feet wide extending from ten feet inside to seventy feet (where feasible) outside the fence line should receive 0.02 footcandle minimum at any point. For the more critical security areas 0.04 footcandle minimum is recommended.

3. Not isolated fenced boundaries are those immediately adjacent to operating areas of other plants or to public thoroughfares where outsiders may move about freely in approach areas close to the fence. Such fence lines should be lighted by a system providing a minimum amount of glare, and the use of conventional street lighting equipment is recommended. A sixty-foot

strip should receive 0.04 footcandle minimum at any point if the strip extends from twenty feet inside to forty feet outside the fence line, and 0.05 footcandle minimum if it extends from thirty feet iside to thirty feet outside the fence line.

Unfenced Boundaries

Where boundaries consist of unfenced property lines located more than twenty feet from on-site buildings, with no outdoor operating areas between the on-site buildings and the property lines, nor off-site buildings adjacent to the property lines, a strip extending at least eighty feet from the bases of the on-site buildings into the unobstructed area should receive 0.04 footcandle minimum at any point.

Building Face Boundaries

Where faces of buildings are on or within twenty feet of the property line, and the public may approach the buildings, a strip extending outward from the base of each building for fifty feet should receive 0.05 footcandle minimum at any point. Since the general public is permitted in the area to be lighted, a lighting system providing a minimum amount of glare should be used and conventional streetlighting equipment is recommended.

Waterfront Boundaries

Where property lines, fenced or unfenced, are on or adjacent to bodies of water a strip extending from ten feet inside the boundary to fifty feet outside the boundary should be illuminated with a value of 0.05 footcandle minimum at any point. (Unnavigable water less than fifteen feet in width should not be considered a waterfront boundary and the approach area should be classified without regard to the presence of the body of water.) The lighting for waterfront boundaries should make it possible to detect the approach of small craft and there should be no shadows on or near the water close to the bank or sea wall. Either conventional streetlighting equipment, or floodlights, at a suitable mounting height to minimize glare, are recommended. Before installation, Coast Guard officials must be consulted for approval of proposed protective lighting adjacent to navigable waters.

Entrances

At those gates, doors or passages used to admit personnel and materials into protected areas, luminaires should be located to provide adequate illumination for recognition of persons, for examination of credentials and for inspection of vehicles and their contents. At pedestrian entrances into fenced areas the entrance road or walkway should be lighted, where feasible, for twenty-five feet inside and twenty-five feet outside the gate with 2 footcandle minimum at any point. At vehicle entrances into fenced areas the entrance roadway should be lighted, where feasible, for fifty feet inside and fifty feet outside the boundary gate with 1 footcandle minimum at any point. Both street lighting and floodlights can be used, but should be located to avoid objectionable glare in the entranceways and surrounding areas.

Industrial Thoroughfares

These include roadways for vehicles, pedestrian walks, intraplant lift truck routes or any such area normally used by vehicle or personnel for movement from one location to another. Those thoroughfares of security significance which are not bordered by buildings should be illuminated with 0.05 footcandle minimum at any point and if bordered by buildings on one or both sides, 0.20 footcandle minimum is advocated. These lighting requirements are the minimum for protective lighting only. Much more light may be necessary for operational needs. However, the minimum should be provided for policing purposes, regardless of the traffic conditions.

Open Yards

Those portions of a security area which do not border boundary limits and are not occupied by structures or vehicular roadways are designated open yards. There are two classifications: material stroage areas, railroad sidings and parking areas, and unoccupied land. Open yards of special security import in the first classification require 0.10 footcandle minimum at any point. All other open yards where lighting is deemed necessary to supplement other protective features, require 0.02 footcandle minimum. A much higher level of illumination may be necessary for operational purposes.

Vital structures most easily and seriously harmed at close range should be lighted to a height of eight feet with an intensity of not less than 2 footcandles minimum at any point on their verticle surface, and the approach areas for a distance of twenty feet from the bases of the structures should receive similar illumination. Vital structures which may be easily harmed from distant areas open to the public should not be specifically lighted for protective reasons.

Vulnerable apparatus should not be highlighted, but the areas that surround it and lead up to it should be well lighted. This will aid in the detection of intruders. The area surrounding the structure is generally considered to mean a twenty-foot wide strip measured from the base of the building outward. Floodlights are recommended when adjacent structures are vulnerable.

SPECIAL USE OF PROTECTIVE LIGHTING SYSTEMS

Certain facilities constitute an extremely attractive target to interests inimical to the national security, and for that reason should be located and operated in a manner to attract a minimum of attention. They may properly be provided with a protective lighting system which, although not normally lighted, is equipped for immediate illumination by manual or automatic means in the event of emergency, intrusion or suspicious activity in or near the area.

These areas should also have an alarm system to overcome the security limitations of guarding in darkness. Although continuous perimeter lighting is essential, provision should also be made that when a guard senses something suspicious along the boundary, he is able to furnish, immediately, a higher intensity of illumination at that point. This can be done by using a searchlight.

In certain areas and at certain times it is very economical to use searchlights to continually sweep areas such as wharves and

PROTECTIVE LIGHTING GUIDES

Area	Footcandle Illumination	Luminaire		Lamp Size	Number	Location of Luminaires (feet)		
		Type	Beam			Inside Boundary	Height	Spacing
Industrial Thoroughfares 50 ft. wide, bordered by buildings on one or both sides	0.20	Street light	Narrow asymmetric	10,000 L	1	2 ft. from side of bldg.	30	135
buildings on one or both sides	0.20	Street light	Narrow, asymmetric	6,000 L	1	2 ft. from side of bldg.	25	80
80 ft. wide, border by buildings on one or both sides	0.20	Street light	Medium wide, asymmetric	10,000 L	1	2 ft. from side of bldg.	30	75
	0.20	Street light	Medium wide, asymmetric	6,000 L	1	2 ft. from side of bldg.	25	50
Open Yards Storage parking areas	0.10	Floodlight	Wide	1,000 W	4		60	125
	0.10	Floodlight	Wide	750 W	4		60	113
	0.10	Floodlight	Very wide	500 W	4		40	95
	0.10	Floodlight	Very wide	300 W	4		40	78
Open Yards Unoccupied land	0.02	Floodlight	Very wide	1,000 W	4		60	225
	0.02	Floodlight	Very wide	750 W	4		60	208
	0.02	Floodlight	Very wide	500 W	4		40	168
	0.02	Floodlight	Very wide	300 W	4		40	135
Vital Structure	2	Floodlight	Narrow	1,000 W	2	40 ft. from bldg.	20	80
	2	Street light	Medium wide, asymmetric	10,000 L	1	4 ft. from bldg. face	30	50

*L = Lumens;   W = Watts

PROTECTIVE LIGHTING GUIDES (Continued)

Area	Footcandle Illumination	Luminaire				Location of Luminaires (feet)		
		Type	Beam	Lamp Size	Number	Inside Boundary	Height	Spacing
Vehicular Entrances 50 ft. inside to 50 ft. outside boundary gate	1	Street light	Wide, asymetric	15,000 L	4		30	25
	1	Floodlight	Very wide	750 W	3		25	50
	1	Floodlight	Very wide	500 W	4		25	25
Industrial Thoroughfares 30 ft. wide not bordered by buildings	0.05	Street light	2-way 4-way	6,000 L	1	Overhang to center of road	25	225
	0.05	Street light	2-way 4-way	4,000 L	1	Overhang to center of road	25	190
	0.05	Street light	Narrow, asymetric	6,000 L	1	Overhang side of road 2 ft.	25	180
	0.05	Street light	Narrow, asymmetric	4,000 L	1	Overhang side of road 2 ft.	25	160
Industrial Thoroughfares 30 ft. wide, bordered by buildings on one or both sid.s	0.20	Street light	2-way 4-way	6,000 L	1	Overhang to center of road	25	90
	0.20	Street light	Narrow, asymmetric	6,000 L	1	2-ft. from side of bldg.	25	90

PROTECTIVE LIGHTING GUIDES (Continued)

Area	Footcandle Illumination	Luminaire Type	Beam	Lamp Size	Number	Inside Boundary	Height	Spacing
Isolated Fenced Boundaries	0.07	Fresnel	Asymmetric	500 W	1	10	20	270
	0.07	Fresnel	Asymmetric	500 W	1	10	20	165
	0.07	Floodlight	Narrow	300 W	2	80	25	225
Isolated and Semi-isolated Fenced Boundaries	0.02	Street light	Medium wide, asymmetric	10,000 L	1	10	30	205
	0.02	Street light		6,000 L	1	10	25	160
	0.02	Street light	Narrow	300 W	2	80	40	180
	0.02	Floodlight	Medium wide, asymmetric	10,000 L	1	10	30	190
	0.04	Street light	Medium wide, asymmetric					
	0.04	Street light	Medium wide, asymmetric	8,000 L	1 (10° tilt)	10	25	145
	0.04	Floodlight	Narrow	500 W	2	80	40	225
	0.04	Floodlight	Narrow	300 W	2	80	40	180
Not Isolated Fenced Boundaries 20 ft. inside to 40 ft. outside fence line	0.04	Street light	Medium wide, asymmetric	10,000 L	1	20	30	215
	0.04	Street light	Medium wide, asymmetric	6,000 L	1	20	25	185
	0.04	Street light	Medium wide, asymmetric	4,000 L	1	20	25	165
Non-Isolated Fenced Boundaries 30 ft. inside to 30 ft. outside fence line	0.05	Street light	Medium wide, asymmetric	10,000 L	1	30	30	215
	0.05	Street light	Medium wide, asymmetric	6,000 L	1	30	25	175
	0.05	Street light	Medium wide, asymmetric	4,000 L	1	30	25	150
Unfenced Boundaries lighted strip 80 ft. wide	0.04	Street light	Medium wide, asymmetric	10,000 L	1	4 ft. from bldg. face	30	190

PROTECTIVE LIGHTING GUIDES (Concluded)

Area	Footcandle Illumination	Luminaire				Location of Luminaires (feet)		
		Type	Beam	Lamp Size	Number	Inside Boundary	Height	Spacing
	0.04	Street light	Medium wide, asymmetric	6,000 L	1	4 ft. from bldg. face tilted 10°	25	145
	0.04	Floodlight	Narrow	500 W	2	At building face	40	225
Building Face Boundaries lighted strip 50 ft. wide	0.05	Street light	Medium wide, asymmetric	10,000 L	1	4 ft. from bldg. face	30	205
	0.05	Street light	Medium wide, asymmetric	6,000 L	1	4 ft. from bldg. face	25	185
	0.05	Street light	Medium wide, asymmetric	4,000 L	1	4 ft. from bldg. face	25	155
Waterfront Boundaries lighted strip 60 ft. wide	0.05	Street light	Medium wide, asymmetric	10,000 L	1	10	30	200
	0.05	Street light	Medium wide, asymmetric	6,000 L	1	10	25	160
	0.05	Street light	Medium wide, asymmetric	4,000 L	1	10	25	155
	0.05	Floodlight	Wide	500 W	2	50	60	260
	0.05	Floodlight	Wide	300 W	2	50	60	180
Pedestrian Entrances 25 ft. inside to 25 ft. outside boundary gate	2	Street light	Wide Asymmetric	15,000 L	1		30	12.5
	2	Floodlight	Very wide	750 W	3	5 at each of 3 locations	25	25
	2	Floodlight	Very wide	500 W	4		25	12.5
	2	Floodlight	Medium wide	150 W	15		25	50

piers. This is necessary, for example, when ships are docked, or under tide conditions favorable for the approach of small craft. Searchlights can then be used to sweep the affected areas.

Where manned guard towers are justified because of the importance of the facility, each tower, whether continuously or intermittently manned, will normally require a movable luminaire or searchlight. Movable lighting units located on guard towers should be installed that they may be focused in all directions in which the guard is expected to render protective observation. This normally will be a full 360°. Further, they should be located in a manner which permits the guard to operate the light without increasing his exposure and without deserting his communication facilities or emergency weapons.

BACKGROUNDS

Indirect interior lighting makes use of the ceilings and upper sidewalls of a room for redirecting and diffusing light given by lamps. This is in part a matter of the light reflecting properties of various colored surfaces. The same is true of exterior lighting. Dark areas are the ally of an intruder. To permit a guard to see an intruder crossing a grass strip outside a perimeter fence or passing in front of a sooted wall of an industrial building, much more light is required than if both backgrounds were light in color. An eight foot strip of white sand outside the perimeter fence or a light-colored wall will permit better visibility with less expenditure of light. In areas where there is little traffic by authorized personnel, a background of alternating light and dark stripes will aid in the detection of movement, but a supplementary light must be available for more detailed examination.

FLARES

Trip flares and parachute flares are expedients adaptable to installation perimeter protection when other types of lighting are not available. Trip flares are equipped with a bracket for attachment to a tree or post. Trip wires to either flare should be placed at least ten inches above the ground to prevent small animals from setting them off. In dry or wooded areas, precautions must be taken to prevent fires resulting from the use of flares.

Chapter 8

ALARM SYSTEMS

INTRODUCTION

At most facilities perimeter protection plus interior surveillance of the area is probably the best combination to use in protecting the facility. This combination gives protection in depth. If an intruder is clever or lucky enough to get by one system, he will be trapped by the others. Such a setup spots not only the crook coming from the outside but also the stay-behind thief, the one who lurks inside until everyone has gone.

USES

Protective alarm systems provide an electrical and mechanical means of detecting and announcing proximity or intrusion which endangers or may endanger the security of a restricted area, a facility, or its components. Protective alarm systems are used to accomplish one or more of the following purposes:

1. To permit more economical and efficient use of manpower by substitution of mobile responding guard units for larger numbers of fixed guards and/or patrols.
2. To take the place of other necessary elements of physical security such as securely locked doors, heavily constructed partitions, frequent patrol or continuous observation by guards, or similar measures which cannot be used because of building layouts, safety regulations, operating requirements, appearance, cost, or other reasons.
3. To provide additional controls at vital areas as insurance against human or mechanical failure.
4. To permit the use of less expensive storage equipment (e.g., transfer cases, open shelves, key lock cabinets or other nonsecurity containers) than would be required without alarm protection.

EVALUATION OF NEED

The use of alarms in the protection program of area or facility may be required in certain instances the critical importance of the area or the facility ar instances, because of situations and conditions pertaining to the location and/or the layout of the area or facility. In some instances, their use may be justified as a more economical and efficient substitute for other necessary protective elements. In determining whether the use of alarms in a restricted area is essential or advisable, the various conditions and situations peculiar to the restricted area or facility will, of course, affect the ultimate decision. However, in general, the following criteria should form the basis for a decision by management to use alarms:

1. The critical importance and vulnerability of certain restricted areas or facilities require the additional control and insurance against human or mechanical failure which is provided by alarm systems. In this group are the following:

   a. Restricted areas or facilities which, because of a concentration of vital components, materials, or data, are attractive, high priority targets for sabotage, theft, espionage, or other criminal acts.

   b. Critical processes and process controls.

   c. Very important restricted areas or facilities where it is desirable to have admission controlled by both guards and operational employees, or where it is desirable for operators to deny access to guards.

2. In certain cases due to restrictions imposed by location, layout, or construction, alarms are necessary to take the place of the more usual protective elements such as fences, lighting, patrols, etc. Included in this group are the following:

   a. Restricted areas or facilities which, because of proximity to adjacent structures, activities, or property lines, require the use of alarms in lieu of physical barriers to limited or exclusion areas.

   b. Restricted areas or facilities which are difficult or impossible to guard effectively due to terrain conditions, personnel hazards, or atmospheric conditions and where other types of protection are not effective or practicable.

c. Restricted areas or facilities, or components thereof, which are small or remote areas requiring more than safe and lock protection but not justifying a full-time guard.

3. Alarm systems, because of their cost, are justified only where their use results in a commensurate reduction or replacement of other necessary protective elements without loss of protective effectiveness. The objective of such use, in most instances, is to reduce the number of guards otherwise required. In determining the advisability of substituting alarms for other protective elements, a careful comparison of relative costs is essential. This should include consideration of the initial cost of the system as well as recurring service and maintenance charges. In this connection it should be borne in mind that many alarm systems have little salvage value and, consequently, the longevity of the activity being protected is an important consideration.

ALARM SYSTEMS

An alarm system is simply a manual or automatic means of communicating a warning of potential or present danger. Electric intrusion detection alarm systems provide an acceptable means of augmenting other methods of perimeter closure, and may be used as a secondary line of defense within the physical perimeter barrier.

Where closure of a perimeter opening interferes with public safety, for instance at fire exits, electric intrusion detection alarm devices with limited guard supervision may be used. If so used, a warning should be conspicuously posted against passage except in an emergency.

When it is not entirely practical to enclose a restricted area, other than one established to safeguard classified security information, or at an entrance to a completely enclosed restricted area, an electric intrusion detection alarm system may be used with limited guard supervision.

ALARM REPORTING SYSTEMS

Central Station System

A commercial agency may contract to provide electric protective services to its clients by use of a central station system. The agency designs, installs, maintains and operates underwriter-approved systems to safeguard against fire, theft and intrusion and monitors industrial processes.

These services generally link detection devices in a plant to a central station, usually located some distance away. This central station is manned by the service's own personnel. When an alarm comes in, the dispatcher sends out a patrol to check on the trouble. He also alerts the local police or fire department.

The central station can also monitor a guard force by checking to see that a guard "hits" each of his reporting stations within a certain time. If he doesn't, or if he fails to respond to a local signal within a given time, an alarm is turned on at the central station. So if he has been attacked, or disabled, help is automatically sent to him.

A twist on the central station idea is for a group of companies to set up their own central station, each paying part of the operating expenses. Each factory selects the security system that best fits its needs, and plugs itself into the central station.

Proprietary System

The operation of a proprietary system is similar to that of the central station system except that it is owned by, and located on, the installation. Control and receiving equipment is located in the installation guard or fire department headquarters. Response to an alarm is by the installation's own fire-fighting or police personnel. In addition, this type of system may be connected with the police and fire departments, and with a commercial central station.

Local Alarm System

A local alarm system is one in which the protective circuits or devices activate a visual or audible signal located in the immediate vicinity of the object of protection. Response is by the guards or other personnel within sight or hearing. This system can also be connected with civil police and fire departments.

When buildings, rooms, safes or vaults requiring alarm protection are so located that a visible or audible signal would be immediately detected by protective personnel or operating personnel specifically designated to take required action upon such detection, a local alarm system may be utilized; otherwise, a central station alarm system should be specified.

Auxiliary System

An auxiliary system is one in which the installation-owned system is a direct extension of the police and/or fire alarm systems. This is the least effective system and because of dual responsibility for maintenance is not favorably considered by many protective organizations.

INTRUSION DETECTION DEVICES

General

There are a variety of commercially manufactured electronic, electromagnetic and ultrasonic devices designed to detect approach, intrusion, or sound. These devices report electrically by means of a light or buzzer to a monitored control panel, usually located in the guard headquarters or central station. Where an audio device is a part of the system, it is possible for the monitor to listen in at the location from which the alarm is received and to determine the cause. Where a closed circuit television is included in the system, the guard may also scan the location of the alarm. These devices are designed to detect not prevent intrusion and do not transform a room or an area into an impregnable vault. The same ingenuity and technical skill responsible for their creation may be applied to circumvent their purpose, and in the last analysis, it is the human factor which makes the difference between adequate and inadequate protection.

Only the highest quality of equipment should be used.

The installation of the system must be made as nearly tamper-proof as possible. Switches and controls should be located in a locked enclosure and should be under the control of supervisory personnel of the guard organizations.

FUNDAMENTAL ELEMENTS

There is on the market a variety of alarm equipment based on different principles of operation and designed to meet various requirements. There are, however, four fundamental elements required in all alarm systems. These elements are the following:

1. A detection device or series of such devices. Most common among these are the following:
    a. Foils, screens and traps which are damaged or disturbed by penetration (usually used for protection of doors, windows, ducts and nonsubstantial walls or partitions).
    b. Photoelectric systems, whereby interruption of a virtually invisible beam of light is detected.
    c. Electromagnetic wave or electronic system whereby entrance of an intruder into the field of the system is detected.
    d. Microphonic devices which detect sound and vibration.
    e. Pneumatic detectors of various types which serve the same purpose as mechanical and electrical detectors.
    f. Thermal detectors, actuated by heat exceeding a predetermined temperature limit or rate of rise.
2. Electrical or electronic circuits for transmitting signals from the protected area to the signal apparatus.
3. An alarm or signal apparatus which will announce by audible and/or visual means any activity which the system is designed to detect.
4. Electronic apparatus which provides visual means of observing activity from a different and remote location.

FACTORS IN SELECTION

Each type of intrusion detection system is intended to meet

a specific type of problem. Factors to be considered in selection of the appropriate system include, but are not limited to, the following.

1. Response time capability of guards.
2. Intruder time requirement.
3. Weather conditions.
4. Ambient sound level.
5. Building constructions.
6. Radio and electrical interference.
7. Operational hours of the installation.

A consideration of these factors readily indicates the advisability of obtaining engineering studies to assist in making a wise selection. Often more than one system is necessary to give adequate protection for an area of structure.

Requirement

To afford the required degree of protection and be accessable as a protective unit, alarm installations should meet the following requirements:

1. The system should be so designed that the interval of time between the detection of activity and the achievement of the objective of such activity is sufficient to permit the application of necessary countermeasures.

2. Central station systems should be specified for all locations where guards are not continually in the immediate vicinity to pick up a local alarm signal and make adequate response.

3. All systems, materials and equipment should meet the Underwriters' Laboratories, Incorporated standards where applicable, for the purpose for which it is used.

4. All installations should be made in accordance with accepted Underwriters' Laboratories, Incorporated standards for alarm systems.

Protective alarm systems, while designed to reduce the guard force, are not sufficiently efficient to replace it. Applications and costs of an intrusion protection system vary with plant needs and type chosen. The engineers of a competent manufacturer of such devices should be called in for a survey of the space requiring protection. Resulting recommendations should be studied by plant engineers from the viewpoint of how well they fit protection needs, how effectively they perform, how they would assist and supplement police and guard forces, and how the cost would be balanced by the savings of installing and operating such protective devices.

PHOTOELECTRIC

A photoelectric system is basically the same setup that opens supermarket doors as customers walk out. Photoelectric detection and alarm systems consist of two major parts, a projector and a receiver. The projector, located on one side of the protected area, transmits a beam of invisible or visible electrically modulated light across the area to a receiver which converts it into electronical energy. Any interruption of this light beam will trigger an alarm.

These devices are designed so that any tampering with the systems—even when it is not in operation—will set off the alarm. Since the beam is modulated, it cannot be "jammed" with a flashlight or another light source.

Visible light is almost worthless in a security installation —an intruder can spot it immediately. An "invisible" beam of infrared or ultraviolet light is better, but there is still a faint glow at the source. An experienced intruder, equipped with special filters, might spot it immediately.

Some of these drawbacks can be overcome by using a flickering beam; one that is interrupted in a fixed sequence. Usually, this is done with a moving disk at the light source, but a better method is to interrupt the beam electronically. Because the receiver is tuned to the same interruption frequency, an intruder cannot simply substitute another light source to bypass the system. The substitute light would not have the same frequency and the alarm would go off.

Since light travels in a straight line, mirrors can be used to reflect the beam over a wide area, or completely around an object, like a safe. But the system can be defeated if the mirrors are knocked out of alignment or get dirty. Photoelectric systems are often used to fill gaps, i.e., doors, gates, windows, in another security system.

ELECTROMECHANICAL

Electromechanical devices are the most common, the simplest, cheapest and easiest to install. They are also the easiest to defeat. An electromechanical system is simply a continuous circuit around windows, walls, skylights and doors of a room or building. Shorting or opening the circuit at any point sets off the alarm.

Because they are so easy to spot and jumper, electromechanical systems are often disregarded; but this very drawback can be turned to advantage. By using both electromechanical devices and another security system, it is possible to trap an intruder. When he bypasses the simple rig, he thinks he is home free, and will blunder through another system he has not noticed.

Another point in favor of easy-to-see systems is that they show the plant is protected. This is often enough to discourage vandals and amateur thieves, who account for a great majority of total break-ins.

These devices give perimeter protection so guards can be on the scene before the burglar gets inside. But they will not trap the stay-behind thief who fleeces you and then breaks out.

At doors, windows and so forth, snap action switches can be used. These permit the door or window to be opened during working hours and secured after hours. Wires, strips, or screen can be laced across permanently closed windows or skylights. Tape and foil can surround specific areas. Sensitive mats can be placed under rugs. Because these are simple devices, they are not susceptible to false alarms—their warnings are genuine.

TAUT WIRE DETECTOR

The taut wire detector system is a rugged, low-cost system providing dependable perimeter protection. It consists of a strand of detector wire, which is virtually invisible, strong along any strategic perimeter, such as the top of a fence or wall around a protected building. The wire is held at a scientifically calibrated tension. A relaxation of the tension (by cutting) or an addition to the tension (by the touch of an intruder) will set off a local or remote alarm. Special equipment assures trouble-free operation in all types of weather by automatically adjusting for changes in temperature and by compensating automatically for rain, sleet, hail and snow.

ELECTROMAGNETIC FENCE

Among the various types of perimeter intrusion detection systems which have been tested, the most satisfactory results have been attained with the electromagnetic fence. This fence, sometimes referred to as an electronic fence, consists primarily of from three to five strands of wire strung horizontally and spaced from nine to twenty-four inches above one another, which serve as antennas, a monitor panel and electrical circuitry. This system establishes an electromagnetic field between and in the vicinity of the wires and between the wires and ground. An intruder approaching within two feet of this fence will disturb the balance of the electronic field and cause an alarm. The fence is zoned to indicate the point of approach on the monitor panel. This fence can be an effective security aid if it is installed and maintained properly and used in conjunction with other security aids. It should be installed inside a chain link fence to minimize nuisance alarms caused by animals, debris and casual trespassers.

CAPACITANCE DETECTION SYSTEM

Proximity Alarm

Another electrical detection system which is frequently used in office buildings is the capacitance or proximity alarm. These devises are designed to protect metal containers, safes, file cabinets, lockers, tool cribs and fences used to protect areas within a warehouse.

A capacitance detection system utilizes two oscillator circuits that are couples together and set in perfect electrical balance. However, one of these circuits includes the antenna (detection wire) which is sensitive to changes in capacitance caused by movement of any kind. When motion occurs, these changes in capacitance are amplified to throw the two oscillator circuits out of balance and effect an alarm signal.

Electrical capacitance can be illustrated by a commonly experienced household phenomenon involving the popular television "rabbit ears" type of antenna. When a person approaches or touches such an antenna, a change in the electrical capacitance of the television circuitry takes place. Often this change is marked enough to actually cause a noticeably clearer television picture temporarily.

The system consists of a control panel and an antenna, normally a detection wire that is run in the protected area. The antenna may be mounted in a variety of ways. The antenna wire can be affixed with tape on hard surfaces such as marble for a temporary installation; it can be stapled to wood surfaces, wallboard, etc., for permanent installations; or copper tubing may be used instead of wire where the installation requires protecting open spaces.

If desired, various objects may also serve as the antenna. For instance, a series of desks, file cabinets or safe may be hooked together by coaxial cable, which is inert. In this way, the objects themselves are "live" and when their electrical capacitance is disturbed by an intruder, the system will set off an alarm.

A person coming within several feet of the antenna causes a change in electrical capacitance. This change is detected at the control panel where computer-type circuitry determines the characteristics of the disturbance. This circuitry eliminates nuisance alarms and activates a fail-safe alarm relay in the event of an attack. The fail-safe feature causes an alarm in the event of a component failure or an attempt to compromise the system. Adjustments within the panel provide a wide range of sensitivity.

After detection sensitivity has been adjusted, the device will automatically compensate for changes in temperature and humidity and retain the required sensitivity. Vibrations, noise or a running mouse will not set off the alarm because these conditions do not disturb the electrical capacitance.

NOISE DETECTION SYSTEM

Essentially, an audio detection system is a long-range hearing aid. It permits a guard at a remote location to hear a prowler moving inside an area or trying to get in. When it "heads" something unusual—footfalls, axes, drills—it turns on an alarm at the guard office. It also "hears" explosions, machinery malfunctions—

any sudden or unusual noise. When a guard gets the alarm signal, he can flip a switch and listen in on the area to find out what's going on.

A basic system consists of microphones in the protected area, a speaker at a guard station, and connecting wires. The number of microphones needed to protect a given area depends on the range and sensitivity of the mikes and the characteristics of the area itself. Filter circuits cut out low and high frequency noise extremes for better detection. Because the system detects unusual noises, it can work only when the background noise level is comparatively low. Otherwise, the sounds of a burglar at work would be drowned out by the racket in the background.

An audio system can be set so it is sensitive enough to pick up even catlike steps on a carpet. But too much sensitivity is a handicap—the alarm might be triggered by a clap of thunder, or the rumble of a passing truck, or an airplane flying overhead.

Special equipment is available which electronically evaluates the significance of sound disturbances to eliminate false alarms while positively detecting intrusion noises. Some systems combine the pickup mike with a loudspeaker. They can be used for audio monitoring and communication, which permits an operator at a remote location to listen in on any protected area and to communicate with personnel in the area. This allows maintenance of control of the area at all times and is especially valuable when guards are dispatched to investigate alarms. The operator at the remote location can be in constant communication with the investigators and can direct and support their efforts, if required. During the day, the system serves as an intercom. Noise detection systems combine perimeter and interior protection; can detect both "stay-behind" thieves and intruders breaking in.

MOTION DETECTION

Motion detection systems, too, use sound to spot intruders, but they use an entirely different principle. Audio-detection discovers burglars by the noise they make; a motion-detection system picks up the thief's movements which cause changes in the frequency of the sound waves. One or more transceiver units establish a sensitive sound-wave pattern in the area to be protected. The transceiver both transmits and receives the basic sound wave pattern. One or more of these units can be installed in a secured area to provide the exact degree of protection required. The transceiver sensing unit may be positioned to provide the exact area of coverage required for any installation. It may be placed in a variety of ways to achieve spot protection or in such a way as to provide a total area saturation wave pattern. Any distortion of the pattern caused by the slightest motion of a person is immediately sensed, relaying an alarm signal to any remote location, such as guard headquarters, and also flashing a local alarm, if desired. The system will detect any form of intruder motion, even that which is extremely fast or painstakingly slow. Nuisance alarms are eliminated by special discriminator circuits that distinguish between an intruder and minor disturbances, such as swaying drapes or a falling piece of paper. External influences, such as thunder, backing firing trucks, etc., are also ignored. Ultrasonic waves do not penetrate walls, floors, or ceilings and are not affected by outside movement any more than they are by audible noise. Sensitivity of the ultrasonic waves can be varied to suit conditions within the protected area. Wave patterns can be set up to provide the exact coverage desired—from a single "zone" or "spot" of protection within an area to total protection of the complete area. If so desired, two-way communication can be added so that headquarters personnel can monitor and communicate with the protected area.

RADIO FREQUENCY SYSTEM

A radio frequency security system saturates an area with invisible electromagnetic waves. Any absorption of this energy changes the antenna loading and sets off an alarm. The waves are not affected by air currents or sound waves.

Radio waves are extremely penetrating, passing right through most materials. This characteristic is both an asset and a liability. On the one hand, the waves permeate an entire room, so an intruder cannot hide from them; but they also go right through walls, so any movement that cuts the wave pattern—even outside—will trigger the alarm. One answer, where exterior movement is unavoidable, is to shield the security areas, but this is expensive. Fortunately, with some devices you can "load" the antenna, to "shape" the wave pattern, and make it fit the area to be protected. In many cases, though, you may want to detect movement outside the protected area.

ELECTRIC VIBRATION DETECTION SYSTEM

Electronic vibration detection system gives excellent protection against burglary, cautious rifling or wanton destruction by detecting vibrations caused by intruders. A highly specialized type of contact microphone is attached directly to objects such as safes, filing cabinets and art objects, or to surfaces such as floors, walls, windows and ceilings. When an intruder begins an attack, slight vibrations occur which are sensed by the electronic vibration detector. Signals are instantly sent to the detector control panel, which acts as a computer center and converts the incoming electronic signals into an alarm. This panel may be equipped with an alarm discriminator unit that eliminates alarms due to incidental vibrations. It is sensitive only to vibrations. It will not respond to airborne noise, making it practically immune to false alarms. A variety of alarms may be used on the premises to frighten off the intruder when he begins his attempt, or a "silent" alarm may be sent to reliable personnel at a remote location so that the intruder can be apprehended.

CLOSED CIRCUIT TELEVISION

An audio system extends the ears of a guard, and a closed circuit television system extends his eyes. The television camera surveys an area and transmits what it "sees" through coaxial cable to monitoring screens. With strategically located cameras, one man maintains constant surveillance over large areas used for storage, employee parking or material movement. Remote controls enable him to direct the camera up or down or from side to side. Using the remotely controlled zoom lens for wide angle or close-up views, he can focus in on suspects and observe their activities.

While the television camera can see, it cannot sound the alarm. Also, guards find that staring at an image of a closed door is pretty boring. The answer is to use some other device to tell them when to look. An electronic fence, for example, could signal the guard when someone approaches. He can check the television sent.

SECURITY SYSTEM SURVEILLANCE CAMERA

The surveillance camera is designed for protection against pilferage, holdup, industrial espionage and vandalism. It is particularly applicable in banks, plants, warehouses, retail establishments, hospitals, trucking companies and military and governmental agencies.

Consisting of a camera and separate control box unit, this system can be used in conjunction with any type burglar or holdup alarm system including ultrasonic, photoelectric or capacitance types. The camera can be set to take pictures automatically when an alarm is set off, or it can be triggered manually by a switch, money clip or foot plate. The silent camera is designed for simple wall or ceiling mounting and the camera is easy and inexpensive to align and install.

GUARD SUPERVISORY SYSTEMS

Guard supervisory systems consist of key operated electric call boxes located strategically throughout an installation or facility. The system can be either proprietary or central-station operated. By inserting the key in the call box a guard can make a routine tour report or summon emergency assistance. Tampering with the transmitting key or the call box automatically locks the latter causing a failure of the signal and an alert for immediate investigation.

This system provides the means for supervision and detection of interference with the guard's normal activities. It must be contrasted with the recorded tour supervisory system which is simply a mechanical means of determining that a guard has covered his assigned route. The latter affords the guard no way of summoning aid and it is adapted to watchman service rather than to guard service.

FIRE ALARMS

In addition to the familiar manual fire alarm box and the heat-activated sprinkler system, there are various automatic devices to report fires either to the central station or direct to the proprietary or civil fire station, or a combination thereof. A sprinkler system that is activated at a predetermined temperature level or rate of rise also may be designed to transmit an alarm by the flow of water through the piping or the change in water pressure. Where water damage is a consideration, spot thermostatic or pneumatic pressure type systems may be used. Photoelectric beams may be sensitized to report an interruption of the beam by smoke. The fire reporting system may be incorporated with other protective alarms and may be monitored at one station and, in some instances, on the same control panel.

INDUSTRIAL MONITORS

For monitoring industrial processes, there are many types of alarms that report any variation from predetermined standards of temperature, humidity, flow, pressure and other factors. These conditions so important to certain processes make them attractive targets for mechanical sabotage. The nature of the alarm and the place where the alarm is reported are matters to be tailored to the particular installation.

RESPONSE

The important part of any such system is that after the report of an intrusion there must be some response which is planned and ready for immediate execution. It may consist of an intrusion alarm system tied in with the plant guard, police, or other protection force headquarters. Such headquarters should be prepared to receive the alarms and initiate immediate response. An alternative system is an electric protective system which sounds a gong, bell, or other audible device and flashes lights or other visible signals in such a manner as to attract the attention of anyone in the vicinity of the facility.

TESTING AND INSPECTION

All alarm equipment and circuits should be tested daily to assure proper functioning and periodically to assure prompt response by protective personnel. Alarm equipment and circuits should be inspected at least annually to assure continuing serviceability.

Chapter 9

COMMUNICATION SYSTEMS

INTRODUCTION

ADEQUATE COMMUNICATIONS are essential to the effective operation of the guard force during normal periods and especially in the event of an emergency or disaster. Protective communication systems will vary with the importance, vulnerability, size, location and other factors affecting a specific installation and must be largely subject to local determination. An industrial facility should have a quick and ready system for communicating with all of its employees. Such a protective communication system should be capable of signalling, alarming or alerting workers at any location throughout the plant, as occasion demands.

The communication system should also provide for employees, guards, or watchmen reporting fires, accidents, or suspicious circumstances without leaving their posts or areas of patrol. Finally, the communication system provides liaison with protective organizations outside the facility.

DEFINITIONS

Communication systems include the following:

1. Facilities for local exchange and commercial toll telephone service.
2. Intraplant, interplant and interoffice telephone systems, with or without switching equipment, using rented circuits and equipment, but not interconnected with facilities for commercial exchange or toll telephone service.
3. Radiotelegraph and radiotelephone facilities for either point-to-point or mobile communication service.
4. Telegraph and teletype facilities for either commercial service or private line operation.
5. Central station supervisory automatic alarms.

6. All other reliable means of electrically or electronically transmitting and receiving a signal, message, or alarm which will be understood as such.
7. Teams of messengers should be organized to be used in case of telephone and radio communication failure.

Interior communications are considered to be two-way communications for the exchange of information between two or more points within a restricted area or facility; and also, one-way communications for the transmission of alarm signals from one or more other locations within the restricted area or facility.

Exterior communications are considered to be two-way communications between a restricted area or facility and an exterior point or points from which assistance may reasonably be expected in the event of an emergency, and for which contingency emergency plans have been formulated.

TYPE

The types of communication systems used and their comprehensiveness will, of necessity, vary with the criticality, vulnerability, size, location and other facts affecting a specific restricted area or facility and must be largely subject to local determination.

EXTERNAL COMMUNICATIONS

Adequate means should be provided for communications liaison with local and state protective organizations. Normally, and especially for off-site facilities, communications with the following organizations should be maintained:

1. Local police and fire departments.
2. County sheriff and/or state police.
3. Local Federal Bureau of Investigation office.
4. Local civil defense organization.
5. Local mutual aid organization.
6. Nearest military establishment.
7. State fire marshal.

Guard forces should have a communication system with direct lines outside and an auxiliary power supply. Although principal dependence is placed on the telephone, the teletype and an automatic alarm system, interior and exterior radio communications play an important part in the protective net of large installations. One or more of the following means of communications should be included in the protective system:

1. Facilities for local exchange and commercial telephone service.
2. Intraplant, interplant and interoffice telephone systems.
3. Radiotelegraph and/or radiotelephone facilities for either point-to-point or mobile service.
4. Telegraph and teletype facilities for either commercial service or private line operation.
5. Central or proprietary station automatic alarm system.

ALTERNATE COMMUNICATION SYSTEMS

Alternate communication systems are always advisable for use in emergencies. The flood of inquiries that follow emergency conditions added to the normal flow of messages may overload the existing system at the very time that sure and rapid communication is vital. The most efficient emergency reporting system consists of direct connections to the guard or communications center from telephones strategically placed throughout the installation. The use of these telephones should be restricted to emergency and/or guard reporting only. The wires of alternate communication systems should be separated from other communication lines, and should be in underground conduits when feasible. For summoning aid from outside the installation, leased wires or a radio adjustable to police and fire department frequencies should be available.

Arrangements can be made with a local radio or television station to make emergency announcements. Such plans should appear in the company manual and be brought to the attention of the employees.

COMMUNICATIONS CENTER

The communications center is the nerve center of the entire installation and should be designated as a restricted area. At a large installation when it is feasible to separate the protective communications center—including the alarm reporting system, the closed circuit television, and the guard telephone switchboard and radio receiver—from the regular telephone and radio center, it is desirable to consider the protective center as a restricted area. Further dispersion of the protective communication equipment, such as separate locations for the telephone and radio receivers, gives additional protection but prevents centralized control.

WIRING

Whenever practicable, the wiring of protective alarm and communication systems should be on separate poles or in separate conduits from the regular communication and lighting systems. Tamper-resistant wire and cable will provide added protection. This is an insulated wire or cable that includes a sheath of foil that transmits a signal when penetrated or cut.

INSPECTION AND TESTING

All alarm and communication circuits should be tested at least once during each guard tour, preferably when the new shift comes on duty. At small installations that do not employ guards, a test should be made immediately before closing for the night. Some commercially manufactured systems have self-testing features but even they should be checked periodically by the guard or operating force. All equipment must be inspected periodically by technical maintenance personnel to repair or replace worn or failing parts. Tests should be conducted at regular intervals and simulated emergency conditions.

Chapter 10

CONTROL OF AUTHORIZED ENTRY
AND MOVEMENT

INTRODUCTION

Controlling authorized entry to a facility includes the screening and identification of employees and visitors, control of trucks and railroad cars, and control of incoming and outgoing packages.

SCREENING OF EMPLOYEES

All businesses whether large or small are concerned with the problem of securing honest, reliable employees. In addition, the screening of job applicants and employees to eliminate potential espionage and sabotage agents and other security risks is important in peace time and is extremely important in time of a national defense emergency. For the results of such screening to be most effective in an emergency, it is desirable that they be incorporated into standard personnel policies for peacetime as well as for times of emergency. Obviously the gravity of the problem will vary widely from one business to another, depending upon the type of work and the amount of trust which must be placed upon the individual, and the vulnerability of the plant or product. Consequently, the amount of time spent in checking an employee's background will vary.

These will range from the simple inquiry designed to obtain the basic information necessary for reporting to government and for plant personnel requirements with no more investigation than a contact with the former employer to the complex security clearance investigation. In either case the first step is to complete a personnel questionnaire. That questionnaire should obtain the following information:

1. Full name, and any other name or aliases used.

2. Date and place of birth and whether the person is a citizen or non-citizen.
3. Complete personal description including height, weight, color of hair, color of eyes, and any other marks, scars, deformities, or amputations.
4. Complete educational history.
5. Statement whether applicant has ever been adjudged insane or has ever been committed to an insane asylum or treated for serious mental disorders.
6. Statement whether applicant has ever been arrested or convicted of a crime. If so, the details.
7. Statement whether applicant has ever used alcohol or drugs habitually or to excess.
8. Statement from the applicant whether he has ever been discharged from any job and if so, the reason therefor.
9. At least three persons who have known the applicant for five years or longer (other than relatives or previous employers) who can testify as to the character and habits of the applicant.
10. Person's employment record for the previous five or ten years with an explanation of any time unaccounted for during that period.
11. Whether he has been bonded and if so, by what company, and whether there is anything in his past record that might prevent his being bonded.

The filled-out questionnaire should be checked for completeness, and obviously undesirable applicants eliminated from further consideration.

Next a sufficient investigation should be conducted to assure that the applicant's or employee's character, associations, and suitability for employment are satisfactory. Emphasis should be placed on an analysis of the periods of lapses in employment, conflicting dates of employment, or unsatisfactory replies to reference checks.

What a Preemployment Investigation Will Cover

Typically a basic pre-employment investigation will probe four facets of the applicant's background.

WORK HISTORY. Information on the applicant's work history is obtained by personal interview with his former supervisor. The report will include, in the supervisor's own words, his summation of the applicant's skills, work habits, attendance, honesty, attitude toward the company and its policies and any other pertinent information.

CREDIT HISTORY. The applicant's credit account history is investigated by means of information obtained from credit bureau or similar agency files. How a man pays his bills is one good measure of his sense of responsibility. Employees who pay their bills on time are more likely to finish a project when they say they will than those who are lax about their financial obligations. Bad credit history also frequently turns out to be associated with alcoholism, emotional instability and other undesirable forms of behavior.

POSSIBLE CRIMINAL RECORD. It is true that relatively few applicants' backgrounds will reveal any criminal record—but the small proportion that does is a highly important one. Perhaps it should be added that it is not necessarily the purpose of the criminal background check to eliminate every applicant who may have such a record. The employer is certainly entitled, however, to know the nature of the crime, what sentence was imposed, and something of the circumstances under which it was committed (in addition to any facts the applicant himself may have volunteered). Convictions as apparently innocuous as traffic violations have enabled some investigators to uncover everything from felonies to clearly psychopathic behavior.

HOME LIFE. Information on the applicant's home life is usually obtained through personal interviews with the neighbors both at his present address and at two or three of his former locations. These interviews can yield a number of pertinent facts about his off-the-job behavior. Thus the neighbor's comments may reveal an unhappy marriage, neglect of the children and so on. On the other hand, they may equally well confirm that the applicant is a thoroughly responsible citizen. This part of the investigation also enables the investigator to observe the condition of the applicant's home and estimate its approximate worth.

The following sources may be helpful in securing employment investigative data:

1. State and local police.
2. Former employers.
3. References (including those not furnished by applicant or employee).
4. Public records.
5. Credit agencies.
6. Schools—all levels.

In requesting investigative data from any of the above sources, it is suggested that the following minimum information be furnished, as appropriate, to identify the applicant or employee properly and minimize errors in identity:

1. Full name and other names or aliases used.
2. Personal description.
3. Date and place of birth.
4. Social security number.
5. Present and immediate previous address with dates.
6. Employment, present and past.
7. Military service (if any) indicating discharge received.
8. Arrests (if any) excluding minor traffic violations. State circumstances and disposition of penalty imposed (if any).
9. List of all organizations (if any).

The period covered by a background investigation should be the last ten years. However, if unfavorable information is disclosed, the period may be greater in order that the unfavorable information may be verified or developed further.

A management official should be responsible for reviewing the results of investigations for such things as the following:

1. Unfavorable information.
2. The need for further investigation to develop information reported.
3. Completeness of investigative coverage.
4. Discrepancies between the information included in the personnel security questionnaire and the investigative report.

5. Insuring that there are sufficient reasons or details to support or provide for proper analysis of information reported.
6. Has the investigating agent or informant been biased in reporting investigative data?
7. Are statements contained in the investigative report subject to more than one interpretation?
8. Is the investigation adequate after considering all of the above?

Fingerprints

Fingerprinting is necessary to assure positive identification, the most important requisite to investigation of personnel. Although the Federal Government is not accepting fingerprint records from private industry for processing purposes, some assistance may be obtained from other law enforcement agencies and private investigators. It is recommended that duplicate sets of prints be obtained.

SCREENING OF VISITORS

All visitors and other personnel granted access to the plant, such as salesmen, servicemen and tradesmen, should be screened before being allowed entry. The type and degree of screening necessarily will depend on the type of plant, the product being manufactured, and the areas accessible to such personnel.

Visitors should be required to identify themselves properly, and normally should be accredited by the managerial staff of the plant. Sales, service and trades personnel should have received prior clearance to the plant from the plant administrative officer, should be required to identify themselves properly and when granted admittance, their movements should be limited to predetermined specific unrestricted areas.

IDENTIFICATION CONTROL

Perimeter barriers and protective lighting provide physical security safeguards; however, they alone are not enough. A positive personnel identification and control system must be established and maintained in order to achieve required compartmentalization, preclude unauthorized entry, and facilitate authorized entry to restricted areas at personnel control points. Access lists, personal recognition, security identification cards and badges, badge exchange procedures and personnel escorts are elements which contribute to the effectiveness of identification and control systems. The best control is provided when systems incorporate all these elements. Simple, understandable and workable identification and control measures and procedures should be used to achieve security objectives without impeding efficient installation operations. Properly organized and administered, a personnel and movement control system provides a man not only of positively identifying those who have the right and need to enter or leave an area, but also of detecting unauthorized personnel who attempt to gain entry. These objectives are achieved by the following:

1. Initially determining who has a valid requirement to be in an area.
2. Limiting access to those persons who have a right and need to be there.
3. Establishing procedures for positive identification of persons within, and of persons authorized access into, areas.
4. Issuing special passes or badges to personnel authorized access into restricted areas.
5. Using access lists.
6. Using identification codes.
7. Using duress codes.

An additional purpose of control is to prevent the introduction of harmful devices, material, or components—and the misappropriation, pilferage, or compromise of installation material or recorded information.

EMPLOYEE IDENTIFICATION

A practical system of positive identification should be used to control the entry of employees to all facilities and special means established to adequately identify employee personnel who are authorized to have access to restricted areas. Employee identification should normally be accomplished by the guards stationed at each entrance to the plant and at such other points within the plant as are deemed necessary to assure adequate security.

Personal recognition as a means of identification of em-- ployees should be used only at small plants or within small areas employing not over thirty persons per shift who are personally acquainted and subject to a low rate of turnover. A personal recognition system is not feasible for use in large facilities. In these facilities, an artificial system of identification should be developed and applied. The most practical method of identification of personnel at large installations is the use of a special pass or badge.

PASS AND BADGE PLAN

The provisions for identification by pass or badge control at a facility should be included as part of the facility physical security plan. The following should be prescribed:

1. Designation of the various areas where passes and badges are required.
2. Description of the various identification media in use and the authorization and limitations placed upon the holder.
3. Mechanics of identification at times of entering and leaving each area, including nonoperational hours.
4. Details of where, when and how badge should be worn.
5. Procedures to be followed in case of loss or damage to identification media.
6. Procedures for the disposition of identification media on termination of employment or as a result of security investigations and flagging actions.
7. A procedure to reissue new identification media when 1 to 5 per cent, depending on the criticality and vulnerability of an area, have been lost or are unaccounted for.

EMPLOYEE IDENTIFICATION BADGES

Definite standards should be set up covering the design, color, size and wearing of employee badges. The design of an acceptable pass or badge must be such that it presents maximum difficulty to any attempt at aleration or reproduction. Laminated or embossed passes or badges are considered to be tamper-resistant if they have the following characteristics:

1. Contain distinctive and intricate background design that is difficult to reproduce by photocopying.

2. Have some of the printing on the insert in an ink or dye which is noticeably affected by any heat, solvent, or erasure that would be necessary to effect alteration.

3. Use an insert of proper size to permit bonding of the plastic edges during lamination. (If the pass is to be worn on a chain, the hole must be punched through the cover and insert before lamination.)

4. Contain a recent and clear photograph of the bearer. Where a duplicate or triplicate pass or badge system is used, all photographs must be from the same negative. The photograph should not be less than one inch in its smallest dimension. Individuals should be rephotographed, when necessary, to reflect any significant physical changes in facial appearance but in no event less often than every five years. If color photographs are used, care must be taken to assure true coloring.

5. Contain the name of the bearer and other data deemed necessary to identify the person, such as signature, date of birth, height, weight, color of eyes and hair, sex and the right and left index fingerprints.

6. Contain the identity of the installation and an individual pass identification number.

7. Include the name and signature or facsimile of the validating official.

8. Indicate the effective dates of the pass or badge.

9. Be of distinctive color or other code designating portion or area of facility and the classification or categories of information to which employee is authorized access. The specific clearance status of the holder may be shown only on badges which are not removed from the facility. Secret," "Secret" or "Confidential" on passes or badges removed from the facility but may be indicated, if de-
It shall not be shown by letters or the words "Top

sired, by color or symbol coding.

10. Where applicable, code denoting area for which a badge is valid.
11. Signature (or facsimile) of validating official.
12. Serially numbered.
13. Sturdy in construction.
14. Resistant to fumes inherent in the industry.
15. Some secret characteristic known only to management. This will serve as an additional safeguard against forged passes or badges, especially on those occasions when it is necessary to inspect them more carefully—during spot checks, or because of guard's suspicions being aroused.

Other features, such as fluorescent inks or dyes, watermarks, and superimposed cross-threads or wires, also aid in preventing alterations or forgeries.

There are commercially available magnetized plastic and metal passes that, in addition to showing all the required identification data of the bearer, may be coded with one or more codes for admission to areas of varying security. The code is read by inserting the pass in a reading device which gives an audible or visual signal when an entry is attempted with an improperly coded pass. The reading device which may be either fixed or portable, decodes and recodes in a manner similar to a tape recorder.

It is recognized that practically any pass or badge can be altered or reproduced by determined individuals sufficiently skilled in printing, engraving and photography. However, passes and badges made in accordance with the above, are sufficiently difficult to alter or reproduce to be acceptable as a deterrent. Additional protection against forged credentials may be afforded by exchange systems and by the rigid control and accountability exercised over the valid media. At locations where credentials which are extremely difficult to duplicate or alter are necessary, the following commercially available materials may be used:

1. Watermarked insert paper prepared with a distinctive watermark specifically for the location. The difficulty and expense of duplicating watermarks make this an effective security feature.

2. Readily visible cross-threads or wires within the plastic cover sheets of the pass or badge which will make obvious any alteration of the device such as superimposing a photograph, signature, number, etc., over the original and then relaminating with a third piece of plastic.

3. Sensitized plastic material on which the photograph is printed, made to the same chemical formula as the transparent plastic covers. When this is used, the photograph panel of the paper insert is cut out so that, in laminating, the front and back transparent plastic sheets and the plastic photograph flow together forming a solid piece of plastic material.

4. A paper for use as insert material which loses its fibrous strength while being heated during the laminating process. Any attempt to separate this paper or to peel off the plastic from it will result in a physical disintegration of the weakened, brittle paper;

5. Ink of a type which bleeds and/or changes colors when exposed to chemical solvents which dissolve the plastic covers.

6. Fluorescent inks which may or may not be normally visible, but which are visible under ultraviolet light.

7. An embossed plastic or intricate design and/or such other similar material as will accomplish the desired purpose of preventing or making counterfeiting difficult.

PASS AND BADGE CONTROL

Since any pass or badge may be altered or reproduced by a person having the time and sufficient skill in printing, engraving and photocopying, the makeup, issuance and accountability of passes and badges must be fully controlled. This control commences with the manufacturer or supplier of inserts. When inserts or complete passes are secured commercially, inquiry should be made as to the control exercised by the supplier. This is especially important where an engraving or a special paper is concerned.

Pass or badge issuance, accountability and control should be accomplished at a central office, preferably the chief guard.

By doing so, a minimum of time elapses between a change in the status of a pass or badge and the notification of the guards. A duplicate of each issued pass or badge and a file on each bearer should be kept, including in addition to the data entered on the pass or badge, the bearer's residential address and telephone number. Strict control must be exercised to insure the return and destruction of passes and badges on termination of assignment or employment of personnel, and to invalidate lost or mutilated passes.

Badges or identification cards should be replaced with a new issuance of distinctively different design, every five years or oftener if necessary. They should be reissued when 5 per cent of a previous issue have been lost.

Records should be maintained of all badges or passes made, issued, lost, recovered, or returned. Such records shall include date of issuance, pertinent identifying data, areas or degree of access authorized and clearance status of the holder. Such data should be available to personnel controlling access to security areas for ready verification of clearance status and operational approval as necessary.

Lost Badges or Passes

Provision shall be made for the replacement of lost badges or passes and notification to appropriate personnel to prevent unauthorized usé. If the badge is lost or forgotten, the man can be routed through the guard house, his identity checked and a temporary badge—clearly identifiable as a one-day badge—issued. This will cause a minimum of delay. If the badge is definitely lost, a new one may be issued after proper investigation.

PASS SYSTEMS

The following identification systems may be used either for passes carried on the person or for passes or badges worn on the outer clothing:

Single Pass System

With a single pass system, permission to enter different areas is shown on the pass by letters, numerals, or colors. For instance, blue may be the background color of the pass for the current period for general admittance to an installation. Permission to enter specific areas of higher restriction within the installation may be designated by specified symbols or colors overprinted on the general pass. This system gives comparatively loose control and is not recommended for sensitive areas. Permission to enter is not always analogous with the need to know, and the fact that passes frequently remain in the bearer's possession during off-duty or off-post hours gives the opportunity for surreptitious alteration or duplication.

Pass Exchange System

This is a system of two passes, containing identical photographs, but having different background colors, or an overprint on one of the passes. One type of pass is presented at the entrance and exchanged for the other which is carried or worn while within the restricted area. This second pass may contain additional symbols or colors which grant further admittance as explained above.

Multiple Pass System

This is a further development of the pass exchange system explained above, except that instead of having specific markings on the installation pass denoting permission to enter various restricted areas, an exchange is made at the entrance of each restricted area within the installation. Exchange passes are kept at each area for only those individuals who have need of entry. This is the most secure and effective pass system.

IDENTIFICATION OF PASS BEARERS

The most vulnerable link in any pass system is its enforcement. When badges are a requisite for entering security areas, they shall be worn conspicuously by the entrants at all times while they are within the areas. All restricted areas must be specifically designated as to type and degree of restriction and the boundary lines clearly set forth.

The personnel concerned with the movement into and out of such restricted areas should understand the method of identification necessary for entering and leaving the area and the limitations placed upon the wearer. The method used for personal identification should be thoroughly understood by employees and visitors so that they will facilitate proper identification by the guards and avoid unnecessary delays in their movements.

Persons controlling admittance to security areas should be required to determine, by examination of identification or other media, that an individual has appropriate operational and security approvals to enter the area before permitting entry. It must be recognized that the perfunctory performance of duty by the guard in comparing the bearer with the pass may invalidate the most elaborate system. The following are some of the steps that should be taken to insure efficient working of any pass system:

1. Guard personnel for duty at entrances should be chosen for their alertness, quick perception, tact and good judgment.

2. A uniform method of handling or wearing passes or badges must be prescribed. If carried on the person, the pass must be removed from the wallet or other container and handed to the guard. If worn on the clothing or necklace, it should be in a conspicuous position to expedite making the comparison.

3. When artificial light is necessary, it should be so positioned that it illuminates the pass bearer.

4. Entrances should be so arranged that arriving and departing personnel are forced to pass in a single file in front of the guard. In some instances turnstiles may be advisable.

5. Guards stationed at control points should compare the photograph in each badge with the person wearing it. Any lack of similarity between the two is cause for investigation.

6. When a badge exchange arrangement is used, the guards should make a three-way comparison of the badge, the pass and the individual. Close supervision of guards when checking identification is necessary in order to prevent

such controls from becoming purely mechanical and the identification system ineffective.

7. Employees should be instructed to look on any unidentified or improperly identified person as a trespasser who should be promptly reported to the nearest supervisor or guard. In restricted areas where zone pass systems are issued, employees should report any unauthorized individual in such areas.

Not only must the access of all persons to the premises be properly controlled, but various safeguards must be set up within the buildings and yards to establish that individuals are not roaming about in restricted areas which they have no right to enter. For this reason the use of roving guards is very valuable in the enforcement of restricted areas. Again, the degree of enforcement required depends upon the secrecy, value and purpose for which the products of the plant are intended.

A system should be maintained for recording the names and pass numbers or departments of all employers leaving the premises after normal business hours. The person may have a valid reason for delaying his departure. On the other hand, he may have been concerned with some illegal activity which is not readily apparent. By recording his name, verified by plant pass or other identification, in the gate register, along with the time and date, the plant guard has established a record of the fact that the employee left at a later hour. He can then be subjected to questioning if it later develops that something happened in the plant during this period of time.

No employee should be allowed to reenter the plant after closing hours without authorization. This authorization should be either in written form or by direct verbal communication between the plant official with authority and the guard. The guard should not accept the employee's word that the proper authority was obtained.

VISITOR IDENTIFICATION AND CONTROL SYSTEM

Absolute security with respect to visitors could be accomplished by the prohibition of all visits, but this is not practicable because certain visits are necessary to the conduct of the business in which facilities are engaged. There are also important values occurring to our economic and technological development and that of foreign countries in permitting visits. Thus, the problem is one of developing a policy which will provide for a judicious balancing of the divergent interests involved; namely, the importance of protecting facilities against espionage and sabotage, and the contribution visits make to economic and technological progress. Management is responsible for making this policy decision. It would appear to be entirely feasible for management to limit visits to those which serve a useful purpose, restrict access to visitors to certain areas of the facility and provide security precautions to control the acts of visitors within critical areas of the facility or installation.

Security checks of visitors by facilities' management is manifestly impracticable. Furthermore, such checks cannot be conducted by the government as the volume would be too great, and the time required to run a security check would be generally disruptive of the accomplishment of the purposes of the visit.

SCREENING

Physical security precaution against espionage and sabotage requires screening, identification and control of visitors. Remember that the mere wearing of a uniform is no guarantee of official sanction to enter a plant. Police and military personnel should be required to show their identification, and be subject to the same regulations as other visitors. Visitors to vital facilities are generally in the following categories:

1. Persons with whom every facility must have dealings in connection with the conduct of its business, e.g., representatives of suppliers, customers, licensors, or licensees, insurance inspectors or adjustors, government inspectors (national, state and local), service industry representatives, contractors' employees, etc.

2. Individuals or groups who desire to visit a facility for a purpose that is not essential to, or necessarily in the furtherance of, the operations of the facility concerned. Such visits may be desired, for example, by business, educa-

tional, technical, or scientific organizations and individuals or groups desiring to further their particular interests.

3. Individuals or groups specifically sponsored by government agency organizations—such as foreign nationals visiting under technical cooperation programs sponsored by the International Cooperation Administration and similar visits by United States nationals.

4. Individuals and groups which the government generally encourages, but does not specifically sponsor, because of the contribution they make to economic and technical progress or to defense production in the United States and/or in friendly countries.

5. Guided tour visits to selected portions of facilities in the interest of public relations and sales promotion.

Through visa, immigration and naturalization, and related procedures, the government endeavors to exclude from the United States foreign nationals whose backgrounds indicate that they might engage in espionage, sabotage, or otherwise present a threat to the national security. The fact that a foreign national has been issued an entrance visa or admitted to the United States should not, however, be interpreted as assuring that no security problem would be presented by his admission to vital facilities. Before allowing any visits by nationals of any countries within the Soviet block, management should consult with the Department of State or the government agency having security cognizance of the facility.

CONTROL PROCEDURE

Arrangements for the identification and control of visitors may include the following:

1. Positive methods of establishing the authority or necessity for admission of visitors, as well as any limitations relative to access. If the visitor is coming to see an official or someone of authority within the plant, then this person should be contacted and his permission obtained for the visitor to come to see him.

2. Positive identification of visitors at the facility or re-

stricted area (to be visited) by means of personal recognition, visitor permit, or other identifying credentials. Employee to be visited should be contacted to ascertain validity of visit.

3. Availability and use of visitor registration forms and records which will provide a record of identity of visitor, time and duration of visit, name, address, citizenship status, name and address of employee, reason for visit and persons and portion of plant to be visited. This record must also reflect the name of the company official who authorized such visit.

4. Availability and use of visitor passes. Such passes should be numbered serially to correspond with the number in the register log and indicate the following:

   a. Bearer's name.
   b. Area or areas to which access is authorized.
   c. Escort requirements, if any.
   d. Time limit for which issues.
   e. Signature (or facsimile) and title of validating official. Visitor passes or badges normally do not contain the bearer's photograph. A badge of easily recognizable and distinctive design worn in a conspicuous place is generally adequate. However, some installations because of their sensitivity issue to authorized visitors passes or badge bearing their photographs. Such photographs may be taken by Polaroid®-Land process when the visitor arrives at the installation.

5. Availability and use of visitor badges embodying applicable features to those established for employee badges; and, in addition, distinctive characteristics which will make obvious that the wearer is a visitor; escort requirements, if any, and any limitations as to access.

6. Procedures which will insure supporting personal identification in addition to a check of visitor passes or badges at restricted area entrances. Each visitor should also carry a written pass containing the bearer's name, the areas to which access is authorized, the escort requirement, the time limit of the pass, the purpose of the

visit, the person or persons to be seen and the signature of the validating official. Admittance and movement of visitors should be rigidly controlled.

7. Procedures for escorting visitors having limitations relative to access through areas where an uncontrolled visitor, even though conspicuously identified, could acquire information for which he is not authorized. Visitors should be escorted to the officials visited. The person being visited should be held responsible for the visitor. Visitors should not be allowed to move unattended throughout the plant. Guards within the plant should check on visitor's movement to assure that visitors do not enter areas for which they do not have the required authorization. Escort personnel may be guards or representatives of the person visited. Escort, when required, must be continuous without being obnoxious.

8. Controls which will recover visitor passes or badges on expiration, or when no longer required. Visitors indicating intention to return at a later time should not be permitted to retain their identification badges.

9. Twenty-four-hour advance approval when possible.

10. Ban cameras, except when previous permission has been granted to take photographs, and supervise the actual photography to make sure that restricted equipment is not included in the picture, whether intentionally or not.

Enforcement of access control systems rests primarily on the facility guards. It is essential that they have the full cooperation of the employees, who should be educated and encouraged to assume this security responsibility. Employees should be instructed to consider each unidentified or improperly identified individual as a trespasser. In restricted areas where passes are limited to particular zones, employees should report movement of individuals to unauthorized zones. Any suspected acts of espionage, sabotage, or other subversive activity should be reported immediately to the nearest field office of the FBI.

Wearing of Identification Badge

Each visitor allowed entrance to the plant should be required to display his identification badge prominently; preferably, it should be attached to his coat lapel.

Access Lists

Added security against unauthorized entrance to a restricted area may be gained by the use of a printed or typewritten list of all persons who are authorized to enter. Regardless of the type of pass carried, each bearer should be checked against the list. Anyone not appearing on the access list should be denied admittance, and detained; the central guard office should be notified. Each time that a permanent addition or deletion is made of those authorized to enter, a new access list must be prepared and the old one destroyed.

REGULAR VISITORS

Visitors who repeatedly return to the plant and/or who remain on the premises for several hours during each visit should be subject of study and special consideration. Such visitors include vendors, tradesmen, utility service men, special equipment service men and outside contractor employees. These visitors may be given a special visitor badge, the color, shape, or number of which indicates that they are restricted to a given area and are provided with a pass to only that area, in lieu of escort.

Such a pass should be issued only upon receipt of a certificate from the visitor's employer to the effect that the visitor is a trusted employee and is, in his opinion, a good security risk. The pass should give the name of the individual, the badge number, the limits of the area in which he is allowed, the time limit for which issued, and it should be signed by the security or other validating officer.

Design of Truck Driver Identification Badges

Badges for truck drivers and helpers should be as follows:
1. Of distinctive size, color and shape.
2. Issued for a single entrance and should be retrieved upon departure.
3. Bear an easily read serial number. This number should be entered on the truck register and meticulously checked upon departure.

CONTRACTOR EMPLOYEE CONTROL

Employees of outside contractors working within the perimeter of the plant should be carefully screened and their movements controlled. Contractor employees who have either limited or direct access to the plant area should be properly screened, identified and their loyalty vouched for by their employer. Such personnel should be checked as carefully as are the normal plant employees.

When a substantial project is carried on for a considerable time, the area of the contract project should be separately fenced and otherwise guarded. The contractor's employees should enter this area without entering the plant proper. When the contractor's employees must pass through the plant area enroute to or from the contract area, guard surveillance should be established to assure that they go directly to their work without straying into the plant or loitering in restricted areas. When the working area is physically separated and adjoining restricted areas are properly guarded, it should not be necessary for the contractor's employees to undergo special screening or investigation by plant management. Such employees should, however, wear distinctive badges so that it is readily apparent that they are not regular employes of the plant.

When fencing is impractical, or when contract work is engaged in at regular or irregular intervals covering a comparatively short period of time and employing individuals not well known to plant management, the contractor's men should be processed in the same way as visitors. In addition, intensified surveillance of the contractor personnel must be effected. If the employment is for a long period of time, the contractor's employees performing continuous service within a restricted area of the plant should be controlled by the same procedures adopted for regular plant personnel.

UTILITY AND MAINTENANCE PERSONNEL

No group of occupations has been used as successfully and as often as a "cover" for unauthorized entry as that of service and maintenance personnel. Appropriate clothing, a toolbox, and a smattering of technical knowledge are the only requirements to pose as a telephone repairman, an electrician, a plumber, or a business machine maintenance man. Legitimate employees of public utilities and some commercial service organizations usually carry company identification. However, they should not be admitted to a restricted area without a telephone check with their home office to establish their authenticity and without a check with the person who purportedly requested the service. Movement within the installation should be subject to the same pass and escort procedures as are other visitors. Maintenance personnel employed by the installation on a full-time basis may be accredited in the same manner as other employees.

PACKAGE CONTROL

The establishment of a definite system for the control of packages entering or leaving the plant's production or manufacturing areas is an indispensable aid in preventing theft, recovering lost or stolen property, and preventing sabotage or espionage.

The first step in establishing such a system is to establish limitations as to types of property authorized, persons allowed to move authorized property, and approved points of entrance and exit. No packages should be permitted to be brought into the plant by employees or others unless they are opened by members of the guard force and thoroughly inspected. Employee lunch boxes should be spot checked from time to time.

As a means of safeguarding essential vital information, production tools and equipment, inspection of all outgoing packages is desirable. In lieu of 100 per cent inspection, frequent unannounced spot checks should be considered.

Package Passes

Such systems usually require employees to check all packages brought into the plant and to display the checking mark or authorized package pass when leaving the premises at quitting time. The system should also cover visitors and salesmen carrying personal packages or samples. These passes may be relatively simple in form. They should contain the following data:

1. Employee's name and department or time clock number.
2. Time and date of removal.

3. Description of package or items being removed from the plant.
4. Signature of supervisor or other person authorized to approve package passes.

It is desirable to have these passes numbered serially. This enables a better degree of control over them as they can be charged, by number, to the person who is to issue them. The number of authorized signatures for such passes should be restricted so as not to be excessive, and so that effective control over the system is not lost.

In a large plant where there are a number of persons authorized to sign package passes, a list of the authorized signatures should be posted with the guards at the exit gates. When in doubt concerning the validity of an authorized signature, the guard may then refer to his authorized signature list. When the package control system provides for the sealing of checked packages, a broken seal should be sufficient reason for a guard to inspect the contents in order to insure there is no irregularity.

A sample package pass procedure is as follows:

1. A package pass must be obtained by any employee who wishes to remove any property from the premises.
2. The original copy of the pass is given to the employee and the duplicate is pasted on the outside of the package, sealing the package from further tampering.
3. These packages are delivered to the exit gate used by the employee prior to quitting time and sufficiently early to allow time for any inspection deemed appropriate by the guard who shall have authority to make any inspection he believes is necessary.
4. The responsibility for the contents of the package rests with the person signing the pass.
5. The original of the package pass, when presented by the owner, is compared with the serial number on the duplicate before the package is released to the employee.
6. No bundles or packages should be permitted to leave the properly authorized package pass. Verbal instructions to premises after closing hours unless accompanied by a

permit the removal of certain packages should be discouraged, for it provides a basis for misunderstanding or a later denial that such instruction was actually given.

VEHICLE CONTROL

A method of removing misappropriated property from the plant is by concealing it in vehicles or by adding it to loads of merchandise legitimately leaving the plant. Goods may also be concealed under cars leaving the area. By the same token sabotage equipment may be smuggled into the facility in the same manner. Thus it is essential to establish a system of close control and inspection of all vehicles entering or leaving the plant.

Parking Areas

Whenever practicable, parking areas for privately owned vehicles of employees and visitors should be located outside the perimeter of the restricted area. By installing employees and visitors' parking lots outside the plant enclosure, the necessity for inspecting such vehicles daily will be eliminated and the ability to use the vehicle in criminal activity will be greatly decreased, if not eliminated.

If it is found impractical for all cars to be parked outside of the restricted area, then only employees should be allowed to park within the enclosure. In case of interior parking, the parking areas should be located away from important processes and separately fenced in such a manner that occupants of automobiles must pass through a pedestrian gate before entering the facility. The parking areas should be fenced, lighted and have separate entrances and exits. The method of parking should be clearly marked and strictly enforced. Security guard personnel should monitor movements of personnel to parking areas to prevent unauthorized removal of property.

Sufficient parking space should be provided for visitors near access control points to prevent visitors from entering areas other than as required to conduct their business. Visitors parking areas should also be under close scrutiny of security guards to prevent unauthorized removal of property.

Vehicle Registration

Vehicles entering a protected installation must be subject to control. Privately owned vehicles of regularly assigned or employed personnel should be registered and should be required to display a tag or decalcomania. In addition, all other privately owned vehicles entering the installation on a continuing basis should be registered and should display a distinctive marking.

Prior to receiving a parking permit, the employee is usually required to submit a parking application form through his department manager wherein he authorizes the plant security personnel to conduct a reasonable search of his vehicle at any time when the vehicle is entering, leaving, or is parked on plant premises.

A list should be prepared showing the following:

1. The name of the driver.
2. The permit number.
3. Name of car.
4. License number.
5. Period for which the permit is authorized.
6. The designated parking area.

Such vehicle registration does not excuse the driver from compliance with the regular identification and admittance procedure in force. Visitors' vehicles should be registered. Visitors should be given a detachable card or sticker bearing the time limit of the visit and information regarding searches. It should be surrendered upon departure.

Truck Control

Incoming trucks should be kept to the minimum essential for the efficient operation of the facility. A definite system should be in use to control the movement of trucks and other goods conveyances into, out of and within the facility area. Insofar as possible, loading and unloading platforms should be located outside the manufacturing and facility operating areas and separated therefrom by controlled and guard-supervised entrances. If this is not possible, turn around areas for loading and unloading should be located at or as near as possible to the truck gates. This will help to eliminate irregularities by dishonest employees and truckers in shipments being loaded or unloaded.

All trucks and conveyances should be required to enter through a service gate manned by plant guards. When such an entrance permits direct access to manufacturing and plant operating areas, truck drivers and helpers and vehicle contents should be carefully examined. The guard check at truck entrances should cover both ingoing and outgoing vehicles and should include the following:

1. Appropriate entries on truck register, including registration of truck, name of truck owner, signatures of driver and helper, description of load, and date and time of entrance and departure and signature of inspecting guard.
2. Identification of driver and helper, including proof of affiliation with company owning truck or conveyance.
3. Check of vehicle operator's licenses of driver and helper.
4. Examination of truck or other conveyance for detection of explosives, incendiary devices, or other hazardous items and for unauthorized passengers and cargo.

The importance of vehicle inspection must be impressed upon security guards. Even though the guard knows the driver, he is inviting trouble when he merely picks up the driver's pass and casually waves him through with a friendly nod of the head. The cabs of trucks and tractors should always be examined and should occasionally more thoroughly spot checked, including an inspection underneath the hood, seat cushions, between the radiator grill, under an apparently carelessly thrown tarpaulin, or in other likely places of concealment. A complete inspection will include the underside of the chassis and body. This can be accomplished by the use of mirrors or a pit from which a guard equipped with a portable floodlight can make a thorough inspection.

Identification badges should be issued to truck drivers and helpers who have been properly identified and registered. Such badges should permit only limited access to the plant, that is, to specific loading and unloading areas. Vehicle escorts should be provided if vehicles are permitted access to the manufacturing, plant operating, or any restricted areas. Guard or other supervision of loading and unloading operations should be strictly adhered to, to make sure that unauthorized goods or people do not enter or leave the plant via trucks or other conveyances.

Check List for Entering Vehicle Clearance

The following procedure should be used in clearing a vehicle for entry:

1. Record license.
2. Record name and license of driver and other occupant.
3. Purpose for entry—pickup or delivery.
4. Inspect vehicle.
5. Complete entry pass and instruct driver that the following procedure will be used to clear the vehicle when leaving:
   a. Stop vehicle for inspection.
   b. Secure pass from driver.
   c. Check pass for proper description and destination of load and authorized signature for proper clearance of material.
   d. Make inspection of vehicle as to contents and check against shipping papers.
   e. Allow vehicle to proceed, and complete information on vehicle record.

RAILROAD CAR CONTROL

The movement of railroad cars into and out of plant enclosures should be so supervised as to prevent the entry of unauthorized personnel or material as well as the removal of property. All railroad entrances should be controlled by locked gates when not in use. Guards should be assigned to the railroad gates while cars are entering or leaving the premises, one to keep a general lookout and another to inspect cars. For good security control, a written record should be maintained of all trains switched in and out of the plant premises. Insofar as possible, loading and unloading railroad car platforms should be located outside the manufacturing and plant operating areas.

Before entry to the facility, the seals on all cars should be inspected for tampering and all unsealed railroad cars should be immediately checked to prevent unauthorized persons entering the plant area therein. Railroad cars should be carefully checked to prevent the conveyance into the plant area of explosives, incendiary devices or other hazardous items. Where railroad co-operation can be secured, and it does not materially interfere with efficient facility operation, railroad switching should be confined to daylight hours.

The numbers of the seals of all sealed railroad cars should be checked immediately upon arrival at the facility against the list of seal numbers which should be requested from the shipper. Broken seals or seal numbers not in accordance with advice from the shipper warrant immediate investigation.

MARITIME CONTROL

Docks, piers and quays should be separated from the main installation by a fence and controlled gates. Passage to and from the area must be limited to authorized and identified personnel. When the inspection of a vessel is necessary to preclude the entry of unauthorized persons, cargoes, or sabotage devices, it should be accomplished by persons experienced in the intricacies of maritime construction.

Chapter 11

IN TRANSIT SECURITY

INTRODUCTION

SECURITY PROBLEMS are increased when property and material are in transit. Loading and unloading procedures, compartmentalization of cargoes in ships, railroad cars, aircraft and movements of carriers all present security hazards of varying degrees.

PROBLEM AREAS

The major problem areas include the following:

INEFFICIENT HIRING METHODS. Pilferage and theft can be stopped effectively by hiring more carefully. Check personal, business and credit references of potential employees more thoroughly, especially the previous employer.

POOR EMPLOYEE MORALE. Causes of low morale are numerous. Whether unhappy because of salary, working conditions, or for any other reason, an employee may satisfy a grievance by stealing from his employer.

INDIFFERENT OR CARELESS SUPERVISION. Because supervisors are part of management, they must be selected with care. Employees are quick to spot and take advantage of a supervisor who is a misfit.

POOR HOUSEKEEPING. An array of discarded cartons offer good camouflage for stolen items and allows employees to retrieve the goods when the "coast is clear." Placing trask baskets and waste containers at designated points can keep these areas clear, make them easier to check.

POOR PACKAGING. Insist on tightly sealed or well-taped cartons and boxes. If opened only slightly, removal of goods is simple. Bad packaging also makes containers or cartons more prone to damage.

INSUFFICIENT SECURITY FOR YARD AND PLATFORM. Most truck

pilferage takes place on the platform, so this is the area to initiate rigid security measures.

POOR LIGHTING. Dim platform and yard lights make a fine cloak for removing goods unobstrusively. They also reduce platform visibility for dispatchers and supervisors, and contribute to an inefficient operation.

INEFFICIENT LOADING OF TRUCKS. Poorly loaded goods make theft easier in transit.

INADEQUATE COMMUNICATION. Departments, executives, supervisors and terminals must all have complete and coordinated records to combat pilfering.

INADEQUATE CONTROL of records, orders and forms.

TRANSPORTATION HAZARDS

Cargoes in transit are vulnerable to both overt and covert hazards, such as pilferage and sabotage. Either may occur while the carrier is at a standstill or while it is moving. Though generally not within the control of security personnel, accidents and natural hazards also present a threat to safe delivery of cargoes. Inclement weather can have a direct bearing on the sucrity of air and sea traffic, and by impairment of railbeds and roads, it can affect rail and highway transport.

LOADING AND UNLOADING

Two of the most vulnerable places during a movement are at the loading and unloading points. The fact that the carrier is at a standstill and the present of laborers and handling equipment present opportunities for pilferage and sabotage. Both loading and unloading should be accomplished as expeditiously as possible. Loading should be done when the cargo is brought to the carrier, and the carrier should be moved as soon as is feasible after the loading is completed. Unloading should commence as soon as the carrier arrives at its destination. Immediate handling of cargo reduces opportunities for loss or diversion.

PREVENTIVE PROGRAM

Once the problem areas have been evaluated, a system to make an employee think twice before he pilfers can do much to decrease substantially the staggering losses suffered each year by fleet owners. Following are suggested methods:

1. Improve personnel selection. A detailed employment application is recommended to obtain a full description of past experiences, background and references. A check on this information will soon turn up any false facts or discrepancies in the applicant's history. Inquire also into the applicant's eligibility for bonding.
2. Intensify training procedures to stress the seriousness and consequences of pilfering.
3. Caution employees to avoid visiting parked cars during working hours and isolate employee parking lot from the loading dock. This will reduce chances of pilfering.
4. Establish the same standards for supervisors as for other employees. A poor attitude by a supervisor in matters of honesty and good conduct affect the action of other employees.
5. Build morale. Keep the employee and his family up-to-date about company activities and policies through such means as newsletters, house organs, bulletins, payroll insertions, posters and other informative material.
6. Take disciplinary action if necessary. Theft violations cannot be ignored.
7. Discourage thievery by using seals and locks wherever possible; also control availability of keys and locks. Periodically change locks in critical areas to prevent unauthorized personnel from entering warehouses or storage areas.

   To reduce theft from trucks, lower the locking point for the rear gates and padlock them about one foot up from the bottom because most thefts from loaded trucks, where the gate is locked at the center, are accomplish by the bending out the upright on the gate.
8. Safety devices, such as guard rails and barred entrances on railroad cars, deter access to unauthorized personnel and prevent damage to equipment and injury to personnel. Tarpaulins, other coverings, ropes and dunnage for securing loads are also among the devices used to

prevent damage and loss.

9. Code markings on containers, railroad cars, and trucks are used to indicate the nature of the supplies contained. Different classes of supplies are marked not only with distinctive symbols, but each of the technical services uses a distinctive color for marking. Marine shipments are usually marked with an additional symbol designating the port of destination. These symbols and markings afford a ready identification to security personnel in making inspections and safety checks and aid them to determine the degree of security required.

10. Maintain both an inside and outside security force on a twenty-four hour basis. Guards should be screened carefully and trained in modern security procedures.

11. Issue identification badges, preferably with pictures of employees. Unauthorized persons are then more easily spotted.

12. Have guards inspect lunch boxes and packages. Let it be known right from the start that management reserves the right to check all packages and containers taken from the plant.

13. Instruct drivers and dock workers thoroughly on procedures for handling at time of pickup or delivery. Require reporting to make sure all goods tally with the shipping record.

14. Establish controls to check freight handling reports and other records which can reveal shortages.

15. Establish close communication. Interterminal communications will notify personnel of imminent arrival of trucks and encourage prompt checking of shipments.

16. Inspect packages frequently for damage. If broken or damaged, reseal and repair containers immediately.

17. Check freight-stowing frequently to make sure freight cartons can withstand stresses and strains of loading.

18. Establish stringent security procedures, especially for value shipments. Keep valuable cargo in locked rooms or in an area where it can be guarded and observed at all times.

19. Request police assistance. Law enforcement agencies and the FBI can help plan a preventive program. They can also outline procedures for prosecuting violators.

20. Plan trailer parking near offices so trailers carrying high-value goods can be watched more easily by guards and supervisors. For weekend security, place loaded trailers back-to-back, or back them against terminal wall or dock doors.

21. Make sure trailers are spotted correctly at docks to avoid possibility of cargo dropping between dock and tailgate.

22. Fence in terminals. A fence, however, isn't enough. Place guards at entrances and install adequate outside lighting for maximum security.

23. Keep a watch on outside servicemen working on your premises. This precaution can save employees from blame and minimize pilfering losses.

24. Consider the use of electronic alarms on trucks and trailers in transit. Such devices offer protection from hijackers and are particularly recommended for transporting shipments of high value.

25. Control the issuance of all orders and establish a system of checks of each step in the processing of said orders to assure that no unauthorized shipment can leave the facility without being discovered.

26. Consult specialists. Helpful data based on wide experience in the prevention of thefts is available from insurance companies, detective agencies, loss-prevention experts, and associations in which a company retains membership.

In addition to the points cited, fleet owners can enlist the cooperation of shippers and consignees. Following are five additional factors which should be called to the attention of shippers and receivers:

1. Full bill of particulars. Papers for all freight should be in order and should include the full address of both shipper and consignee. Bills of lading should give complete routing instructions and a description of shipment.

2. Prompt reporting of losses. Fast claims filing gives everyone the opportunity to stop possible pilferage promptly.
3. Printed tapes to combat pilfering. Specialty tapes show immediately whether anyone has tampered with seals.
4. Check of goods received. Until an entire shipment can be accounted for, do not sign anything.
5. Controlled deliveries. Do not let outside delivery men into stockrooms and storage areas unless supervised. Whenever possible, have outside men leave goods on the platform.

ACTIVE SECURITY MEASURES

Active security measures may supplement preventive measures or be used separately. To make certain that maximum benefits are derived from security devices, trained personnel should supervise the proper installation, utilization and maintenance of such devices. Specially trained guards may be used to safeguard critical or sensitive supplies or equipment in transit. Cargoes which are moved by rail or road may be additionally safeguarded by integrating dispatch operations with security operations.

HIGHWAY TRANSPORT

Loading

The method of loading can contribute to security by the placement of sensitive and pilferable items where they are not easily accessible. Equitable weight distribution is a consideration in loading, but sensitive and easily pilferable containers can usually be placed forward or in the middle of the load. When this is not possible and van type trucks are not available, one or more large items across the rear of the load will offer some protection. If a flatbedded vehicle is used, the load should be covered with a tarpaulin and lashed diagonally.

Convoys

When practicable, trucks traveling to the same destination or in the same general direction should be grouped into convoys for protection of drivers and cargoes. Armed escorts may accompany the convoys.

Routing

Specific routes should be designated for convoys. Arrangements should also be made for alternate routes, refueling points, parking and billeting areas. Whenever possible, routes having steep grades, obstructions, one-way defiles, or heavy traffic are avoided to reduce the vulnerability to pilferage occasioned by slow speeds and halts. Where such routes cannot be avoided and attempts at pilferage are expected, extra precautions may be taken by assigning police or other patrols to the critical portions of the route.

Security Personnel

A convoy traveling alone is vulnerable to attack and should be preceded, accompanied, or followed by security detachments capable of protecting the convoy. Police or tactical units may be assigned as escorts.

RAILROADS

Inspection of Railroad Cars

Prior to loading, each railroad car should be thoroughly inspected. Cars which have holes or damaged places in their floors, roofs, or sides do not provide effective security, and their use should be avoided where possible. If a damaged car must be used, minor repairs should be made, and notice of car condition should be given to the conductor and the commander of the train guard. The latter can then give particular attention to those cars which present additional hazards to the security of their contents.

Loading and Sealing

Careless, loose loading of freight contributes to loss and damage caused by the movement of the train and invites pilferage. The standard method of sealing a railroad car door is by means of a soft metal strap and a numbered seal. Its purposes are to indicate that the car has been loaded and inspected and to reveal tampering, but it offers little protection. Better protection is provided by heavy duty padlocks, or by tightly twisting a length of heavy wire through the locking eyes and closely snubbing off the wire ends.

Grouping Cars

It is standard railroad practice in making up trains to group the cars according to their respective destination. If, however, cars containing pilferable freight can be grouped within the train it will permit economy in the use of guards. This method is particularly adaptable when the entire train has the same general destination.

Open Railroad Cars

When it is essential that train guards be employed, they are placed to permit continuous observation and protection of open, flat, or gondola-type cars used for shipment of sensitive or easily pilfered items.

Mission

The mission of the guard force is to insure that the train arrives at its destination with freight intact. The strength of the guard force is dictated by the sensitivity of the cargo, the priority of its need, and the circumstances prevailing in the area to be traversed.

Where a few guards are sufficient to secure a car or cars containing sensitive freight, they may ride either in the specific car to be protected or in the caboose. If the guards ride in the caboose, they must be able to observe the sides of the cars, and to dismount and observe the cars from positions in the roadbed when the train stops. The guard force should not be spread throughout the length of the train, but concentrated in one or two positions.

There should be radio communication among personnel deployed in two or more locations. When possible, there should be radio communication between train and police units and tactical units in the area.

Vulnerability of Railroads

Railroads present many targets for the saboteur. These targets range from a switch thrown the wrong way to the demolition of a trestle. The destruction of switches, signals, or track may be only a harassment, or it may trigger a chain reaction of a larger scope. The destruction of a bridge or a tunnel may disrupt a whole railway system and may require a long time for repair or replacement. Each individual bridge and tunnel must be considered as a separate security problem.

Bridges

A railroad bridge, because of the weight which it must support, may be rendered unserviceable merely by being weakened. A bridge approach or abutment is extremely vulnerable to attack. For a single span bridge, the destruction of an abutment is usually sufficient, as not only is the bridge wholly or partly wrecked but also the destruction of an abutment makes it difficult to obtain a footing for the foundation of a new bridge on the same site. For a multiple span bridge, the demolition of an intermediate pier usually has the same effect as the destruction of an abutment.

SECURITY MEASURES FOR A BRIDGE

The security measures required for a bridge are determined by its sensitivity. The sensitivity is determined by a bridge's location, its relation to other structures and alternate routes, and its proximity to populated areas. Usually the most effective security measure is a stationary guard force, although mechanical aids may be used as a supplement. Guards should be so placed at both ends of the bridge that they observe its understructure as well as its roadway. The guard force should be quartered at a safe distance from the bridge but near enough for personnel not on duty to be readily available in an emergency. The full length of the bridge should be inspected at irregular intervals. Guard dogs may be used to supplement personnel.

In addition, the following protective steps may be taken:

1. Light up bridges where possible.

2. Floodlight river above (upstream) bridge.
3. Place booms across water above and below bridge. Attach barbed wire to bottom of booms to prevent swimmers diving below booms to attack bridge supports.
4. Drop small depth charges or grenades into water at irregular intervals. This will cause any underwater swimmers to come to the surface as they will not be able to stand the pressure on their eardrums.
5. Have small boats patrol above, below and under the bridge.
6. Assign additional guards at water level to detect swimmers or floating objects.

TUNNELS

The most vulnerable point of a tunnel or tube is the place where it passes through loose or shifting earth, sand, or other unstable material. At such a location, a saboteur may attempt to destroy the lining by placing explosive charges along the crown or haunches. It may be sufficient to destroy one side of an arch ring in this manner, for the pressure of the overburden may bring down the roof and fill a section of the tunnel. This type of destruction is not possible in firm soil or solid rock without the use of large breaching charges, which saboteurs usually avoid due to the difficulty of placing them surreptitiously. However, a similar but not as serious a result may be obtained by derailing a train in the tunnel. Ventilating shafts are also vulnerable points.

Security Measures for Tunnels

In general, the security measures which apply to bridges also apply to tunnels. However, special precautions must be taken to protect ventilating shafts or other outlets that extend above the ground. In addition to the guards, an alarm system may be used to detect tampering with ventilating equipment or power units. If motor vehicles use the tunnel, a wrecker should be stationed at each end.

MARINE TRANSPORT

General

The avoidance of manmade hazards is a matter of preventive security and safe practices, in addition to those protective measures peculiar to normal waterborne traffic.

Pier Security

The landward side of a pier can be protected by fencing and pass control, but the part of the pier which protrudes over the water cannot be protected in this manner. Not only is this part of the pier accessible from the sides and end, but also from the underside. Methods for securing the pier along its water boundaries are as follows:

PATROLS. Patrols in small boats should be used in pier areas to prevent unauthorized small craft from operating in adjacent waters and to recover jettisoned cargo. Patrol boats should be sufficiently narrow of beam to enable passage between the pilings when inspecting the underside of piers. Patrols walking along the end of the pier may be used separately or in conjunction with boat patrols.

LIGHTS. The lighting of the working area of piers should be adapted to the particular construction and work needs. The slips and underpier areas should be lighted sufficiently to give night protection. Lights under a pier can usually be affixed to the piling close to the pier flooring. Wiring and fixtures in this area should be waterproof to insure safety in the event of unusually high tides.

BOOMS. Under certain circumstances it may be advisable to close off the water side of a pier area by the use of booms. A floating boom will prevent entries of small boats. To deny underwater access, a cable net must be suspended from the boom.

ANCHORAGE SECURITY. When a port lacks sufficient pier space to accommodate traffic, ships may be required to anchor, or even to load or unload, offshore. Positions of the ships in anchorage are assigned by local port authorities. Cargo is loaded or unloaded by means of lighters, which also transport the stevedores to the ship being worked. This type of operation has advantages and disadvantages with respect to security of the ships. The trips to and from anchored ships give added time for the inspection or surveillance of the laborers, but it is difficult to control the movements of small boats which bring provisions to the ships. Such craft may be the instruments of pilfering, smuggling, or sabotage. Police water patrols and alert supervision of the stevedoring offer the most effective protection. Police water patrol boats should have sufficient power to overhaul other small craft.

Other Transportation—Air Transport

The hazards to air travel can result in even more disastrous consequences than those at sea. The protection of an aircraft against all hazards is primarily the responsibility of its crew. The crew may be augmented by guards for surveillance while awaiting shipment and loading. Every precaution must be taken to prevent the introduction of a sabotage device into the cargo. Cargo receiving points should be controlled to prevent unauthorized entry, and sensitive cargo should be kept under guard surveillance while awaiting shipment. Packages and containers should be checked against shipping papers for irregularities, and any item that excites suspicion should be removed from the cargo area and examined by competent personnel before it is approved for shipment.

PIPELINES

The location of a pipeline and its sensitivity determine the security measures required. From a security standpoint, the ideal location is parallel to a highway or road and close enough for observation from the road. When local conditions or terrain prevent such a convenient location, other means of protection must be devised, which include air surveillance or mounted and/or foot patrols. Initial security is provided by operational and maintenance personnel, but where their strength is insufficient for the duty, guard patrols may be required. Particularly vulnerable sectors of the pipeline, such as isolated areas and pumping stations, may be protected by guard detachments. Sentry dogs may also be used to advantage in such locations.

An industrial monitoring system that can detect tampering and localize the area of interference can save manpower and insure more rapid repair.

Chapter 12

EMPLOYEE SECURITY PARTICIPATION

INTRODUCTION

STRUCTURAL AND MECHANCIAL aids to security are valueless without the active support of all personnel. Thus the management of each facility should encourage employees to participate voluntarily in plant protection activities. Employees may be of invaluable assistance in the promotion of plant security by supplementing the efforts of regular plant guards and watchmen in the detection of sabotage, thefts and other acts and conditions which threaten a plant's production.

Employees should be on the alert to detect all security deficiencies, and attention should be particularly focused on those deficiencies which may permit unauthorized entry to the plant or plant area and subsequent actions of espionage or sabotage. Such security deficiencies should be immediately reported to the plant protection officer.

At the outset we must recognize and realize that one of the biggest problems to overcome is apathy. Once we recognize this apathy, we can launch our campaign to overcome it. All companies have something to sell whether it be a service, electronic computers, or lead pencils. To stay in business you must develop a demand for it. This same principle holds true in successful security planning, you must create a demand for it.

Most advertisers will agree that the best way to create this demand is to demonstrate the need for the product. Make the customer demand your product to fulfill his need.

It should be brought to the attention of each employee that his own life and livelihood, as well as the welfare of the country, depend on the security and continued production of his plant. Any suspected act of espionage, sabotage or theft should be reported directly by the observer to the plant security officer. Too often in the past, honest employees have overlooked or failed to prevent and report dishonesty among their fellow workers because management has failed to impress upon them how industrial thefts hurt all employees. The fact that thieves cut profits and thereby increase layoffs comes under this heading. Workers can be encouraged to "police" themselves through this program, by giving them examples of how thefts hurt them personally. Union leaders and the officers of other employee groups, the company newsletter and the regular training program may be used successfully to get this message across. Keeping adequate loss records that are reported to the employees annually, awarding an employee bonus as losses decrease, and making specific use of the employee suggestion system to elicit "stop-loss" recommendations will demonstrably reduce losses of applied intelligently and consistently.

EDUCATIONAL PROGRAM

The keystone of sustained employee security participation is a continuous and forceful employee education program. It assures constant employee awareness of the problem and need for their help. The minimum requirements of an effective program should include the following:

1. Direction and guidance at management level, and active support by top management.
2. A mandatory indoctrination for all personnel at time of assignment or employment.
3. An advanced and continuous program to selected audiences on timely and applicable topics to develop and foster a high degree of security consciousness.
4. The distribution to supervisors of a monthly kit consisting of posters, placards and suggestions pertaining to the physical security of the installation and the participation by all personnel in support of the system.

OBJECTIVES

The goal of a security education program is to acquaint all personnel with the reason for security measures and to secure their cooperation. The assumption by employed personnel that they are not concerned with security unless they work with classified matter or in a restricted area must be overcome. This can be done by emphasizing the loss to each of them individually in terms of pay increases and job security. It must be impressed upon them and continually reiterated that a locked gate or file cabinet does not constitute an end in itself, but is merely an element in the overall security plan. At any facility, the security officer should attempt to instill a willingness on the part of all personnel to learn and comply with all security systems. Likewise employees must be persuaded to personally remain honest. This is best done by impressing upon them the hazards involved in stealing.

One graphic illustration may be made by showing the income loss of an employee discharged because of theft and comparing it to the value of the goods stolen. Another method is to place the most frequently stolen tools on a bulletin board and attach a price tag to each tool under the caption, "Isn't your future worth more than this?"

Let the employee know what he is up against. Publicize the fact that a well-trained security force is in the plant for the protection of all concerned and for the apprehension and exposure of pilferers.

PROGRAM OF INSTRUCTION

The security officer is responsible for planning an effective program of instruction. Profitable use of the limited time devoted to receiving instruction demands the technique of a competent instructor. The security officer should give the more important portions of the instruction. Other competent instructors may be used for less important phases.

SECURITY INDOCTRINATION

All newly assigned personnel must be given a security indoctrination. The reading of printed security regulations is not sufficient to insure complete understanding. Indoctrination should consist of a general orientation on the need for and the dangers to security, and the individual's responsibility in preventing infractions. It should include a discussion of those hazards common to all personnel, with emphasis on the dangers of loose talk and operational carelessness. It should define in detail the general security measures, such as the pass system, private vehicle control and package inspection. The security indoctrination is an introduction to the subject as applied to the particular installation. Further instruction should be applicable to the individual's duty assignment.

AUDIENCE PLANNING

After the indoctrination, continuing security education should show a direct relation to an individual's duty in order to create and maintain interest. Segregation of personnel by the level of security education they require will usually conform to their general academic background and make it possible for the instructor to select a level of instruction best suited to the audience. Separate instruction programs for each group should present subjects within the group's security interest. Suggested groupings and the general area of their security education follow:

1. Supervisors of classified activities and security control officers should receive instruction in the general security responsibilities of their positions, the physical aids available as safeguards and the enforcement policies and procedures of the installation.

2. Technicians, craftsmen and others having access to classified information or material should be instructed in explanations of the terms *chain of custody* and *need to know* and in telephone security and destruction procedures.

3. All personnel whose normal duties do not require access to classified information or material should be reminded constantly of the dangers of unrestrained talk about the duties they perform or materials they handle.

SCHEDULING AND TESTING

Frequent short periods of instruction are more effective than less frequent, long periods. The ideas contained in four well-planned, weekly, fifteen-minute classes are more readily absorbed than those contained in an hour lecture once a month, regardless of how well the latter is planned and delivered. Instruction that infringes on the free time of the audience is seldom well received. Short periods of instruction to selected groups are easier to schedule without disrupting the operation.

In any form of instruction, testing serves the dual purpose of keeping the audience alert and indicating the efficiency of the presentation. Tests do not necessarily involve written answers. In fact, skits and hypothetical situations tend to enliven the instruction and by audience participation in giving the consequences or solutions accomplish the same result.

PUBLICITY CAMPAIGN

A good publicity program can cut theft losses by a staggering percentage. It is by far the easiest and most effective way to combat crime in the plant. All methods of communication, plant papers, bulletin boards, meetings, even the grapevine, should be utilized.

Here is what to emphasize:

1. Crime does not pay. Prove it by publicizing discharges for thefts. Let employees know you mean business.
2. Prompt reporting. Urge employees to notify the plant security people immediately when articles are missing. The sooner a loss is reported the quicker management can take action.
3. Crime costs plenty. Displays, with price tags, can be made showing the high cost of stealing and why it cannot be tolerated.
4. The plant is secure. Stories on cooperation from local police, the plant guard force, and the plant's detection methods will show an employee what he is up against—in a subtle way.
5. Self-protection. The employees should be informed what they can do to stop thefts of their own tools.

The selection of the proper media requires careful planning. The best written prose in the world is of little value if it is not read and digested. There are many different methods available to tell employees, and others, about the company's plans. Naturally, some will adapt themselves more readily to a particular plan than will others, but all are methods that have been used successfully for decades by businessmen intent on selling their products. Select the lines of communication which seem to be the most sought after by the group you wish to reach. A company's advertising and public relations experts may be of considerable help in preparing effective copy and in selecting the best media.

In the search for the proper media the first step should be a complete appraisal of the company's established information pipelines. Study these sources of information since everybody is already familiar with them and naturally turn to them for news on company happenings. Most of us like to do things the easy way and our "audience" is going to find it easier to learn about the plans if the information comes from an old familiar source.

The number one item on this list may be the employee paper or magazine which is normally read by all and is usually considered to be a reliable report of the recent activities of the company and its employees. Many organizations regularly use this publication for announcing changes in company police, new employee benefits and future plans.

Daily newsletter, bulletins, bulletin board postings, sales magazines and any other regular publication should be evaluated as a possible channel for the dissemination of information. In looking at these items, check carefully the normal makeup of the reader complex. To be effective, the medium must carry the proper message to its specific group of readers; for example, information appearing in bulletins that are normally directed to management will be prepared differently from that contained in a general employee announcement.

Because of their regular duties, supervisory personnel are considered spokesmen for company policy and employees naturally turn to these individuals for information. It is vital that all executives, supervisors, managers, foremen and section heads be thoroughly acquainted with the plan and give it their full support.

Routinely, almost every company holds meetings and seminars for its supervisory staff. The security should be given the opportunity to talk with the groups during one of their regular sessions, explaining and selling the program to them. It is usually best to work with and gain the support of these people before any attempt is made at general employee education. Provide the "bosses" with pamphlets and schedules so that they will be prepared to answer the questions of their people when the grapevine starts to function.

It should be pointed out to these supervisors that it will be an advantage to everyone concerned if they will hold regular meetings with their people to discuss the plan. Supervisors and foremen should be prepared to provide general instructions about what to do in an emergency and should be sure that everyone in their charge is acquainted with their part in the plan.

Poster, Placards and Leaflets

The example of modern commercial advertising leaves no room for doubt on the effectiveness of constant repetition of a message. The armed services and many private industries make use of posters, placards and leaflets to enhance their security education program. However, some thought must be given to their display in order to reap the maximum benefit. Little good results from circulation and initialing, and from display above eye level or in places where few people will see them.

Posters are usually large, depicting an eye-catching illustration with a brief and pointed message. They are designed to import their message at a glance. They should be displayed in locations where the maximum number of people pass or congregate, such as at entrances, gates, at the head of corridors, or on lunchroom walls.

Placards are usually small, may or not be illustrated, and carry a more lengthy message. They are designed for use where attention is ready made, and where people must normally stop or are expected to loiter, and have time to read. Examples of possible display locations are bulletin boards, counters, telephone booths and the vicinity of vending machines.

When posters and placards are issued in series, the method of display is dictated by the individual situation. Where display space is limited, they may be used one at a time and changed at intervals. Where the operation is scattered over a large installation, the posters and placards may be rotated among all the locations. When some of the posters or placards are particularly apropos to some phase of the operation, they can be given special display and kept posted for longer than normal periods of time.

Leaflets or pamphlets are usually pocket size and may or may not be illustrated. They are given to each individual, frequently as an enclosure in a pay envelope, and contain a brief and pointed message.

PUNITIVE ACTION

Occasionally, the employee education program may include strong punitive measures for serious offenders. When this is necessary, it must be done in a way to gain the respect of other employees, not their disgust.

When an employee is released for stealing, let it be generally known around the plant what has happened. In such cases it is best not to publish the name of the employee, even though it will, in fact, be known to all.

SECTION II
DISASTER ORGANIZATION AND PROCEDURE

Chapter 13

EMOTIONAL ASPECTS OF
DISASTER CONTROL

INTRODUCTION

IN ADDITION TO THE SECURITY provided by the protective measures already discussed, each installation must have a plan to
minimize the effects of a major disaster which could strike any
facility. This emergency plan should be designed to handle
situations such as fires, explosions, riots, major accidents, black-
outs and air raids. It must include the entire strength of the
organization in joint action to combat the common danger. Only
a carefully organized and rehearsed plan can assure a fast and
effective response to any situation that may arise.

A comprehensive industrial disaster control program for each
facility will, of necessity, vary greatly with the complexity and
physical characteristics of the facility itself. Emphasis is placed
on those items which enhance the security of the facility without imposing an undue and intolerable financial burden.

The objects to keep in mind are (1) The protection of
lives and property, and (2) the preservation of the corporate
organizational structure. In all action management must keep
uppermost in its mind the thought that the human factor pre-
dominates. The state of mind of the people is often the decisive
factor which determines policy and procedure.

PERIODS

In the pattern of individual reactions to an acute disaster,
at least three periods can be distinguished:

Period of Impact

This corresponds to the duration of the direct initial stress
of emergency. The behavior of the people may be confused,

disorganized and panicky. It may also be of a semiautomatic nature. In the face of an emergency many people calmly proceed to perform more or less aimlessly or to do otherwise useless acts. Well known is the tendency for an occupant to take, say, an umbrella from a burning building and leave money or other valuables behind, or to pass a fire extinguisher while looking for a bucket to carry water. Some persons have manifestly inappropriate responses such as confusion or paralyzing anxiety, inability to move or think, or uncontrolled crying or screaming.

Period of Recoil

This period follows immediately after the impact period and may last several hours to a day or more. Persons may show apathy, childish dependency, or hostility. Women may have typical alternate periods of crying and laughing, with some disturbed overt behavior. For the majority of survivors there will be a gradual return to self-consciousness and some awareness of what they have just passed through. There may be emotional outbursts and usually there is a need to talk out one's experience, to express anger and other typical feelings that result from the experience.

Posttraumatic Period

It is during this period that full awareness of the results of disaster develops and the individual is faced with an altered environment. This period is also characterized by the development of the more easily recognized types of psychoneurosis, psychosomatic symptoms, or delinquent behavior patterns. They include persistent anxiety states, fatigue states, recurrent traumatic dreams, depressive reactions and rage.

REACTION TO DISASTER

Normal

Some people are able to remain remarkably calm, at least for a time, even under the most extraordinary circumstances. Most, however, show obvious signs of disturbance. They will perspire profusely, tremble, or even feel weak and nauseated for a little while. Clear thinking may be difficult for a time. Fortunately, many can regain their composure, fairly well, soon after the first impact of a devastating experience. It would be misleading to classify all of these natural, and largely transient, reactions to stress as abnormal.

However, even very stable people are occasionally so overwhelmed by an event that they are unable to recover a semblance of emotional balance for quite a while and may require help from someone else. Calm reassurance, without pity, is usually best.

Abnormal

The four abnormal reactions to stress are individual panic (blind flight), depression, overactivity, and bodily disability. The victim may suffer from one of a combination, simultaneously or progressively.

INDIVIDUAL PANIC (BLIND FLIGHT). This reaction occurs far less frequently than one would expect, but its notorious contagiousness gives it an importance far out of proportion to its initial frequency. A very small number of individuals in true panic can easily precipitate the headlong mass flight of a crowd. The outstanding characteristic of panic is its blindness. All judgment seems to disappear and to be supplanted by an unreasoning attempt to flee.

In other instances, individual panic may be manifested by pointless physical activity, such as uncontrolled weeping or wild running about. Situations most conducive to this behavior (often called hysterics) are those where known escapes actually are, or are believed to be, threatened by progression of the danger. Sheer horror may also precipitate a very similar reaction. The sight of close friends or family who have suffered grusome mutilations may bring on wildly disorganized behavior in certain persons, especially if they themselves have suffered little or no physical injury.

Four main factors are characteristic of the panic-producing situation: partial entrapment, a perceived threat, partial or complete (or imaginary) breakdown of the escape route, and front-to-rear communication failure. The best general antidotes for most abnormal reactions are information, communications and calm reassurance.

Mass panic springs from the panic of a few. It is best avoided by keeping a watchful eye out for signs of individual panic; then isolating or calming the sufferer at once. However, use discretion and be calm, lest you create a worse condition than you cure.

DEPRESSED REACTIONS (SLOWED DOWN, NUMBED). In disasters many people will act for a time as though they were numbed. They may stand or sit in the midst of chaos as though they were alone in the world. Their gaze vacant. When spoken to they may not reply at all, or simply shrug their shoulders and utter a word or two. Unlike the persons in panic who seek physical escape at any cost, they appear to be completely unaware of the situation and devoid of emotional reaction to it. They are unable to help themselves without guidance. Fortunately, a great many of the less severely disturbed persons in this group can be salvaged rather quickly and with fair success.

OVERLY ACTIVE RESPONSES. In contrast to the victims just described, others will explode into a flurry of activity which, at first glance, may seem purposeful, but will soon be recognized as largely useless. They will talk rapidly, joke inappropriately, and make endless suggestions and demands of little real value. They will jump about from job to job and seem unable to resist the slightest distraction. They appear to flee into an unreal confidence in their abilities, which causes them to be relatively intolerant of any ideas but their own. Consequently, even one such person can become a disturbing nucleus of opposition to sounder procedures than he himself can actually propose.

BODILY REACTIONS. Some bodily reactions have already been mentioned as the normal temporary responses of many of us to unusual stresses. Although these "normal symptoms" are troublesome while they last, they generally do not interfere seriously with a person's ability to carry on constructive activity in a difficult situation. Certain of these reactions, especially weakness, trembling and crying may not appear until after a person has effectually met and surmounted the immediate danger.

Most serious and sustained bodily reactions can be truly

disabling, however, and it is important to recognize their emotional significance. Severe nausea and vomiting are particularly common. Because these symptoms may also result from serious exposure to radiation or to certain biological and chemical warfare agents, remember that many individuals not so exposed may suffer from these symptoms for purely emotional reasons.

Another type of bodily disability is basically different from the disorders just mentioned. This is technically called conversion hysteria (not to be confused with the condition of wild excitement usually designated by the popular term hysterics). In conversion hysteria, the person unconsciously converts his great anxiety into a strong belief that some part of his body has ceased to function. For all practical purposes he may be unable to see, or to hear, or to speak. All feeling or power may disappear from one or more limbs. The development of such conversion symptoms reduces the overwhelming anxiety previously felt. It should be emphasized that such casualties are not "faking" (malingering). They are completely unaware that no physical basis for their symptoms exists, and they are just as disabled as though they had a physical injury.

MENTAL DANGER SIGNALS

It is management's responsibility to spot the persons who need help and to see that they get help from a doctor or psychiatrist before it is too late for speedy recovery. Here are some of the danger signals run up by the mind that has had too much punishment and strain:

1. Look for anything that makes a man stand out in an awkward or queer way from the others in this group. Does he stay by himself too much? Does he go for long periods without speaking? Does he find conditions intolerable that other men get along under?
2. Look for any sudden change in the person's own personality. Has he become moody and depressed, boisterous and noisy, dirty and disheveled? These factors are signs of mental trouble.
3. Look for a marked increase in the following: absenteeism, absent rate, alcoholism, anxiety, arguments, arro-

gance, carelessness, complaints, dissatisfaction, loud talking, mental vagueness, "quits," resentment, rumor (and belief), scrap rate, malingering, suspicion, theft, and tool damage.

4. Look for a marked drop in the following: cheerfulness, dexterity, enthusiasm, food purchases, general efficiency, job interest, judgment, personal cleanliness, plant housekeeping, production rate and product quality.

INDICATIONS OF PSYCHONEUROSIS

When one has been through a particularly trying experience, more acute signs of psychoneurosis may show up. All persons should know that these signs are the natural result of fear and strain. They mean that the person needs care, rest, food, medical attention and psychiatric first aid. These signs are the following:

1. Inability to sleep.
2. Inability to eat.
3. Buzzing or humming in one's ears.
4. Shakiness, weakness in certain parts of the body, as the knees or wrists.
5. Dizziness, peculiar feelings in the heart.
6. Difficult breathing.
7. Restlessness combined with a feeling of being penned in, an overwhelming desire to push people and walls out of the way.

FIRST AID

Mental first aid is just as important as physical first aid for preventing casualties and losses. To be of maximum service to an emotionally disturbed person, certain general principles must be reasonably understood and believed. A brief discussion of them is provided here. With only basic training, the layman can be of considerable benefit to the victim as well as expedite recovery from a disaster situation. All emergency workers should be given the principles, and special training should be provided for wardens, auxiliary police and medical first aid people. Here is a summary:

1. Accept every person's right to have his own feelings. Do not blame or ridicule a person for feeling as he does. Help him cope with his feelings, do not tell him how he should feel. Realize how impossible it is for human beings to make a conscious choice of their deeper feelings. Freedom of choice and responsibility depend upon what is done about feelings, to relieve the tensions they create. Nothing is gained by trying to deny that the initial feeling existed just because it conflicted with what one may think a self-respecting human being should feel. Do not overwhelm a casualty with pity. This will only confirm his worst fears about himself. Just do enough to renew his confidence.

2. Accept a casualty's limitations as real. When a man's thigh is shattered, no one (including the patient) expects him to walk for a time. When a man's ability to cope with his feelings is shattered, many (often including the patient) are inclined to expect him to function normally again almost immediately. "It's all in your head." "Snap out of it." "Pull yourself together." Such goading, scolding forms of "reassurance" have no place in psychological first aid. These people do not want to feel as they do. The thing to do is to help them regain as much effectiveness as they can as rapidly as they can by accepting the handicaps they present, and by helping them rediscover quickly a few of their assets which they can use at once. They need your help—not your resentment.

3. Size up a casualty's potentialities as quickly as possible. While allowing for a disturbed person's limitations, be on the lookout for skills and other assets that can be revived and utilized. Inquire gently into what happened to him. Find out if he is particularly concerned about family or friends. Ask brief questions about his normal occupation. Let him talk for a few minutes about his own experience. Do not let him ramble, but even a few minutes will relieve some of his feelings of despair and helplessness. Give him as honest an estimate as possible of where, when and how he can reasonably contact those dear to

him. If he is too depressed to talk, some statement of what may have happened to him (based on your general knowledge of the disaster) can increase his confidence in you to the point where he can talk a bit. Show sincere respect for his feelings. Help him to find a way to utilize his skills. This is the best cure.

4. Accept your own limitations in a relief role. Do not try to be all things to all people. There will be much you would like to do in a disaster beyond your strength and skill. Establish a set of priorities in your thinking about what you will allow yourself to undertake. Quickly form a mental program of action and stick to it. Don't allow yourself to deviate in all directions and, utlimately, you will accomplish much more. Your first responsibility will be to whatever emergency job you have previously been assigned. Even under pressure to do some of the jobs assigned to others who may be casualties, do not accept more responsibility than you can handle efficiently.

FATIGUE

Mild disaster fatigue normally will soon pass without treatment and with no ill effects. It is not usually a concern to emergency personnel except as it somewhat reduces a person's effectiveness. However, if it does not disappear or becomes slightly more severe we may classify it as moderate disaster fatigue. If the fatigue does not pass or if it tends to become progressively more severe, treatment is indicated and persons so affected should have medical attention. In the case of severe fatigue, medical attention is mandatory. Unless medical service is promptly available the person may be past susceptibility to treatment and quick recovery. Immediate medical treatment may return persons to normal, prevent increased fatigue and loss of effectiveness, and avoid unusual problems of prolonged care and treatment.

The impact phase is a critical one because of its disruption of the community and its normal activities. Power may be lost, transportation may break down, and entire geographic areas may be physically isolated or destroyed. Persons will have disaster fatigue in all of its manifestations and degrees. Big problems will demand attention and action. As the postimpact phase is entered, recovering from the effects of the disaster will move toward the resumption of normalcy. At this point special attention must be given to problems of disaster fatigue, prompt treatment provided for those in need, and accommodations provided for those who may require extended care. Many who survive the critical emergency period may now have delayed emotional reactions, and attention must be given to identifying these people and providing them with necessary care.

If an individual can be relieved from his work for a time, given a rest when he needs it urgently, he can usually be counted upon to come back to the rehabilitation work with fresh vigor If, however, he is allowed to go on past his mental breaking point without letup, if he is permitted to wait until he collapses or until urgency of his needs requires medical attention, the chances are much smaller for his rapid recovery.

Chapter 14

PANIC

INTRODUCTION

Panic is the child of fear and ignorance. Panic is a violent fright and disturbance in a crowd, caused by mass fear resulting from collapse of buildings, fire in closed area such as a movie or night club, the occurrence of a crisis—any unusual event which signifies danger. It is an overpowering unreasonable terror with or without cause inspired by fright, creating highly emotional behavior, violent action, paralyzing immobility, or blind, unreasoning, frantic efforts to reach or secure safety by flight.

There is another type of panic that is not restricted to people who are physically together as a group. It seizes a person who, for one reason or another, is particularly susceptible to some suggestion, when face-to-face with a situation that seems critical. Panics generated by shocking news, false rumors, terror propaganda are of this variety.

It is impossible to convey in mere words a true perception of what a panic is. Words cannot convey the horror of waves of mankind breaking upon themselves, shattering and crushing each other in a blind, wild scramble of self-preservation.

CAUSES

The origin of panic lays in the fear of being exposed to immediate or imminent danger, such as death, or ruination physically, financially, or socially. To put it another way, panic is due to abnormal circumstances—such as fire in a theater, a ship sinking, earthquake—that are not anticipated, for which the people are unprepared, and which suddenly occur.

Danger and fear alone are not enough to generate it. There must be the possibility of escape and also the possibility for en-

192

trapment, for panic behavior is escape behavior—an attempt to avoid, to flee.

Characteristics of a Panic Producing Situation

The characteristics of a panic producing situation are as follows:

A PERCEIVED THREAT. The threat, real or imaginary, may be physical, psychological or a combination of both, and it is usually regarded as being so imminent that there is no solution except frantic efforts to escape and survive. The perceived threat may be a rumor such as one that a dike, levee, or dam has been blown up and that a tidal wave is rapidly approaching the crowd or mob; or that the area is dangerous due to radioactivity, deadly gas or germs.

It may also be an actual disaster, i.e., large fires, earthquake, or explosions caused by rioters or natural causes.

PARTIAL ENTRAPMENT. There is only one, or at best, an extremely limited number, escape route from a situation dominated by the perceived threat.

PARTIAL OR COMPLETE BREAKDOWN OF THE ESCAPE ROUTE. The escape route becomes blocked or jammed or it is overlooked.

FRONT-TO-REAR COMMUNICATIONS FAILURE. The false assumption that the exit is still open leads the people at the rear of the mass to exert strong physical or psychological pressure to advance toward it. It is this pressure from the rear that causes those at the front to be smothered, crushed, or trampled.

How large a crowd does it take to make panic possible? The answer is precise, but it is not the kind of answer we might have expected; it is not a fixed number. The answer is, it takes a crowd big enough to produce the entrapment dilemma in the given situation. The numbers must be great enough so that the exits are not completely adequate, so that there is a possibility of entrapment.

During the second World War on a transport which was plying between the United States and England there were, in a compartment below decks, eight members of the crew who had sailed together for many months. While they were resting during their off hours, unusually high waves struck the ship, cascading a torrent of water down the open hatch and into their compartment. Fear that the ship had been struck and was sinking seized them, and these eight companions simultaneously made a mad dash for the door, fighting and clawing with each other to get through. The first through jumped on the ladder, but before he had climbed more than a rung he was dragged down by his companions and replaced by another, who attempted to climb to safety. He, too, was dragged down by his hysterical friends, all now completely seized by panic. It was some time before they realized that there was actually no danger, but by then one of the men had suffered a fractured skull, another a broken arm, and all of them had sustained minor injuries. This clearly underlines the principle that any number of individuals can make a panic crowd.

PANIC RIPENESS

We find panic occurring in individuals whose resistance to fear has not been adequately strengthened, in those who are lacking in self-confidence, in those who have too little feeling of security, and in those who have a feeling of isolation and of not being a part of a group. Anything that makes persons tense, on edge, jittery and oversensitive to slight noises, half hidden sights, or sudden movements will make them easy victims of panic. For this reason prolonged anxiety makes one panic-ripe. So does overfatigue, too much beer or liquor, or a hangover. So does lack of proper food, especially a deficiency of B vitamins. Prolonged exposure to the noise and alarm of modern warfare may produce the jumpy state of mind from which panic arises.

Insecurity, whether actual or imaginary, sets the stage for panic. An emotionally insecure person feels himself helpless against overwhelming odds. This feeling may be purely imaginary or realistically justified, and the feared danger may be existent or nonexistent. In both instances, the immediate reaction is one of reaching out for help, either directly or indirectly. A feeling of emotional insecurity is augmented by a feeling of inferiority, inadequacy and a lack of affection, which increases the individual's sense of vulnerability in a hostile world. In other words, the individual lacks inner self-confidence and his basic approach to a situation is "I can't," instead of the confident "I can" of the emotionally secure individual. Prolonged tension before an incident may predispose a group to panic by amplifying unconscious anxiety related to interpersonal conflicts. When the group is tense and insecure and when imagined danger has not been dispelled, the stage is set for the introduction of defensive attitudes. These attitudes manifest themselves in a loss of initiative, a search for means of protection and escape, a secret fear and overestimation of the enemy and defeatism. When states of insecurity and tension become particularly marked, actual panic reaction may set in.

Another source of the mental anguish of panic is the realization that one is helpless to defend those one loves. Not only are the group ties broken, but the capacity to undertake any constructive action in the face of danger is lost. The loss of inner resources against disaster renders one powerless to stop destruction, and there is a tendency under these conditions to become panic-stricken. The lack of information or knowledge of some phenomena may contribute to panic and create a conducive atmosphere for rumor and suggestibility.

PSYCHOLOGICAL FACTORS

Let us now look more closely at the pattern of action that evolves in a panic. Take as an example a fire in a theater. Everyone knows what smoke is and how it smells. However, if the smell of smoke occurs in a theater, the reaction is totally different than if it is smelled outside. The audience may have been preconditioned about fires and panics in theaters. The subconscious mind instantly sounds an alarm. Everyone is more or less startled and aroused to action, but is incapable of achieving immediately a pattern for adjustment to the new set of circumstances. There is at first a sense of shock.

Psychological shock interrupts the normal course of action, causing the shocked person to freeze momentarily in his tracks. At the same time, it provides in him an irresistible urge to violate action. This is quickly followed by a comprehension of the danger with resulting great fear and terror. Individuals driven by overwhelming fear react emotionally and irrationally to avoid the source of the danger. Their basic instinct of self-preservation will assert itself and the result will be that of panic with everyone struggling to save himself and disregarding the safety of the whole. "Each for all" gives place to "each for himself." This reaction is automatic, even when common sense tells us that an orderly exit would probably save everyone.

All persons involved in escape crowd are not motivated by panic. Some keep their heads and react intelligently. No collection of reacting individuals can, however, long refrain from interaction with one another, if for no other reason than that they are likely to come into physical contact with one another. Inevitably, therefore, the period during which the members of a group react as individuals is brief and is followed by some form of collective behavior.

Panic interaction is a direct consequence of the mimicry by many panic-stricken individuals of the overt behavior of some one of them. The behavior of the one who sets the example may be of almost any order. Often it is a pattern of behavior which might be expedient for him alone but becomes highly destructive when it is taken over by the others. Thus the members of the panic group may act together, but the action may be illogical, irrational, fruitless and even dangerous to the group members. Each individual takes a "me first" attitude. There is ruthless disregard for others in the crowd. Indeed, there is the tendency on the part of each individual to fear the others and to believe that he is threatened on all sides.

This pathetic cycle need not occur. If, at the onset of the panic, firm leadership springs up which brings the crowd to its senses, the maladaptive and dangerous escape behavior may be avoided.

PANIC PREVENTION

The effects of panic can be minimized and controlled by adequate preparation a nd planning, good organization and practical training in the handling of panic situations. Every person must be indoctrinated beforehand so that he will know how to conduct himself and not become panic-stricken when a crisis arises. The following are suggested preventive measures that can be taken:

1. Do everything possible to keep down anxieties and to provide security for all employees. A basic rule for minimizing anxiety is to give people a routine. Once tasks, habits and expectancies have been set, do not make any more changes than are absolutely necessary. The possible anxiety that may be caused in a population by a shift in the way of doing things should itself be an important consideration in the decision whether or not change is practicable. If change is decided upon, give plausible reasons for it. Do not let anyone get the impression that you do not know what you are doing, that you are whimsical.

2. Teach obedience to orders. Create a respect for discipline and an ability to follow it.

3. Panic thrives on ignorance; full and appropriate information is the normal antidote. Keep in constant and close touch with the people to discover the points on which they want and need information. Do not allow any prolonged bewilderment or perplexity to bother any section of the people. Keep people fully informed on all subjects around which panics might start. The effect of insecurity and uncertainty is particularly severe if individuals are not told the character of a situation. Back up your information output with demonstrable, plausible facts and events.

4. Control rumors. Watch out for rumors, note their frequency and nature. They often are instrumental in creating a panic. The following steps will prove helpful in controlling rumors:

What an Individual Can Do

How can an individual check up on rumors and stop the spread of dangerous rumors? There are no sure answers, for rumors are as complicated as the human mind, but the following are some suggestions that will help keep you rumor-wise:

GET THE FACTS. This is basic. Knowledge is the best way to offset rumor. If someone tells you something about which you have the facts, and the facts are at odds with his story, you can set him straight and squelch the rumor.

Getting the facts means getting all the facts. It means hearing more than a few words of a newscast and reading more than the headlines of a newspaper's account. Headlines are written to catch the eye and make you read what follows, but they are rarely able to give a complete story in themselves. Getting the facts also means going to the source of facts, whenever possible. Moreover, getting the facts sometimes means waiting until the facts are announced by the proper authority.

KEEP A SKEPTICAL ATTITUDE. Be "from Missouri" on the tall tales you hear. Ask yourself, "Does that sound true?" "Is it likely, in the light of other known facts?" Asking a few questions of the storyteller may help. If you can make the rumormonger question his own story, he may cooperate with you in trying to find out the facts.

FIND OUT WHO BROUGHT THE NEWS. If you hear something that sounds as if it could be a rumor, ask the person where he heard the news and how he knows. Does he have reason to know what he is talking about? Did he hear or see the evidence himself or did the story come from a source of questionable reliability? If a startling piece of news can be traced to a well-known rumormonger, you will have good reason to doubt it.

LAUGH IT OFF. Ridicule helps kill rumor. Laugh down a foolish and dangerous rumor.

What Management Can Do

Experience indicates that efforts to track down the source of inflammatory rumors are seldom productive. When it is possible, trace false rumors to their source and discredit the individual who started them. Point out the fallacy of listening to unofficial information.

Even though it is impossible to find the source of the rumor, management can and should be on the lookout for rumors, for officials must regard them as symptomatic. They can become an effective barometer of the state of mind of the employees and can indicate the likelihood of trouble in any particular area.

The first step in countering rumors is the development of faith in management. People will listen to and obey those whom they can trust. Also develop and maintain confidence in the official announcements or other communications. Loss of faith in the reliability of official reports tends to spread rumors.

The following procedure should be used in handling rumors:

1. Do not repeat a rumor. Never repeat a rumor, even to deny it, as rumor is usually so much more sensational than fact that the latter is likely to be overshadowed.

2. Stop the spread of the rumor. Check the spread by questioning the person attempting to pass it on and insist upon verification and proof. Make plain to the carrier of the tale the implications of his acts. Stopping a rumor at any point can be of great significance in preventing its spread.

3. Counteract the rumor. The most reliable antidote for poisonous rumor is fact. When people cannot get the facts, they will accept rumors.

4. Keep everyone occupied. There should be adequate work for everyone; if not, create tasks in order to prevent idleness. Idle minds have busy tongues with which to spread worries and untruths.

5. An effective program of teaching personnel what is expected of them in an emergency is to train persons in thinking how they would react in any sort of danger-prone situation and in how to respond to all types of emergency situations which may arise. People work better when they understand clearly what they are to do and what is expected of them.

6. Develop situations such as the following:

   a. Acquainting them with the actual danger if there are effective means of combatting it.

   b. Gathering self-confidence through the knowledge and experience of others.

   c. Associating past conditions of security—presence of others, loved ones and acquaintances; being in a well-lighted room at home in a familiar place; performing acts associated with security and pleasure, such as singing, dancing, etc., reading, etc.; being in a place one can get out of readily; being in a low place (basement or cellar) rather than in a high building.

7. The example of strong, competent leadership by officers of the organization will go far toward preventing panic.

8. Pre-emergency preparations should include arrangements to facilitate control of people during disasters, such as having enough exist, clearly marking routes to be taken in evacuating the building or going to shelter, and locating organization personnel where they can take command and give calm, decisive instructions at places where groups are likely to congregate. In an emergency the organization should be prepared to remove the injured and the dead from general view, clear away debris which appears to cut off escape, quickly control fire, and approach any disturbance with calmness.

   It is important that a leader be able to communicate with his group if he is to control them efficiently. Therefore a communication system should be established which is capable of providing accurate information or directions under any and all conditions.

   Public address systems and electric megaphones are particularly effective. The persons announcing over such systems of communication should have a calm, flat, unemotional voice. The firm authoritative voice of the announcer may be extremely effective in reducing confusion and emotional excitement.

9. In severe cases use severe treatment to avoid panics. Threaten punishment or enforce strict discipline—but remember this is a last resort and not a permanent solution.

PANIC CONTROL

In certain circumstances, despite pre-emergency preparations, an unorganized group may be on the verge of panic. People may be tempted to join a fleeing crowd; the fright of those in motion is enough to suggest the presence of something to fear. When this stage is reached, it may become difficult to control the group. Attempting to reason with such a crowd may be futile, but it may be possible to control the group by assuming leadership or distracting key members of the group. In any case, corrective action should be taken before the movement stage, if possible.

Organizational personnel should be prepared to deal with this in terms of the following principles:

1. Provide assurance. Exert positive leadership. Reassure the group by giving information and instructions calmly.
2. Eliminate unrest. Dispel rumors. Identify troublemakers and prevent them from spreading discontent and fear.
3. Demonstrate decisiveness. Suggest positive actions. Indicate what to do, rather than what not to do.

The following are some of the procedures that have proved effective:

1. Remove the cause of the panic if that is possible. If one person panics or loses control of himself while in the company of others, there is a tendency for his behavior to communicate itself to those about him. There is, therefore, an immediate need to identify anyone who shows signs of such behavior. Two courses are then open. First, the person may simply be isolated. Second, he may be given treatment that may involve the use of sedatives on a doctor's orders, other medical measures, use of a psychiatrist if circumstances permit, and supportive efforts of lay persons who can assist the disturbed person to reorient himself. For example, this may sometimes be accomplished by giving him work to do which occupies his mind and reduces his apprehensions.
2. Try to divert attention. It may be necessary to work out quickly some dramatic way to get attention, since people will be absorbed with their own problems.
3. Once attention is diverted, provide information quickly. Tell people "I have just received word that . . . " Repeat your information over and over again in as many ways as possible. Keep your information simple, direct. If people are not physically together and if the message must reach scattered individuals, use radio, telegraph, press, messengers—everything. Have your information

come from authoritative, well-established sources whose integrity is beyond question.

4. Humor and singing may sometimes be used as diversions. They can occasionally snap people back to objectivity, give them insight and promote cooperation. However, in such cases you must be sure of the appropriateness of these devices. They are most likely to be successful if a panic has occurred in a fairly homogeneous group where the same thing will appear funny to people or the same song will catch hold.

5. Keep avenues of escape open and keep the crowd moving to reduce panic. Continual reconnaissance of the flow of people and traffic is very helpful.

6. Channel the crowd through areas where destruction of property has been minimal to prevent further panic.

It is often necessary to remove individuals from the danger area during a fire, or following an explosion or other disaster. Wrongfully executed, efforts to evacuate personnel may further intensify the feeling of panic. It is therefore vital that whenever an evacuation of personnel is undertaken, it should be accomplished under competent leadership operating under a well-thought-out plan.

Some of the points to remember in the evacuation of personnel from a building or other danger area are as follows:

1. Have a simple plan which may be easily understood and carried out by all persons concerned.

2. Designate certain individuals to assist in carrying out the plan. Messengers will be needed to relay orders. Guides will be necessary at various control points and exits. Searchers will be needed to insure that no one is left behind in the washrooms, offices or other parts of the building.

3. Inform all personnel of the evacuation and the need for cooperation of all persons involved.

4. In putting the plan into action, emphasize that the greatest benefits to all will be derived from strict conformity to the directions given, and that by fair play and reason-

able behavior each person can be of the greatest help to his fellow workers.

POSTPANIC PROBLEMS

Sometimes people hold in their feelings and exhibit their disturbance in a delayed reaction. Delayed, severe reactions may present more of a problem than is generally expected. On September 8, 1900, a hurricane followed by a tidal wave and flood devastated Galveston, Texas. Approximately 6,000 people were killed. Personal demoralization and social disorganization followed. A week after the disaster some 500 of the survivors went "insane" almost "in unison."

It is not sufficient to accept the fact that an individual seems to be unaffected at the moment of physical or psychological impact. Military experiences indicate that there is a considerable latent period before the stunned person reacts in retrospect to the dangers he has escaped or the frustrations he has experienced. The fact that a person is apparently psychologically unchanged immediately after the shock should not deter you from employing vigorous therapy and administering practical and emotional support.

Chapter 15

PLANNING

INTRODUCTION

N̲o̲ ̲p̲l̲a̲n̲t̲ ̲i̲s̲ ̲i̲m̲m̲u̲n̲e̲ to catastrophe. Emergency action plans, therefore, should be ready to meet all eventualities—both peacetime and wartime hazards. The goals of emergency planning are defined as follows:

1. To minimize effects of any incident of disaster proportions upon plant and community personnel.
2. To keep property and equipment loss at a minimum.
3. To ensure cooperation of all plant departments charged with specific activities in time of an emergency.
4. To ensure appropriate cooperative action by and with outside civic and governmental agencies.

The benefits of careful planning are many. With an emergency plan well thought out and put on paper, a company is much better prepared to have competent, trained men carry out established procedures in handling an emergency than it would be if frantic efforts were made to bring order out of chaos during and after a catastrophe. Indeed such plans may benefit management in evaluation of current operating problems and result in increased efficiency or lower cost.

It will be advantageous in formulating a detailed plan to consider first the following questions:

1. Has a facility defense coordinator been designated?
2. Has liaison been established with the local civil defense director?
3. Has an advisory committee been set up to assist in the development of the facility protection plan?
4. Has a written protection plan been prepared for the facility?

5. Has the facility plan been coordinated with the civil defense plans of the jurisdiction in which the facility is located?

6. Has a self-help organization been formed for the protection of life and property in and around the facility?

7. Have the coordinator and the chief officers of the self-protection organization been trained in civil defense schools?

8. Have arrangements been made for receiving and disseminating a warning?

9. Have a control center and a communications system been established for use in emergency?

10. Have emergency shutdown procedures been developed?

11. Has a fallout shelter been provided for building occupants and the public?

12. Have plans been made for the movement of building occupants to a fallout shelter within the building or elsewhere?

13. Does the plan provide for enlarging existing protective groups (e.g., guard forces, firemen, etc.) for use in an emergency?

14. Have sufficient building occupants been trained in civil defense skills (shelter management, radiological monitoring, first aid and medical self-help, decontamination, rescue, fire fighting, utilities control, etc.)?

15. Has the facility fallout shelter been marked and stocked?

16. Are employees informed about the facility protection plan?

17. Are drills or exercises held to test the plan?

18. Have preparations been made for assessing and reporting damage after attack?

19. Have arrangements been made for emergency repair and restoration?

20. Has information about personal and family survival been distributed to all building occupants?

21. Has protection information been included in em-

ployee publications?

22. Are protective measures presently available for personnel and equipment?

23. Will present measures provide the maximum protection?

24. Should additional protective measures be provided?

25. How seriously will production be affected by the loss of a department?

26. What alternatives are available for minimizing loss if a disaster occurs?

27. Can partial production be maintained?

28. Can subcontracting be utilized?

20. Can alternate equipment be used?

30. What conditions determine the minimum time required to fully restore the department to full production?

31. In what manner and by what means can these conditions be altered to reduce the time?

KEY STEPS IN PLANNING

1. Get in touch with your local civil defense authorities. They have pooled information and experience from many sources and can give you valuable guidance. If you tie in your program with theirs, and standardize your equipment in accordance with their recommendations, you stand a much better chance of coming out ahead in an emergency than if you operate alone.

2. Visit neighboring plants. Find out what they are doing and discuss their programs and your plans with their management. Try to coordinate your activities with theirs.

3. Survey your plant for possible hazards and take immediate action to lessen or eliminate them. Your emergency program should be based on essentially sound safety and protection practices.

4. Appoint a disaster director or defense coordinator, who will organize a central system of control and protective services. Provide adequate protection equipment and a centralized control room.

5. Early in the planning stage, present the program to your employees and enlist their active support. If you have few employees, you may want to call them together in a general meeting. When employment is large, consider presenting the program to employee representatives or department heads, who will relay the plan to their respective departments. Also, consider letters to all employees at their homes, and bulletin board notices. If your plant is unionized, discuss your plans with the union president or steward before presenting it to the employees. Workers, and most unions, will cooperate in carrying out their part of the program if they participate from the beginning and understand the problems involved.

6. Call an organization meeting of the heads of services, employee representatives and key personnel. Outline the purposes of a self-help disaster control program and explain how the plant should organize for protection. Make it clear that disaster control benefits everybody—initially in protecting life and limb, later in protecting the job.

   This meeting will be important to the success of your program. If you conduct the meeting forcefully and effectively, cover the important subjects, explain all the details, and ask for the cooperation of those present, your program will be off to a good start.

   Here is a suggested agenda:

   a. Announce the purpose of the meeting.
   b. Explain the program and its aims.
   c. Describe the reason for the plant survey and its results; then discuss your plans for eliminating hazards and correcting deficiencies.
   d. Explain the functions of the local civil defense organization, the relationship between it and your company, the impracticality of expecting full protection in a major disaster from outside services, and the thinking behind "mutual aid."
   e. Introduce the person you have appointed plant defense coordinator. Then let him take over chairmanship of the meeting, and do the following:

(1). Appoint the heads of protective services.

(2). Supply each service head with information, manuals and an initial procedure guide covering the whole plan and especially his particular service. Instruct service heads to study carefully.

(3). Make the service heads responsible for organizing and training their units. Limit the time for organization and arrange for progress reports to be submitted periodically. Explain who will advise on organization and training, if advice is needed.

(4). Point out to the service heads that they should recruit mature persons who have presence of mind and are calm in emergencies, and who can act as advisers and guides to others.

(5). Arrange for representatives of management and labor to meet with each service head.

(6). Allow time for specific questions.

(7). Announce date and time of next meeting.

(8). Adjourn the meeting.

7. Define the program. As soon as this part of the program is underway, the plant defense coordinator should begin work on a disaster control manual. This is the policy and procedure manual for the program. He will need the help of most executives and many department heads during its preparation. This manual should be authoritative, clear, precise, specific, and should cover every eventuality. It should provide for modification and improvement in both policy and procedure, as well as for flexibility of control.

THE PLAN

To be effective, the plan must be detailed and explicit. It must be in writing. The plan must be reviewed periodically, and lists of names and addresses of key personnel must be kept current. Sufficient copies of the plan must be made availabale to all departments in the facility for wide distribution.

"Dry runs" must be made under realistic conditions. Co-operation between departments and units within the company must be assured by top management.

The local civil defense director can be of great help to the company in preparing its disater plan. It is his job to provide guidance and assistance, and to coordinate the emergency planning activities among the various departments of local government. This includes also the development of plans to utilize fully and to coordinate the nongovernment leadership and resources into community emergency preparedness.

The planning may include, but is not limited to, the following:

1. Dispersion of new facilities and major expansions of existing facilities to locations away from target areas.
2. Dispersion of movable supplies and equipment, i.e., inventories, spare parts, vehicles, etc., which are not continuously needed in target areas.
3. Relocation of key production of critical items to other existing facilities not in target areas.
4. Arranging for alternate sources of supply.
5. Protective construction.
6. Provision for facility air raid shelters.
7. Provision for the evacuation of facilities.
8. Continuity of management.
    a. Selection and equipment of alternate headquarters.
    b. Establishment of personnel succession lists.
    c. Protection and duplication of vital records and documents.
    d. Review of all legal documents such as charters, by-laws, etc., to assure that surviving directors and officers have authority to continue operations.
    e. Development of emergency financial arrangements.
9. Disaster planning for emergency repair and restoration.
10. Organizing and training employees for self-help.
11. Establishment of industrial mutual aid associations for civil defense.
12. Preparation of emergency shutdown procedures.
13. Preparation of corporate and plant disaster plan manuals.

14. Preparation of handbooks for employees.
15. Support of and assistance to the community civil defense efforts.
16. Establishment of a facility warning system.
17. Establishment of a control center.
18. Establishment of a communication system.
19. Establishment of a medical plan which covers the following:
    a. Rescue of the injured.
    b. Provision of first aid.
    c. Triage (sorting) of the injured.
    d. Dispatching to a hospital those most seriously injured.
    e. Provision of such care as may be required for the less seriously injured.
    f. Availability of hospital facilities.

WRITTEN PLAN

A plan that is not reduced to writing and circulated to those who must follow it is useless. It is of vital importance that the corporate and plant emergency plans be put in writing and organized in manual form. The manual should, if possible, cover every eventuality. It should be designed so that policies and procedures can be revised because of changes or when improvements in the program need to be made.

The success of a disaster plan depends on the quick understanding and the immediate cooperation of each person in the affected installation. A disaster plan locked in a safe and known only to a limited number of supervisory personnel will not effect the desired result. When the nature of an installation is such that certain areas or processes carry a high security classification, the portions of the emergency plan providing for the security areas or materiel should be contained in an annex to the general plan. Such an annex can be disseminated separately on a need-to-know basis while the general plan can be disseminated to all concerned.

Over and above that, it must never be forgotten that the manual is only a record of the plan and a guide to emergency action. Much training is necessary to assure that action will be proper and automatic when emergency occurs.

Some companies have made their disaster plan manual distinctive by using a special color for the cover, or a size different from the usual company manuals. Several companies have used bright red or yellow covers for their disaster manuals to conform to similar colors used for helmets or other emergency equipment.

GENERAL OUTLINE

The following is a general outline for such a manual:

Administrative Directives

1. Protection of personnel and production.
2. Master plan director.
3. Pre-emergency committee directory.
4. Postemergency committee director.
5. Plant protection unit director.
6. Communications unit directory.
7. Engineering unit directory.
8. Medical unit directory.
9. Welfare unit directory.
10. Mutual aid directory.

Operational Procedures

1. Fire and explosion.
2. High wind and tornado.
3. Flood.
4. Strike or riot.
5. Sabotage and/or espionage.
6. Enemy attack.
7. Chemical or radioactive fallout.

Each section should describe control organization, task organization, general situation, mission, implementation, administration and function, and command and communications, as they are affected by that particular source of danger both before and after the disaster. In addition, each chapter should cover activity responsibilities, as annexes to the plan, for each service unit as they relate to the specific danger, plus a section describing location of emergency equipment, evacuation plan, shelter provisions and any other pertinent information.

OUTLINE OF SUBJECTS FOR A DISASTER PLAN MANUAL

The following is a more detailed outline of the subjects for a disaster plan manual:

1. Company policy statement regarding disaster planning.
2. The purpose of the plan.
3. Authority for the plan.
4. Types of disasters expected.
5. Physical layouts—maps, blueprints.
6. Data on adjacent areas.
7. Assessment of vulnerability to enemy attack.
8. Relationships with local government, including civil, defense authorities.
9. Relationships with disaster and welfare agencies, including American Red Cross.
10. Principles observed in disaster planning.
11. A message from management to members of the emergency organization.
12. A message from management to all employees, emphasizing individual responsibility.
13. Name of the plant and disaster control director.
14. Names of disaster advisory committee members.
15. Organization chart.
16. The plant warning system
    a. Receipt of warning.
    b. Warning to employees.
17. The plant control center
    a. Location.
    b. Equipment.
    c. Operation.
18. The plant communications system
    a. Internal.
    b. External.
19. Emergency shutdown procedures.
20. Emergency evacuation routes, including directional

signs from the workplace to outside of building.

21. Evacuation routes from the plant to outside the city or to reception and mass care centers.
22. Shelters for employees.
23. Location of shelters, including floor markings and directional signs from workplace to shelter.
24. Location of hazardous areas which should be avoided in going to shelter.
25. Care of visitors during emergencies.
26. Organization and composition of self-help or protective groups
    a. Fire.
    b. Police—weapons, apprehension and restraint, traffic control.
    c. Rescue.
    d. Medical and first aid—location of first aid stations.
    e. Chemical and biological defense.
    f. Radiological monitoring.
    g. Warden.
    h. Welfare—housing, clothing, financial assistance, transportation, counseling.
    i. Reconnaissance parties.
    j. Reporting damage.
    k. Disaster equipment.
    l. Uniforms and helmets.
    m. Disaster corps identification.
27. Protective construction.
28. Protection of equipment and machinery.
29. Deployment plans and procedures.
30. Measures for prevention of sabotage and espionage
    a. Physical security—critical area protection.
    b. Investigation of applicants and employees.
    c. Employee responsibilities.
31. Protection from delayed or unconventional weapons effects.
32. Participation in industrial mutual aid associations for emergencies
    a. Membership.

    b. Personnel.

    c. Equipment.

33. Plan for continuity of management.
34. Amendments to bylaws and administrative regulations.
35. Alternative company headquarters in emergencies
    a. Location.
    b. Operation.
36. Employee reporting centers.
37. Recall of personnel.
38. Registration of personnel.
39. Rotation of personnel.
40. Personnel utilization.
41. Inventories of employee secondary skills.
42. Plan for protection of vital records and documents
    a. Duplication method.
    b. Storage location.
43. Emergency financial procedures.
44. Emergency repair and restoration of plants and equipment
    a. Alternate sites.
    b. Alternate sources of supply.
    c. Stockpiles.
    d. Alternate production methods.
    e. Subcontracting.
45. Utilities repair and restoration
    a. Gas.
    b. Sewage.
    c. Fuel.
    d. Water.
    e. Electric.
    f. Communications.
46. Policy on deconcentration and dispersion of production.
47. Methods of informing employees about the company disaster plan.
48. Disaster plan testing and exercises.
49. Program for informing and educating employees in civil defense preparedness at home.
50. Policy regarding utilization of employee publications

and organizations to inform and encourage employees in disaster preparedness.

51. Policy for informing stockholders and the general public regarding company civil defense and emergency plans.
52. General support and assistance to local government in community survival planning efforts.

DISASTER PLAN OUTLINE

In order that the scope of the plan and its organization may be better understood, one will now be set forth in detail. This disaster plan outline is presented as only one possible format for a plan for a facility. It is by no means the only form recommended. However, it is one which has evolved over the years as the most practical without being so detailed as to discourage prospective planners by the length and complexity of the project.

The *Annex* concept is used, permitting the circulation of applicable parts of the plan to persons having a limited responsibility and "need to know." This will permit the issuance of the Annex to the master plan within affected departments or sections for individual testing and training activities without unnecessarily involving the entire organization.

Considerations suggested in this plan, in conjunction with modifications required by a particular industry, its organization, resources, physical plant and geographical location, should act as organized reminders of risk factors requiring careful thought and practical solutions.

DISASTER PLAN

(Facility Heading)   (Date)

1. *Purpose.* This paragraph should include statements comparable in scope to the following: "To establish a continuing program of preparation for the protection against disaster from all causes; to insure survival of this facility through effective organization, coordination, and operational programs designed to assure maximum protection, continuity and recovery of personnel, premises, products and services."

2. *Planning Factors*

a. *National assumptions.* This paragraph should include a statement to the effect that potential enemies of the United States possess nuclear and thermonuclear weapons in sufficient quantity, and delivery means to enable them to attack, with little or no warning, selected targets in the United States. These potential enemies are capable of widespread sabotage action against United States industry.

b. *Local considerations.* In this paragraph include an analysis of sabotage effects, and the types of natural disaster to which the area is subject.

c. *Facility and geographic configuration.* Outline pretinent information pertaining to facility location, layout on the ground, organization and location of subsidiary facilities in relation to the parent organization. Discuss productive processes which are particularly conducive to hazards and include maps and sketches of facility and area as required. Employee patterns of residence may be included for casualty planning purposes.

d. *Operational data*

(1) Personnel. Indicate the total number of employees of the facility and specify the number of contractual or vendor personnel present daily.

(2) Shift operation. Indicate the total number of employees and contractual personnel, male and female, assigned to each shift. Indicate the capability of expanding to two- or three-shift operation.

(3) Production data. Discuss pertinent information pertaining to items produced, mobilization production plans, alternate end item production, and other information essential for planning purposes. ,

3. *Organization*

a. *Organization chart.* Attach a schematic chart of the organization for normal operations, and a chart indicating organization for disaster planning and control purposes.

b. *Alternate headquarters.* Indicate location, operation and staffing of the alternate headquarters, if any.

c. *Responsibilities.* In this subparagraph a statement should be made pertaining to management responsibilities for disaster planning and disaster control operations, as well as employee responsibilities to the facility. This statement should indicate that maximum support is required to the facility by both management and operational personnel.

4. *Implementation*

a. *Authority.* Designate the authority to place the disaster plan in effect.

b. *Actual and practice alerts.* The policy on types of alerts and their implementation should be discussed here.

5. *Training and Testing*

a. *Training responsibilities.* Indicate the type of training to be conducted, responsibilities for training, scheduling, etc.

b. *Supervisor responsibilities.* Discuss responsibilities of supervisors to ensure that their subordinates are familiar with this disaster plan and the individual responsibilities outlined in the plan. Discuss supervisor reporting responsibilities upon observance of inadequate action, or insufficient planning.

c. *Tests, evaluation, review and revision.* Frequency of tests may be discussed, and responsibilities assigned for evaluation of tests, frequent review of the disaster plan, and responsibilities for timely revision based upon experience gained in tests and through other means.

ANNEXES

1. Operations
2. Coordination
3. Protection
4. Continuity
5. Recovery

_____

Authenticating Official

ANNEX 1

OPERATIONS

I. *Disaster Control Organization.* Organization Chart should be included in the basic plan, or as an Appendix to this ANNEX.

A. *Control Centers*

1. *Location.* Indicate the local of an adequately protected site within the facility to be designed as the primary disaster control center; an alternate should be selected. Include schmatic drawing of internal layout of control center, to include location of equipment, communication, supplies and personnel.

2. *Equipment.* Indicate equipment to be permanently maintained in the control center, e.g., communication equipment, public address system, emergency power source, maps, food, blank forms, office supplies, etc.

3. *Operations and Staff.* Indicate when the control center will be activated and by whom it will be staffed, either by position in the organization chart, or by name of designated individuals.

B. *Alternate and Emergency Headquarters*

1. *Location.* Indicate a selected off-site location where facility business can be continued temporarily and plans made for restoration of operations or production.

2. *Equipment.* Indicate equipment to be maintained at the alternate headquarters, e.g., communications equipment, production information, office supplies, maps, food, etc.

3. *Operations and Staff.* Indicate when the alternate headquarters will be opened, its mission, who will staff it, and other pertinent information which will allow for smooth and rapid transition to the alternate headquarters.

C. *Relocation Site.* Indicate location, operation, equipment and staffing of the relocation site, if any.

D. *Assembly Areas and Reporting Center*

1. *Location.* These should be designated areas, away from the facility or critical target area, for reassembly of employees after the facility has been evacuated or in the event of a disaster to the facility during nonoperational hours. In selecting the site or sites, consideration should be given to using employee's homes, subsidiary offices or other predesignated areas, and also the location of planned evacuation routes. These areas should be connected by communications to the control center and alternate or emergency headquarters. Coordination must be made with the Civil Defense organization.

2. *Equipment.* Minimum equipment should be employee registration cards, arrangements for financial assistance, and welfare services.

3. *Operations and Staff.* Designate personnel responsible for supervision and administration of each site. Designate personnel who have authority to implement emergency financial procedure. Establish procedure for registration and reporting of personnel to control center and/or alternate or emergency headquarters.

II. *Communications and Warning System*

A. *Internal*

1. *Normal Operations.* List present system and extent of use such as telephone, radio, public address, etc.

2. *Emergency System.* Show here the communications to be used during emergencies only, such as radio, public address and messengers. Also indicate primary and alternate power sources to be used.

3. *Warning System.* Designate the alarms to be employed for fire, natural disaster, enemy attack, evacuation. Alarm signals should be different for each type of impending disaster and clearly audible to all employees. Signals for enemy action should be the same as those used by the local civil defense organization to familiarize employees with the signals wherever they may be located in the emergency. Arrangements should be made to receive all warnings from the local civil defense organization by "bell and light

systm," telephone, radio, or messenger, or locally devised means.

B. *External.*

1. *Normal Operations.* Same as internal normal operations.

2. *Emergency System.* Enter the extent of system planned for emergencies such as transistor radios to receive EBS messages, telephone, broadcasting and shortwave radio. Outline procedure for activation of communications in control center, alternate headquarters, assembly areas and reporting center.

3. *Warning System.* Enter here the system which is tied in with the local civil defense organization, weather stations, local law enforcement agencies and utilities.

C. *Emergency Notification.* This should list means of notifying key personnel. Consideration should be given to the use of "group alerting," messengers, chain of progressive altering and other locally devised means.

III. *Security*

A. *Utilization of Guard Force.* Outline the organization and responsibilities of the plant security force at present. Indicate additional guards necessary in an emergency, how they will be obtained and trained, and where they will be assigned. Guard orders for the additional posts should be prepared in advance and attached to this Annex as enclosures.

B. *Identification and Control.* Indicate the identification and control system to be instituted in an emergency. This should include visitors, employees, contractors and vendors. Provisions should be made for special package control, and control of vehicular traffic in and out of the facility.

C. *Critical Area Protection.* List vital or critical areas that may require special protection. Indicate specific instructions relative to protection of these areas.

D. *Emergency Procedures*

1. *Weapons.* Indicate authority for use of weapons and when they may be used.

2. *Apprehension and Restraint.* Has this been coordinated with local law enforcement agencies? Under what conditions will apprehension or restraint be effected? Designate place of temporary detention.

3. *Reporting of Incidents.* How, when, where and to whom will incidents be reported?

4. *Unexploded Ordnance Reconnaissance.* List closest military Ordnance Disposal Teams. Indicate actions of guards upon discovery of unexploded weapons.

5. *Emergency Shutdown.* What will security personnel do

until maintenance staff arrives? Indicate exact procedures.

6. *Safeguarding Classified Material.* Specify procedure for safeguarding or removal of classified material in an emergency or disaster. Security personnel should know how to contact designated custodians of classified material.

7. *Guard Force Implementing of Emergency Notification System.*

ANNEX 2

COORDINATION

I. *Civil Defense*

A. *Liaison.* In the initial stages of planning, contacts must be established with the appropriate civil defense and other governmental agencies.

B. *Emergency Identification.* Coordinate with civil defense and other interested agencies identification procedures that will be used to authorize entrance to damaged areas post disaster.

C. *Transportation.* Provide for civil defense vehicle identification, and coordinate with Military Traffic Management Agency and other government agencies to ensure proper road clearance in the event of a widespread disaster.

D. *Manpowr Utilization* Indicate coordination effected to ensure that personnel designated for key civil defense and other governmental jobs are not included in key personnel positions within the facility.

E. *Communications.* Ensure coordination has been effected to tie in the appropriate civil defense communications.

F. *Testing.* Coordination with civil defense personnel in the testing of the facility disaster plan will assist in effected testing of the disaster plan.

II. *Mutual Aid*

A. *Organization.* List the name and location of each facility of the mutual aid association. Indicate who in each facility can approve the implementation of aid. Also include any other mutual aid associations with which you may have unilateral agreements.

B. *Communications and Control.* List the primary and alternate methods of communications that will be used to alert the mutual aid pact members and your facility. Include methods of alerting during normal working hours and nonworking hours. Include the methods that will be used in controlling personnel at the disaster scene, including direction of police and emergency vehicles and crews. Coordination must be made in advance for use of facility security personnel, state and county police, as applicable.

C. *Facility Responsibilities*

1. *Personnel.* List by job title the various skills that you have agreed to furnish the mutual aid organization. Maintain a current roster of these personnel by name, with alternates. Include supervisory responsibilities when aid is required.

2. *Equipment.* List here the material and equipment that your facility will have available for mutual aid. Establish a method of having the material and equipment delivered as needed.

D. *Other Participants Responsibilities*

1. *Personnel.* List here by job title or skill, the personnel to be furnished by other mutual aid participants. Indicate procedure for their reporting, utilization and control. Indicate responsibility for control and supervision of each group.

2. *Equipment.* List here the material and equipment that may be obtained from other mutual aid members. All items should be listed by location and include procedure for obtaining them.

E. *Legal Aspects.* Insure that legal personnel assist in developing the plan, and are available to assist in fixing legal responsibilities.

F. *Operational Procedures.* List special limitations, legal aspects, feeding and transportation of personnel, prorating cost and use of any special items not covered above.

III. *Local Liaison.* Liaison should be effected with the agencies listed below, as well as other interested agencies, to ensure the proper coordination of the disaster plan.

A. Local government.

B. Police.

C. Fire.

D. Medical

E. Industrial associations and professional organizations.

ANNEX 3

PROTECTION

I. *Personnel Protection*

A. *Shelter Requirements.* List total shelter requirements based upon the maximum number of personnel at the facility at any one time. It is recommended that an allowance be made of fifteen square feet a person.

1. *Location and Capacities.* Indicate location of shelters, and indicate capacities if desired. Protection factors of each shelter may be shown in plan if desired.

2. *Marking and Stocking.* Discuss continuing plans for marking and stocking of shelters, and for inspection of shelters and equipment.

3. *Organization and Operations.* Designate shelter managers, alternates and committees, and assign functions and operations as required.

4. *Monitoring Teams—Internal, External.* Designate composition of monitoring teams for internal and external monitoring of fallout intensity. Prescribe equipment and duties, as required.

5. *Stay Time.* Indicate level of radiation at which occupants may be moved to shelters of lower protection factor, be allowed to exit the shelter for brief periods of time, and for final exist time.

B. *Evacuation*
1. To shelters as result of nuclear attack.
2. From the facility location as result of natural disaster.

C. *Emergency Welfare Services.* Discuss in detail services that are to be planned for post disaster, such as financial assistance, emergency housing and mass feeding, shelter requirements for families, etc.

D. *Public Information Services.* Designate public information officer and prescribe information to be supplied to him, and information which he is authorized to release pertaining to damage, identity and extent of casualties, etc.

II. *Facility Protection*
A. *Functional Areas.*
1. *Criticality.* List functional areas, in order of priority, most critical to overall facility operations and/or production. This should include considerations for both enemy and natural disasters,
2. *Protection.* Functional areas most critical to the overall operation and/or production should be given priority of protection prior to, during and after disaster.

a. *Buildings.* Include considerations for reenforcing walls, roofs, floors and protection of wall openings such as windows and doors of existing buildings. These protection factors should be considered in new construction.

b. *Machinery.* Factors to be considered are dispersal, protection of one piece of equipment by use of another, blast walls, blast roof construction and parts removal. Location of overhead cranes should be indicated.

c. *Hand tools.* Indicate individual action and responsibilities for protection of hand tools. Include tools crib dispersal.

d. *Special equipment.* Indicate methods to be used or

used to disperse on- or off-site parts, subassemblies, completed items, jigs, dies, patterns, molds and other critical items.

e. *Transportation.* Indicate dispersal location of transportation equipment and specify those that will be utilized to protect machine tools.

f. *Utilities.* Indicate protection afforded utilities and include location and protection of electrical transformers at load centers.

B. *Shutdown Procedures.* Specify shutdown procedures to include methods and sequence for individual sections within the facility and the facility as a whole. Designate title positions of individuals responsible for implementing shutdown procedures.

C. *Fire Control.* List responsibilities and training requirements of fire brigade. Indicate equipment available within the facility and that which may be available from other sources.

D. *Dispersion.* Consider the dispersion of facilities, machinery, material and personnel.

E. *Other Measures.* List other measures peculiar to your facility that may be necessary to minimize damage.

ANNEX 4

CONTINUITY

I. *Personnel*

   A. *Emergency Succession*

      1. *Legal Aspects.*

         a. Have state and local laws been examined to determine the legality of the management succession list?

         b. Have company by-laws been adopted or revised to provide adequate authority and necessary quorum for surviving board members?

         c. Military installations should follow normal command succession.

      2. *Succession List*

         a. Has a management personnel succession list been developed to provide alternates or successors for key positions? The plan should provide for at least two or three successors for each position.

         b. Have provisions been made for succession or emergency utilization of key operational personnel?

         c. Geographic employment location and residence data should be carefully considered in preparing succession lists for both management and operational personnel.

         d. In preparing succession lists consideration should be

given to mobilization assignments of personnel who may be members of the executive or military reserve.

B. *Personnel Utilization*

1. *Employee Registration*

a. Prepare registration card on each employee for file at alternate headquarters, assembly areas and/or reporting centers. Registration cards should contain information regarding secondary skills, dependents, coded pay data and personnel identifying data.

b. Registration of personnel at assembly areas and/or reporting centers after a disaster.

2. *Recall of Former Employees.* This should provide for recall of personnel still available who have left the facility under satisfactory conditions, and retired persons.

C. *Casualty Estimates.* This information should reflect the maximum number of casualties which the facility may expect to sustain in the event of a nuclear explosion. For planning purposes, it should be assumed that the facility is in a light-to-moderate damage area. Estimates should be made for day and night attacks, with and without warning, during operational and nonoperational hours.

D. *Medical Requirements.* Based upon existing medical organization and casualty estimates the following should be taken into consideration in preparing for emergency medical requirements:

1. Is there a physician on duty at all times?

2. Alternately or additionally, is there a nurse on duty?

3. Have plant emergency first aid stations been established?

4. Have first aid teams been organized?

5. Have litter-bearer teams been organized?

6. Have ambulance services been organized?

7. Have other health service personnel been organized and traind in sanitation, radiological monitoring and personnel decontamination?

8. Has the plant health service plan been coordinated with the local civil defense health program?

9. Has the American National Red Cross first aid course been offered to plant employees?

10. Is the plant health service organization a part of a coordinated mutual aid organization of several plants?

11. Have emergency first aid supplies been stocked in sufficient quantities for a major disaster?

12. Have employees been blood-typed?

13. Does the plant have a blood program?

14. Has a policy been established on medical evacuation?

II. *Records*

A. *Classes of Records*

1. *Administrative*. Indicate those records needed by the administrative functions of a facility, to include as a minimum payroll, accounting, personnel and sales records.

2. *Operational*. Indicate those records needed by the operations, engineering, or maintenance sections, and production records.

B. *Reproduction Methods and Priority*. Indicate the methods that will be used to reproduce administrative and operational records. Protection considerations should be given to microfilming, use of film sort cards, carbon copies, photocopying and duplicate records. Specify the records in order of priority for reproduction. Administrative and operational records will have to be considered together. Reproduction of classified material must be coordinated with the issuing agency.

C. *Safeguarding of Records*. Indicate the location of reproduced or duplicated records. Consideration for the location of duplicate records should be given to the use of alternate headquarters, small town banks, commercial depositories, the homes of key employees living out of the probable damage area, and vaulting in special circumstances. If classified material is stored, suitable clearance from the issuing agency must be obtained for the location where the documents are to be stored. Special instructions should be included for safeguarding records in the hands of employees at the time of disaster.

D. *Operations*. List special instructions for handling and storage of records. Indicate the person or persons charged with records protection responsibilities and establish his definite authority.

ANNEX 5

RECOVERY

I. *Damage Assessment*. Multiplant corporations, and military installations having a national mission, should establish damage assessment centers located at the alternate headquarters and/or relocation site. Other facilities may establish damage assessment sections in conjunction with the disaster control centers.

A. *Internal Reporting*. Indicate procedure for reporting damage within the facility to the disaster control center. The damage reported will be assessed for overall effect on the facility and as a guide for restoration.

B. *External Reporting*

1. Multiplant corporations and military installations having a national mission. Indicate procedure for reporting damage

from facility to damage assessment center and from damage assessment center to higher headquarters, OCD or other designated
agencies.

2. Facility. Indicate procedure for reporting damage to
damage assessment center, if applicable, and for reporting to higher
headquarters, OCD or other designated agencies.

II. *Recovery Operations*

   A. *Radiological*

      1. Designate facility vital areas.

      2. Determine maximum permissible doses for normal and
emergency operations.

      3. Designate recovery and mission personnel.

      4. Establish radiological monitoring capability and survey
of radiation.

      5. Plan for possible utilization of a staging site.

      6. Establish methods of vital area and personnel decon-
tamination.

      7. Establish radioactive waste disposal areas.

   B. *Chemical and Biological Agents*

      1. Establish facility vital areas as outlined above.

      2. Establish methods to detect and identify agents and
to locate contaminated areas.

      3. Esablish chemical and biological decontamination capa-
bility to include protective clothing and masks.

      4. Establish methods to be used to decontaminate vital
areas or personnel.

III. *Restoration Measures*

   A. *Relocation Sites.* Consider other locations in areas of relative safety that may be used to rebuild or begin operation if present area is destroyed or heavily contaminated.

   B. *Alternate Sources of Supply.* List the names and addresses
of those firms which can be used as a source of alternate supply. List
agreements that have been made with them.

   C. *Stockpile.* Cover information concerning stockage of essential raw material, component parts, parts for machine tools and
maintenance and critical machinery.

   D. *Alternate Production Method.* Indicate those processes
that lend themselves to alternate methods even though they may
be slower and more costly. Outline the alternate methods and indicate conditions under which they will be put into effect.

   E. *Subcontracting.* Indicate those facilities or installations
with which subcontracting agreements have been made.

F. *Utilities.* List the requirements of each subsection of continued operation. Include agreements with local utilities and others having facilities for furnishing electricity, water, gas, sewage and fuel.

G. *Salvage Procedures.* List procedures for salvaging and rebuilding machinery, equipment and buildings.

H. *Transportation.* Based upon anticipated loss of transportation and remaining capability, determine additional requirement if any. Coordinate transportation requirement with the Civil Defense organization.

IV. *Review and Analysis.* Prescribe methods to assure timely review of actions taken to recovery from disasters.

V. *Reconstruction.* The final phase of any disaster control activity is the actual reconstruction of the damaged facilities. Certain planning can take place before the disaster to minimize the time lost in planning, etc.

MUTUAL AID

Few plants can provide all the services and equipment needed in time of disaster. By joining with other large facilities in the neighborhood and through proper coordination with departments of local government, assistance can be provided to one another in the form of equipment, materials, or personnel in time of disaster.

In mutual aid planning, someone must take the initiative to bring the necessary people together. While many variations are possible, the program can start by having an informal meeting to which are invited people from all of the plants, large and small, in the community plus the heads of the various community services. The latter include fire chief, police chief, heads of hospitals, safety director and the people in charge of utilities such as electric power, water, gas, telephones, public transportation and the city engineer. Also invite your local civil defense director.

From each plant there should be at least three people: the president, the plant defense coordinator and the plant engineer or plant manager. Each of these individuals is essential to cover the key considerations for his plant: executive policy and personnel, internal disaster control program and available facilities, equipment and supplies. The first meeting should be kept informal and under two hours. Before the meeting detailed presentation should be prepared. At the meeting the need for interplant coordination should be stressed, the basic information required from each plant should be detailed and the time and place for the next meeting established.

The next step is the organization of a Council for Interplant Coordination. People from each of the plants belonging to the council will attend regular meetings. They will pool ideas, recommend joint action, establish the location of a central machinery pool and a central employment office for disaster recovery—if this is not handled by civil authorities; set up joint standards for materials' identification, fasteners, bearings, lubricants, standard engineering parts; also exchange facilities lists, and develop a master area facilities list. In other words, do all of the things which will, by mutual assistance, provide for rapid production recovery.

Reciprocal agreements should be made well in advance in order that the maximum benefit will be realized when the need arises. In developing a mutual assistance plan the following points should be considered:

1. Each participant should develop a list of those items necessary for repair of machinery, buildings and utility services.

2. Prepare a joint inventory of equipment specifically for disaster control, such as emergency trucks, fire engines, ambulances, mobile cranes, tractors, portable compressors and generators; also fire hose, pulmotors, stretchers, breathing oxygen, rope and cable, wrecking tools, etc. Describe equipment by name, list pertinent specifications and give exact location.

    Standardize hose couplings so that fire hose from any plant is usable in all the other plants. It is not necessary to change outlets. Simply provide adaptors for use when hose is borrowed.

3. Exchange facilities lists, including service equipment, so that each plant knows what is available for emergency operation and repair. Include air compressors, pumps, generators and all vehicles. Make available to

each participating company a list of production machines and equipment. Indicate amount of time which might be available for emergency production work, indicating type and size of such work. These lists should be reviewed periodically and kept up-to-date.

4. Reconcile the shortages between that equipment which is available and that needed, in order to ascertain how it can be obtained.
5. Exchange information as to the extent to which machine repair work can be undertaken.
6. Determine what personnel, both skilled and unskilled, can be made available for emergency repair and clean-up work.
7. Establish liaison with local utility companies. Plan with your utilities people resources of emergency power, water and communications. Coordinate activities to keep power factor and telephone traffic minimized when emergency lines are overloaded.
8. Plan for the movement of essential personnel and necessary equipment during emergency periods.
9. Solve mutual employee vehicle traffic problems by better planning of vehicle movement and parking lot operation. Frequently a daily headache, this is critical in a disaster.
10. Promote a working relationship with local and state civil defense organizations. Be sure to coordinate your activities with those of the community civil defense, as well as with the local protective services such as fire and police. Discuss with your local civil defense director the need for additional warning devices, fire-fighting equipment, ambulances, etc. Take advantage of Federal "matching funds" to obtain needed equipment. Use your civil defense director's help to set up joint training programs.
11. Ask the head of your local hospital to help you plan for emergency medical treatment, perhaps beyond the hospital's normal capacity. What additional supplies will be needed? How about emergency ambulance service?

12. Plan control procedures for combating a major disaster in an individual plant; then for combating an area-wide disaster.
13. Develop an area-acceptable emergency-pass system. Thus you will assure the entrance to a disaster zone of those people who are necessary; you will keep out unnecessary spectators.
14. Set up specialist groups—communications, public relations, fire, rescue, medical, police, engineering, welfare—in the mutual aid organization, just as in your own plant disaster control program. Get local experts to direct the specialist groups.
15. To keep peace in the family, ask someone such as a local lawyer, not directly connected with any individual member, to head the organization.

Chapter 16

PLANNING FOR CONTINUITY OF MANAGEMENT

INTRODUCTION

INDUSTRIAL DISASTER PLANNING involves every department head in a company: the security officer, the purchasing agent, the treasurer and controller, the personnel director, the general counsel, the secretary, the production manager, the chief engineer, the head of research and, of course, the board of directors. Each department head must examine the functions for which he is responsible and work out answers to problems involving continuity during and following a disaster. Unquestionably, many normal functions would become unnecessary while some would become extremely complicated and vitally necessary.

Efficient emergency administrative operation requires five key elements: (1) communications; (2) availability of records; (3) living and operating facilities; (4) assigned personnel, with alternates, and (5) available cash—financial freedom.

CONTINUITY OF MANAGEMENT

The key to management continuity is advance planning. Here are some guides for planning management survival:

1. Avoid assigning, as alternates for the same key position, people who reside in the same neighborhood.
2. Avoid the traveling together in the same vehicle of top management, or all of any management level or specialty group.
3. Develop organization in depth.
4. Introduce executive and supervisory development programs.
5. Do not assign all key people to the same shelter area.
6. Deconcentrate executive and administrative offices.

7. Insist on yearly, or semiyearly, medical examinations for all key people.
8. Tell all employees of the plan, and who will assume responsibility.
9. Set up a policy for emergency operation; establish degrees of emergency.
10. If you company has plants at more than one geographic location, exchange key personnel between plants to orient them in other operations.
11. Prepare a job classification file showing interrelated skills, etc.
12. Keep a daily record of important transactions, agreements, policy decisions, payments, etc., at a safe location, away from the plant.

One of the first steps in planning is to review each function of the various offices and departments of the company to determine whether the function should be continued in a disaster, and alternate solutions listed. It must be ascertained if legal authorization is necessary in order for these officials to act in event of emergency. Bylaws should provide for emergency action by the board of directors. It should be authorized to act in event of an emergency with the statutory minimum for a quorum. This minimum number, in turn, should be empowered to augment their membership to the extent necessary to reconstitute the industrial corporation and get back into business following an enemy attack or major disaster. The Bylaws should also provide authorization for establishment of succession lists and for reestablishing the company and continuing production under conditions caused by disaster.

After the legal problems have been settled, the next step is the development of an emergency executive-succession chart or list. This chart described the emergency functions of surviving successors and lists these personnel in sequence. Thus, in an emergency, the available survivor highest on the list assumes temporary direction of the organization or function. At least two alternates are provided for each key emergency position. It is important to understand that the emergency succession list if not always exactly the same as the normal "chain of command." The two are based upon different factors and different sets of conditions. The list should not and cannot be regarded as a promotion list nor the subject of any other normal personnel action. It is to be clearly understood as a list for emergency use only.

In setting up the chart, it is important to select people who are usually in the plant, who know the personnel, and who have the physical stamina to keep going under adverse conditions. For example, an executive vice-president who spends most of his time on the road is not necessarily the logical emergency successor, if the president becomes a casualty, even though he may be the logical successor under normal conditions. The same rule applies in setting up the chart for individual departments. However, the people selected for leadership should have the respect of the personnel who will answer to them and should be kept informed of plans and policy developing in the positions they may fill.

Executive vulnerability is a primary factor in planning succession. In other words, there is little merit in making the next door neighbor to an executive his successor simply because they are neighbors, they are friends and they have similar specialties. If disaster strikes while they are both at home, both may be victims and the plan would fail. Appraise both the geographic and the volume concentration of your executives. A high percentage of them may be in one area, either business area or a desirable residential area, but it may not be a critical target area. Conversely, all of your executives may be widely dispersed, but with many or all of their individual locations in critical target areas. Select successors, if possible, who reside at some distance from each other yet who can reach the plant or emergency reporting center.

The following type of form may be used for all levels of management succession planning. Names should be listed for each position in the sequence of desired emergency replacement. Each name may appear in a number of columns, and with varied priority of succession.

EMERGENCY EXECUTIVE SUCCESSION LIST

Confidential			Confidential
Position number and title	Position number and title	Position number and title	Position number and title
President	Exec. Vice-Pres.	VP-Manufacturing	Treasurer
1			
2			
3			
4			
5			
6			
etc.			

This list can be prepared and kept confidential until such time as it may be needed, or it can be openly discussed with the executives involved. Companies using such lists claim that when succession is fairly and logically determined, there is no sign of jealousy or envy. After all, the assignments are temporary, subject to the emergency condition, and based upon the ability to take over an empty position quickly.

In decentralized companies with headquarters in major cities, the plan may provide for the autonomous management of outlying plants and divisions during the early restoration period. This would provide for continued legal operations of various plants and divisions as independent entites until central corporate headquarters could be reestablished.

Under major disaster conditions, it is possible that the management of a parent company might be replaced by various surviving top executives of dispersed affiliated or subsidiary companies. Thus there never is any doubt as to the source of leadership, and there is little cause for arguments to develop over who has authority.

If a company has two or more decentralized plants, it is extremely valuable to set up an executive exchange program to familiarize the executives of each plant with the other plants.

An emergency management team can be sent from one plant to another on short notice.

Another approach to the problem is to develop supervisory teams chosen from employees at various distant plants. These teams will report immediately to an assigned plant in the event of a major disaster at that plant.

RESPONSIBILITY

The organizing for continuity of management is one of the responsibilities of the disaster control director but, due to the importance of continuity of management, this task should be assigned to a designated individual. It is essential that he be from upper management levels and be able to wield the necessary authority and have sufficient time allotted to perform this duties.

Basically his duties are as follows:

1. Directing an advisory committee on the continuity of management in developing the company or plant continuity of management plan.
2. Promoting and helping formulate continuity planning in the companies and plants of the corporation.
3. Running a periodic audit of the overall planning and progress under such planning.
4. Keeping various units of the company advised of changes that should be made to keep their continuity plans up-to-date.

Some organizations may not desire to put the responsibility for continuity of management in the hands of one individual. In this situation an advisory committee can be established to decide the order of management succession and to develop all necessary action for planned continuity management. This committee should be composed only of top management officials. Its makeup should be varied. It should include production, manufacturing, medical, personnel, financial, marketing and legal representation.

The executive succession planning should be spread downward to the plant level with local subcommittees to do the plant level organizing of the succession list for the plant.

For larger, decentralized organizations, this plan can be developed for executives within each autonomous unit. The separate lists can then be integrated with the main headquarters list.

ALTERNATE HEADQUARTERS

If continuity of management is to be effective, managers must have a place to assemble and work which is equipped and furnished for carrying on corporate operations during and after the disaster. Thus, it is advisable to designate a specific location as emergency or alternate company headquarters. It should be furnished with the minimum equipment and facilities needed to permit effective operation of the reconstructed executive team.

Before selecting a location, consider these requirements: security, accessibility, communications and accommodations. Construction, rental, or purchase of facilities may be involved. The objective should be development of a normal capacity to manage the corporation from any one of several different plant locations in less vulnerable areas by transfer or delegation of management and command authority during the survival and early recovery period.

The following is suggested criteria for selection of alternate company headquarters:

1. It must be outside the critical target area.
2. The neighboring community must be large enough to provide facilities for adequate transportation and housing for employees.
3. A desirable location is at a point where the company already has an operating installation.
4. The location should be within reasonable traveling distance from the main headquarters.
5. The location shall be accessible from some nearby business center.
6. It should be possible to reach the emergency headquarters by public transportation.

PROTECTION OF RECORDS

Certain records in every company are essential to continued operation. Their destruction could result in the following:

1. Slowing or stopping production.
2. Impeding financial and/or physical rehabilitation.
3. Complicated dealings with suppliers and customers.
4. Harm to employee and stockholder interests.

What Records Are Important?

A partial list of important records is as follows:

Accounts payable	Notes receivable
Accounts receivable	Patent and copyright
Audits	authorizations
Bank deposit data	Payroll and personnel data
Capital assets list	Pension data
Charters and franchises	Policy manuals
Constitutions and bylaws	Purchase orders
Contracts	Plans; floor, building, etc.
Customer data	Receipts for payment
Debentures and bonds	Sales data
Engineering data	Shipping documents
General ledgers	Service records and manuals,
Incorporation certificates	machinery
Insurance policies	Social Security receipts
Inventory lists	Special correspondence
Leases	Statistical and operating data
Legal documents	Stock certificates
Licenses	Stockholder's lists
Manufacturing processes data	Stock transfer books
Minutes of director's meeting	Tax records
Minutes of stockholder's meeting	

Program

There is no single solution nor any standard plan applicable to the protection of records by every kind of business. Differences in nature, size, location, organization and variations in the kinds of records maintained prevent adoption of one best way for all. However, certain general principles are established.

Development of a protection program requires certain basic knowledge about the records. Thus, the first three steps to be taken are (1) an inventory of paperwork; (2) an appraisal, and (3) a classification.

To determine what records should be protected, the program should be established under the direction of a top level management committee. Members of the committee should have company-wide, rather than departmental, responsibility.

Classification of Records

In planning, differentiate clearly between essential records and those which are merely desirable. Unless this delineation is made, your protection program may include so much material that it becomes excessively costly, cumbersome and disruptive to administrative procedures. Similarly, the system is designed to permit reconstruction of essential business information after a disaster, hence excessive material might make this task lengthy and extremely complicated.

Thus the first step is to establish categories of records. Generally business records can be divided into the following classes:

1. *Vital Records.* Irreplaceable; records in which a reproduction does not have the same value as the original; those needed to recover monies promptly; and those necessary to avoid delay in the restoration of production, sales and service.

2. *Important Records.* Those which would be very expensive to reproduce, either in money or time. Many statistical and operating records are in this class and the number is increasing.

3. *Useful Records.* Those whose loss might be inconvenient but which could be readily replaced. That is, records which are not classified sufficiently important to require special forms of protection.

4. *Nonessential Records.* Those which, upon inventory and examination, are ready for destruction.

Records classification implies considerations of importance and composition. Accordingly, each company record should be considered in descending order of importance. Each record should be labeled "vital," "important," "useful," and "nonessential." Just the number may be used in sequence of priority. Or, colored index stickers and even colored paper will classify.

Methods of Protection

Records protection is accomplished by one, or a combination, of four basic methods: by office personnel, by duplication, by dispersion and by vaulting.

OFFICE PROTECTION. Train all office personnel in the steps to be taken for protection of working records in the event of a disaster. Place an "emergency duty card" at each desk detailing each employee's responsibility for moving to safety any vital records upon which he is working. Possibly some records can be placed in a vault when an office is evacuated, with a monitor responsible for vault sealing. If time is insufficient for this procedure, records should be placed in steel desks or filing cabinets, or, they may be carried to safety, if not too bulky, by designated personnel. Avoid leaving records out of the vault or protective storage overnight.

DUPLICATION. Records needed in day-to-day operation should be duplicated and the duplicates stored at a safe distance from the business establishment or plant; or placed in disaster-proof vaults. Duplication is most practical and widely used because of its adaptability and the number of processes available, such as hand copying, typing, making carbon copies, machine duplicating, or photocopying. Choice of process depends upon (1) physical size of records; (2) volume to be duplicated; (3) intervals at which duplication is required, and (4) cost.

Handwritten notations, if legally identifiable, may be the cheapest and easiest method of duplicating such essential records as accounting books and production reports. Similarly, a pocket "diary" of essential factors can be very useful.

Carbon copies are not adaptable to some types of records and have the disadvantage of requiring a large amount of protective space. Accumulations of carbon copies may be avoided by establishing new basic records periodically.

Photocopying records through such processes as photostating, microfilming and, in some instances, offset printing, is used extensively. Some of the newer processes permit very rapid copying on a continuous basis and to a variety of scales. Remember that cutting the linear dimensions of a reproduction in half can reduce the volume of paper 75 per cent.

Microfilming is preferable for duplicating large volumes of

material because of the space saving. About 5000 sheets of paper, eight and one-half by eleven inches, can be recorded on seventy-five feet of film, or a pocket-size roll. Also, copies are exact images; thus are of legal value. Full-size copies are readily viewed from the microfilm enlarged image on reader equipment. Large quantities of recorded documents can be quickly reviewed to locate a specific piece of information. Permanent, full-scale photoenlargements can be made from the film image.

Principal disadvantages are as follows:

1. Records correction may require refilming.
2. Special equipment is needed.
3. Cost is higher for small quantities.
4. Skilled personnel and an efficient record-handling system are needed.

Dispersion

Storage duplicates of essential records in one or more locations is a relatively easy method of protection for companies having a number of offices and/or plants. Some firms protect vital records by depositing them in out-of-town banks and bonded storage warehouses. Other companies have built specially designed depositories outside potential target areas or hazardous locations; some are using caves or abandoned mines. The latter, of course, must be conditioned to protect against dampness, chemical gases, etc., and some people now make a business of providing underground storage for a fee.

For smaller companies the homes of key employees should be considered for records storage. However, beware of basement storage anywhere. Use only a place on high, well-drained ground, dry, and reasonably fire resistant.

Costly methods of protection are seldom required under a dispersion program because many files, especially those less than "vital," can be stored on open shelves or in any type of storeroom free of water and fire hazards. Equipment, storage and handling costs can be further reduced by systematic, periodical destruction of records no longer useful. Facilitate the system by affixing a retention date on each record or document at the time of storing.

Nevertheless, under this system original copies should not be destroyed until a thorough examination has been made of the rules of evidence in the states in which the company transacts business, and of laws concerning the use of duplicates in the states and counties in which the company operates.

Broadly speaking, vital records to which reference is infrequent should be safeguarded by dispersion. In selecting the dispersal site, consider adequacy of transportation and communication between site and office or plant.

Vaulting

As discussed here, records storage in vaults, safes, or storerooms on the premises, rather than the dispersed storage of duplicates, is considered vaulting. Generally, dispersion is more protective.

During normal use, provide good air circulation and a dehumidifier. Store records in compartments that can be seggregated if a fire develops within the vault. Use carbon dioxide rather than sprinklers, for fire protection. Keep all records at least two feet above the floor. Build a good safe into the vault for storing valuables subject to theft. Determine the method of locking the vault by the content. Keep the interior of the vault, as well as the records, clean. Dust and dirt will hasten the deterioration of record materials.

Records Reconstruction

Records are protected so that after a disaster strikes you will have sufficient information, knowledge, proof of ownership and legal status to reconstruct the business. Most of the necessary material is in two places: the records and the brains of the personnel. If a plant suffers a major disaster, the loss of both records and qualified employees familiar with them would result in the loss of know-how required for the continuity of management and the resumption of business. In order to overcome that gap, plans must be made for the reconstruction of the records. Instructions for reconstruction should be simple and, most important, easily understandable by persons unfamiliar with the particular business and its operations. Further, these instructions should be filed in several places to insure recovery of at least one copy. Grant of legal authority to a number of people for the task of reconstruction should be filed with the instructions.

In addition, alternate methods of reconstruction should be developed wherever possible. With the instructions file a list of secondary sources of information that will expedite the program, sources such as the records of accounting firms retained by the company, banks, lawyers, tax offices, insurance companies and the company's suppliers, customers, contractors, or subcontractors.

Materials, machinery and equipment necessary for reconstruction should be available at the records depository. These things may be impossible to obtain following a disaster.

Salvaging Records

In salvaging records after a disaster, the crime laboratories serving police departments in most large cities can be of considerable help. It might be well to discuss the problem with them as part of your planning. For example, images on charred paper can sometimes be brought out by chemical treatment and infrared photography. Proper handling may save apparently ruined documents.

Water soaked papers and cards are best dried by placing them in warm sunlight and gently moving air, or in a warm room with good air circulation and a dehumidifier. Do not try to dry them too rapidly. Be very cautious about unfolding or separating very wet papers. There is less danger of smearing, peeling, or destruction if they are handled after drying or when only damp.

Most important, do not be too hasty after a disaster to open files, ledgers, books, folders and charred or soaked bundles of papers. Even if charred and crumbled, the pieces will be in the proper relationship to each other and some careful reconstruction might be possible. Once they are disturbed, the task becomes infinitely more difficult, if not impossible. Similarly, thoughtlessly opening a soaked ledger may do irreparable harm. Do not be too sure that important papers are useless until an expert looks at them.

EMERGENCY FINANCIAL PROCEDURES

A simple emergency accounting and audit system must be established to assure that adequate cost records will be maintained, the financial status of the organization protected and conserved and current profit or loss reflected. This system will be the basic foundation in taking care of the fundamental financial requirements during the period immediately following disaster conditions until time again permits a more formal approach to accounting and auditing.

Certain preventive steps are also advisable. They call for a study of the methods of safeguarding records. Consideration should be given to duplicate billing daily and its remote record storage. Avoid use of a hazardous area for the accounting department or record storage. Try to arrange vital files in readily portable units.

With relation to United States Government and marketable securities, conduct a survey to determine the adequacy of custodian's procedures and records. Find out whether broker or bank maintains a duplicate record. Be sure that the storage place is safe from fire, flood and explosion. Insist that certificates be placed in storage, rather than in or on a desk, when not actually being used.

Inventories should be analyzed with regard to size, character, value, safety of records, adequacy of storage relative to inherent hazards, practicability of dispersion or subdivision. Avoid using areas for storage simply because they are inadequate for any other purpose.

Money will be needed for prompt payment of wages, cash advance to employees and payment of bills. There must also be funds for necessary cash purchases of emergency medical supplies, food and equipment. This can be achieved by maintaining bank accounts of an unrestricted nature at scattered locations and establishing lines of credit at a variety of places.

Emergency procedures for drawing company funds should be developed in advance. Banking arrangements should be adequate and corporate bylaws should have ample provisions for withdrawal of funds. A list of individuals who are authorized to withdraw funds must be established. Also there should be established a list of persons at various and widely separated locations who are authorized to countersign checks. A current list of all depositories, all people authorized to withdraw company funds and authorized to countersign checks should be available in all company security storage vaults.

One company has simplified the problem by preparing checks, drawn on out-of-town banks, in unit sums of $25, $50, $100, and $500 and signed by the treasurer. To cash one of these checks it is only necessary to fill in the name of the payee and have it countersigned by one of the people whose names are listed on the organization disaster plan. Both a copy of the plan and the checks are kept in the company security storage vault.

In addition to the emergency check system, it is advisable to have a supply of actual cash, as a reserve fund. This money can be kept in the storage vault at the emergency headquarters—with instructions for use in the disaster plan, or distributed to three or four safe locations, such as banks, in towns no nearer than twenty-five miles away and well dispersed. Ten dollars per employee is a reasonable figure for an emergency cash reserve.

COMPANY EMERGENCY HEADQUARTERS PLAN

The following is the outline for a company emergency headquarters plan:

Organization

SCOPE. The purpose of such a plan is to establish the action to be taken in case of a major disaster which requires evacuation of the company headquarters facilities. It should provide for reestablishment of the corporate offices at a different location, on a temporary basis. A separate emergency headquarters plan should be maintained by each plat site. These plans should provide for the re-establishment of group and division headquarters organizations following a major disaster.

ACTIVATION. In the event of a widespread disaster (such as a nuclear enemy attack), the entire company emergency headquarters plan would be put into immediate effect. In case of a local disaster (such as a major fire) designated officers should determine whether the company emergency headquarters plan is to be implemented, either in whole or in part.

STAFF. As soon as travel is possible, predesignated company executives should assemble at a temporary headquarters location. Their immediate mission will be to do the following:

1. Establish communications with local and/or federal disaster agencies.
2. Assess the extent of damage.
3. Take appropriate steps to implement a prearranged plan to recover and protect government and/or company property and records.
4. Evaluate operating capabilities.
5. Determine marketing potentials and develop a course of future action for the company.

A designated officer or officers should be designated to assume charge upon arrival at the emergency headquarters. Those who will staff the company emergency headquarters should be predesignated and their specific functions clearly defined.

SUPPORT. Provisions must be made to provide support facilities, equipment and services during the period of emergency headquarters operations.

Location

GENERAL. The selection of an emergency headquarters site will depend on actual disaster conditions. Such conditions can never be accurately predicted. Tentative sites should be selected for the emergency headquarters. These selections will be subject to change if actual disaster conditions require or warrant a more logical choice of location.

Services

Support services for the emergency headquarters must be provided by a designated unit. Specific responsibilities of this organization are as follows:

RECORDS. The records coordinator will be responsible for the following services:

Vital Records. The key records essential to reestablishment of the company's operations must be immediately available for use. Therefore special provisions must be made to designate them and see that they are protected and that copies are readily available from an alternate offsite storage area. Those records will include many of the following:

1. Minute books.
    a. Board of directors.
    b. Management executive committee.
    c. Policy committee.
    d. Finance committee.
2. Lines of management succession.
3. Real estate acquisitions.
4. Deeds and deeds of trust.
5. License and royalty agreements.
6. Contracts.
    a. Prime.
    b. Sub.
    c. Purchase orders.
7. Leases—property and buildings.
8. Invention assignments and disclosures.
9. Options on products and property.
10. General ledgers.
11. Accounts receivable detail ledgers and banking operations.
12. Insurance policies.
13. Notes receivable.
14. Non agreements
15. Company policies and practices and organization manuals.
16. Payroll registers.
17. Sales summaries.
18. Current personnel lists.
19. Retired, skilled personnel lists.
20. Approved vendor lists.
21. Facilities maps and prints.
22. Annual company property listing.
23. Marketing projections.
24. Engineering and technical.
25. Electronic data processing.

Forms. A supply of key company forms must be maintained so they are available for use by company emergency headquarters personnel.

Reproduction Services. Make reproduction equipment available.

Procedures. During operation of the emergency headquarters the records coordinator will develop, coordinate, publish and distribute all required organization charts, policies, practices, procedures and bulletins for internal headquarters distribution (except for the news bulletin to be issued by public relations and advertising).

Information Center. A chart room will be set up as part of the public relations and advertising information center, for the emergency headquarters. Displays will include topographical relief maps, road, rail and air route maps of the United States, a geographical world map and detailed facility maps of each plant site.

Classified Document Control. Control of classified documents will be accomplished as a part of the records center operation.

Historical Record. The records coordinator will maintain an historical record of the establishment, personnel actions and principal occurrences at the company emergency headquarters.

Support. The support coordinator will be responsible for the following services:

Personnel. Interim personnel records will be established for all company emergency headquarters personnel. Forms will be obtained from the records coordinator. Support personnel will be obtained from or through the plant site organization. This includes stenographers, chauffeurs and clerical personnel.

Office Services. Office space, furniture, equipment and supplies will be obtained from the plant site organization and allocated to each company executive and his staff.

Cashier. A temporary cashier's office will be established to serve the headquarters. Operating funds will be obtained from the plant site finance organization. Forms will be obtained from the records coodinator. The cashier's office will disburse payroll funds to the headquarters staff, on authorization from the treasurer or his alternate.

Security. The headquarters area will be given special

twenty-four-hour security protection, in anticipation of the extensive use of classified information. Security guards will be obtained from the plant site organization and, as necessary and available, from ther plant site organizations. Arrivals, departures and current location of all headquarters personnel will be recorded in a master log. Signs, provided by the records coordinator, will be posted to identify the headquarters area and each company executive's office.

General duties of the security force will include controlling access of employees and visitors, protecting facilities and property, monitoring security communications equipment, safeguarding classified information during nonworking hours, and generally preserving order and performing other security responsibilities.

COMMUNICATIONS/TRANSPORTATION. The communications/transportation coordinator will be responsible for the following services:

Telephone. Telephone service will probably be the principal means of external communications. Calls will be governed by the availability of long distance circuits following the disaster. Efforts will be made to obtain emergency priority ratings from the state disaster control authorities.

Radio. Radio service will consist of the transmitter and receiver units available at plant sites, including the equipment in emergency-purpose vehicles, and the AM radio receivers which are maintained in each company building.

Mail. Distribution of incoming and outgoing mail will be governed by the current operations of the United States Mail service.

Teletype, Telegraph. Local teletype and telegraph facilities will be utilized, as required, to the extent possible.

Messenger/Courier. Messenger service will be made available within the headquarters/plant site area. Courier service to other plant sites and customer organizations may be provided, depending on the extent of need and current travel conditions.

Automotive. Company cars, driven to the emergency headquarters site by executives, will be maintained in a pool and dispatched on a priority-use basis. Additional vehicles may be obtained for the pool, depending on need and availability of equipment and fuel. Efforts will be made to secure needed priority ratings for fuel supplies.

Rail. Schedules will be maintained of all available rail service into and out of the emergency headquarters area.

Air. Commercial air transportation should be arranged, as available, by the communications/transportation coordinator.

SUBSISTENCE. The subsistence coordinator will be responsible for the following services:

Housing. Arrangements should be made with nearby hotels and motels for required accommodations. Sleeping space can be arranged in the office facility in case hotel space is unavailable or travel between the office and hotels is not practical. For emergency sleeping arrangements at the alternate emergency headquarters facilities, spare medical cots, lounges and the sleeping bags can be carried in the trunk of each company executive's assigned vehicle.

Feeding. If possible, arrangements will be made to have prepared meals delivered to the headquarters on schedule, or for reserved space in a nearby eating facility. Emergency food supplies, to cover one month's requirements, can be stored at the emergency facility. In addition, a two-week supply of emergency rations can be carried in the trunk of each company executive's assigned vehicle.

Medical. Emergency stocks of medical supplies and equipment must be stored at each site. In addition, a first aid kit should be stored in the trunk of each company executive's assigned vehicle.

Lighting. An emergency-use supply of flashlights with batteries and auxiliary lighting equipment should be maintained.

Procurement. Purchase requests for required supplies and equipment will be processed through the plant site purchasing office.

Maintenance. Facilities maintence and rearrangement services will be secured from or through the plant site organization.

Special Requirements. Special needs connected with establishment and operation of the emergency headquarters not here specified will be handled by the subsistence coordinator.

Public Relations

Public Relations. The director of public relations and advertising and his staff will do the following:

1. Obtain, evaluate and disseminate information on disaster conditions and the current situation at each plant site. News bulletins should be periodically distributed to all emergency headquarters offices. The assistant director of administration and material will aid in the collection and interpretation of disaster information.
2. Communicate all formal announcements relating to the disaster to company organizations.
3. Communicate all announcements to outside news organizations for the headquarters.
4. Operate the headquarters information center. A full time employee will maintain files in the information center containing all available intelligence on conditions at each plant site and the status of each major organization. Maps put up in the information center by the records coordinator will be posted to reflect current information on disaster conditions.

Reports

DISASTER SITUATION REPORTS. Each major organization will forward a disaster situation report to the company emergency headquarters as soon as possible after a major disaster. Reports originated by divisional organizations will be channeled through their group headquarters. Additional and changed information will be promptly dispatched in the form of report revisions. All such reports will be logged in by the emergency headquarters information center and routed to concerned offices. Disaster situation reports will reflect the organization's current status, including the following:

1. Area conditions—extent of damage to utilities, roads, and structures in the surrounding community, and estimated time required for restoration of basic public services and facilities.
2. Facilities—condition and estimated work, time and cost

to restore each damaged plant site building and the outdoor facilities.

3. Plant services—condition and estimated work, time and cost to restore basic services, including the telephone, water, power and light systems.
4. Working conditions—estimated length of time before general outdoor and inside work can be commenced at the plant site.
5. Personnel—summary casualty reports and an estimate of the organization's available work force by job classification.
6. Equipment—condition and estimated time and cost to repair or replace damaged items of vital capital equipment.
7. Material—items and quantities of important materials on hand in usable condition.

PROGRAM STATUS REPORTS. Each major organization having product line assignments will forward a program status report to the company emergency headquarters as soon as possible after a major disaster. Reports originated by divisional organizations will be channeled through their group headquarters. Additional and changed information will be promptly dispatched in the form of report revisions. All such reports will be logged in by the emergency headquarters information center and routed to concerned offices.

Program status reports will reflect the current status of each of the organization's major programs, including the following:

1. Product line code.
2. Reference number.
3. Customer.
4. Contract number.
5. Program title.
6. Phase (study, development, implementation, production or follow-up).
7. Delivery status, if applicable.
8. Personnel situation.
9. Material situation.
10. Equipment situation.
11. Facility situation.

FUNCTIONAL STATUS REPORTS. Each of the company executives, constituting the emergency headquarters staff, will begin communications with the operating organizations as soon as possible. Information reports will be solicited from each functional area to support the company executives' postdisaster responsibilities.

Operations

For a period of time in the postdisaster era, the company's headquarters organization will exercise highly centralized management control. Particular emphasis must be given to coordinated planning and the utilization of all available company facilities, material, personnel and other resources on a composite basis.

A management executive committee should be designated. As the company's senior committee, it will have overall responsibility for management and welfare of the company. It will reestablish over-all company objectives, assess the company's resources, develop and evaluate alternate courses of action, determine the company's future activities and take decisive action in any and all matters.

The duties and responsibilities of all designated officers should be clearly defined. Thus, in a major industry the following functions might be assigned.

EXECUTIVE VICE-PRESIDENT. He will do the following:

1. Provide overall direction of operations of the emergency headquarters, except for those activities which customarily report directly to the general manager.
2. Plan and coordinate preparations for all management executive committee meetings.
3. Direct the implementation of decisions made by the management executive committee.
4. As a continuing, normal responsibility, direct and coordinate the company's operations.

THE SECRETARY will do the following:

1. Serve as a member of the board of directors.
2. Serve as a member of the management executive committee.

3. Continue to serve as secretary to the company.

THE GENERAL COUNCIL will do the following:

1. Provide overall surveillance of company legal matters.
2. Continue to administer law, patents and licensing functions.
3. Provide counsel and advice regarding the company's legal rights, liabilities and responsibilities.
4. Determine the status of suspension of contractural obligations in the postdisaster period by governmental edict.
5. Provide legal guidance with respect to alternate courses of future action for the company.

ENGINEERING AND MANUFACTURING. The head of the engineering department will do the following:

1. Furnish counsel in scientific, engineering and technical matters.
2. Assess the company's postdisaster engineering and manufacturing situations. Provide counsel as to short-term and long-range capabilities in both fields.
3. Develop alternate, possible assignments of new and old product lines.
4. Provide guidance with respect to alternate courses of future company action.

FINANCE. The treasurer will do the following:

1. Serve as a member of the management executive committee.
2. Determine the company's financial position in terms of cash position, accounts receivable (and probability of collection), inventories and accounts payable.
3. Estimate the company's financial liabilities due to the disaster.
4. Develop estimates of future revenue and expenditures.
5. Determinine the mechanics of government price controls, wage and salary stabilization measures, and emergency rationing; announce their effect on the company.
6. Determine the workings of the banking system and controls established by the Federal Reserve Board and (if operative) the Asset Validation and Equalization Corporation; announce their effect on the company.

7. Reestablish the company's credit and banking arrangements.
8. Determine the possibility of reimbursement by the Federal government for capital losses suffered by the company as a result of enemy attack.
9. Reestablish the corporate financial records system.
10. Develop a plan for reactivation of the company's accounting structure.
11. Plan and coordinate reactivation of the company's data processing activities, giving priority attention to reestablishment of the corporate records function.
12. Initiate insurance claims for the company.
13. Monitor the emergency headquarters payroll system.
14. Provide financial counsel with respect to alternate courses of future company action.

ADMINISTRATION AND MATERIAL will take the following actions:

1. Assess the company's total position as to usable, repairable and destroyed facilities. Compile data as to possible utilization, by kinds of activity, of the usable and repairable facilities. Develop alternate courses of possible action in reactivating operations, from a facilities standpoint. Compile supporting cost and time estimates.
2. Arrange the acquisition of needed, additional facilities.
3. Negotiate settlement of damage to leased facilities, with counsel from the treasurer and the vice-president, secretary and general counsel.
4. Advise and assist the director of marketing research in the planning and analysis of major sales proposals, in order to guide "make-or-buy" decisions, source reviews and supplier selections. Assist in negotiations, and prepare any contracts, subcontracts, basic agreements and procurement contracts to be issued by the emergency headquarters organization.
5. Determine the company's total available manpower, by skills and geographical location.
6. Develop possible courses of labor relations action.

7. Explore possible sources of replacement manpower.
8. Prepare any company-wide communications to employees, in coordination with the director of public relations and advertising.
9. Reestablish the company's security system.
10. Prepare budget estimates for anticipated company payroll and benefits, based on alternate courses of possible future action.
11. Ensure the application of a uniform compensation and benefits program, based on postdisaster conditions.
12. Ensure company-wide uniformity in the procedures to be followed in effecting any layoffs.
13. Plan for possible transfers of employees between plant sites.
14. Determine the status of life, group and disability insurance program settlements. Develop appropriate courses of action.
15. Develop and publish all required company organization charts and statements of authorities and responsibilities.
16. Advise and assist operating organizations in plans for training employees in new kinds of work, as may be required.
17. Provide industrial relations guidance with respect to alternate courses of company action.
18. Develop estimates of stocks on hand and location of vital materials.
19. Establish a control, priorities and allocation system to govern the conservation and utilization of all company stocks of equipment and materials (special efforts will be made to insure adequate control of gold and of radioactive materials).
20. Determine governmental plans, methods and priorities for allocating both production and reconstruction materials.
21. Devise alternate courses of action, both for reconstruction work and production activities, based on present and anticipated availability of materials. Compile sup-

porting material lists and acquisition time estimates.

22. Coordinate the company's contacts with governmental bodies concerning material matters. Ensure uniform compliance with the Federal Defense Materials System. Furnish required reports concerning material activities to governmental agencies.

23. Develop sources of supply and purchasing agreements for major material requirements.

24. Ensure that procurement termination proceedings are initiated, in applicable cases.

25. Prepare budget estimates for restoration work, new facilities, new equipment and materials based on alternate possible courses of future action.

26. Compile records of major government and company property (including property at suppliers), reflecting item descriptions, valuation, condition and location.

27. Negotiate settlement of losses of government property, with counsel from the vice-president, treasurer, secretary and general counsel.

28. Advise and assist operating organizations in obtaining replacement and new government and company property, when specific operations are to be commenced.

29. Determine the effect of government controls in the fields of traffic and transportation. Secure needed priorities and allocations.

30. Develop the best means of transportation between plant sites for company personnel on authorized business.

31. Assist the flight test division in arranging clearances for earliest possible use of operative company aircraft.

32. Compile records of the company's usable vehicles and materials handling equipment by location.

33. Develop plans for transporting personnel, materials, property and equipment between plant sites in accordance with alternate plans under consideration by the management executive committee.

34. Compile supporting cost estimates.

RESEARCH will do the following:

1. Furnish advice on the current status of research work

being performed in the company.

2. Develop alternate courses of possible action for the research areas.

MARKETING RESEARCH. The director of marketing research will do the following:

1. Furnish advice on market research, proposals, product exploitation and overall marketing management.
2. Plan and prepare all sales proposals issued by the emergency headquarters organization.
3. Collect and analyze customer requirements, plans and potentials, and prepare projections applying to group and division capabilities and needs.
4. Develop alternate possible plans for forceful exploitation of the company's most recent product and service lines, and potential goods and services.

COMMERCIAL PROGRAM DEVELOPMENT. The director of commercial program development will do the following:

1. Furnish advice on commercial program diversification possibilities.
2. Participate with research and the director of marketing research in the integration of technical and marketing considerations in proposed plans of future company activity.

Chapter 17

TESTING AND TRAINING

TEST THE PLAN

Testing the industrial defense plan is required in order to determine flaws, weaknesses and requirements for revision of the plan and its parts. Reasonable and practical time and space factors must be tested for flaws which might render the plan ineffective or impossible in actual performance. Testing may take the form of exercises involving only a few people in the plant; it may be done by sections in order not to disturb more people than necessary, or it may take the form of command post exercises.

The following are steps that can be taken in testing the plan:

1. Make arrangements for the alert warning to be received at the plant. In some instances this may be accomplished by special direct alerting systems; however, most plants will receive the warning by the community outdoor warning system.

2. Relay the alert warning to employees.

3. Man and activate the plant emergency control center.

4. Test the communications system. Periodic reports can be received from all departments on their progress in meeting the emergency. New information can be transmitted to the emergency service groups, all such information can be correlated, and reports can be made to the control center.

5. Put the plant disaster control organization into action to save life and property. This is an opportunity to practice a full schedule of activities in saving lives of employees and protecting plant property by the self-help teams.

6. Simulate a plant shutdown. Trained personnel in each department of the plant should be required to carry out a simulated plant shutdown procedure by going through the motions of pulling switches and closing valves.

7. Conduct fire drills and fire fighting exercises.

8. Conduct rescue drills and simulate the rescue of employees "trapped" under fallen rubble or from upper stories of buildings.

9. Plan for mock injuries to employees and arrange for industrial medical groups and first aid teams to rush to the aid of "injured victims." Such victims can be rushed to the emergency hospital or first aid station in a token ambulance run.

10. Arrange for plant police to check for "sabotage and espionage," direct traffic and report "unexploded ordnance."

11. Have emergency repair and restoration teams make simulated repairs to essential utilities. Such teams can make simulated emergency repairs to power and communications facilities, provide emergency water resources and team up with local power and communications companies to assist in restoring services to the community.

12. Test industrial mutual aid measures by practicing life and property saving activities under simulated attack conditions.

13. Activate the alternate company headquarters or "remote emergency location" and work at solving appropriate postattack problems such as the following:

    a. Assessing damage to the company plants and losses of personnel and reporting such data to local authorities.

    b. Testing communications with company plants in other areas and with local government.

    c. Arranging for continuity of management by temporarily filling vacancies due to loss of directors or other key personnel.

d. Simulating the use of duplicate records stored at safe locations or at the company emergency headquarters.
e. Restoring company operations.
f. Moving emergency stockpile or equipment, raw materials and finished products as directed by appropriate agencies of government.
g. Simulating use of substitute raw materials and equipment for production.

Management personnel should go to the emergency headquarters and practice working under simulated emergency conditions. During tests certain key personnel should be declared casualties. Individuals that appear on the succession list as replacements for these casualties should be allowed to carry out their emergency functions until the company can be reconstituted in the normal course of business. They should report to the alternate headquarters, make use of the essential records stored there and gather vital information from secondary sources to get the operation back on its feet for immediate limited production and into full production in the shortest possible time. Only by such testing can weaknesses of the plan be brought to light and necessary action be taken to strengthen and improve it.

TRAINING

Training is the key to the success of any plan. For that reason, a training program which will reach everyone in the organization should be undertaken. Intensive training promotes prompt and efficient reactions to emergencies; but, further, the training must be so thorough that it ingrains responses so that they are habitual or automatic and take precedence over more primitive forms of automatic behavior.

The training program should reach three levels: training for the program disaster control director, his administrative heads and the disaster committee; general training for all employees, and specialized training for those responsible for services such as police, fire and work. The program should be designed so that each employee will be aware of the responsibility he will be required to accept in case of a disaster.

It is desirable that as many as possible in each group receive full training, but where group training is not possible, the leaders should receive the instruction and pass it on to their rank and file. It is recommended that all training be on company time, or that compensation be made for personal time devoted to training, provided attendance at classes is regular and progress satisfactory. The defense coordinator should take courses in basic disaster control and civil defense organization and obtain current pertinent information as it becomes available.

EMPLOYEE TRAINING

All employees should be given instructions in self-protection. Most of this can be done by the printed word. Ask your civil defense director for available literature, use your house organ, post instructions on bulletin boards, prepare posters and send information to employees' homes. When meetings are held for other purposes, make this one of the topics. Try to create acceptance of the program so that self-protection is a part of everyday activity.

All employees must be told what to do in an emergency. This may be limited to drills in shutting off their machines and proceeding in an orderly manner to their assigned areas or to the evacuation point. Preferably, they should be encouraged to take instructions in elementary fire fighting, first aid, psychological first aid, rescue, etc. This participation will help prevent a stampede in time of actual disaster and will stimulate enrollment of volunteers for specific duties, as well as help to keep interest in the protection program alive. Shutdown and exit drills should be held frequently and include all employees. Announce the first few drills in advance; then provide them at irregular intervals and without warning. Drills, properly planned and studied afterward, will uncover weaknesses in planning, coordination and communication. An additional advantage to such training is that each employee has the training and information to be helpful in the survival of his family at home. The community-wide protection is also increased because trained employees can assist the local police, fire and other groups on a volunteer basis during a disaster.

SERVICE GROUP TRAINING

Specialized training for the service groups within the organization such as police, fire and medical is of paramount importance to the success of any disaster program. It is desirable that these units be supplemented with volunteers from other organizations within the plant who can help to enlarge the effectiveness of the service groups. The head of each service unit should take the responsibility for the training of the regular members of the unit as well as the volunteers who will assist during an emergency.

Both the office of the civil defense director and the head of the appropriate municipal department can help with the planning and training of the service unit. As the municipal organization and the plant service unit will work closely together in case of an emergency, it is essential that these units closely coordinate their activities.

Training for the service groups should be handled in four stages. First, the members of the service organizations should receive training as individuals. Second, team training then follows in which the individual's duties within the team and the team functions are stressed. Team training also provides for an interchange of duties within the unit. The third phase is the training of all service units together to assure coordination of activities and to assure the smooth functioning of all the units participating in the event of disaster. A full-scale exercise not only offers excellent practice and training, but will give the members of the unit an opportunity to get an idea of conditions during an emergency. The fourth phase of training for the service units is training exercises designed to train the industrial groups with those of the local civil defense organization, the groups of neighboring plants or with the municipal service organizations.

Handbooks should be prepared containing basic information regarding the company plan and information necessary for self-protection at the place of work, including description of the warning signals, floor plan drawings showing shelter areas and hazardous points within the plant, maps showing evacuation movement routes within the plant and to safe areas outside the city. Such information should be given to all employees.

EXAMPLE OF INSTRUCTIONS

The following is a sample of instructions for employees:

1. General. The Self-protection Organization, for the (*name of facility*) has been established to protect YOU in case of emergency. This Organization will direct and supervise the evacuation of your building and your movement to shelter areas in the event of an attack. It will conduct drills to familiarize you with what you are to do in emergencies. You are asked to help prevent accidents and fires in your building, and to volunteer your assistance in the handling of emergency situations. This program is for your protection—your cooperation is requested. Although emergencies arise most frequently from fire, other emergencies should be anticipated. You should be familiar with all of the following instructions:

2. FIRE. Emergencies in buildings may arise from fire

   a. IN CASE OF FIRE YOU SHOULD KNOW

      (1) The fire alarm signals for your building.

      (2) Where the nearest fire alarm box is located.

      (3) How to operate the fire alarm box.

   b. IF YOU DISCOVER A FIRE

      (1) Proceed immediately to the nearest building alarm box or city fire department alarm box and turn in an alarm. Do not hesitate to telephone the guard office or the city fire department, or to follow other established procedures to report a fire. Always be sure to give the exact location of the fire to avoid loss of time. The fire department would rather turn out for a very small fire than have you delay and give the fire an opportunity to spread. The first few minutes are vital. In buildings without guard offices, call the building manager.

      (2) Contact your building warden and give him the location of the fire.

   c. WHEN YOU HEAR THE FIRE ALARM

      (1) Close all windows in your office immediately.

      (2) Leave the building in an orderly manner by your assigned exit, but do not run.

      (3) Obey the instructions of your floor warden and monitors.

      (4) Avoid crowding or undue haste.

      (5) Descend the stairs in an orderly manner as directed by the Stairway Monitor, a fall might injure you and those who follow.

(6) Stay in formation until you emerge at the ground floor exit.

(7) As soon as you are out of the building, move at least 100 feet away from the exits.

3. FIRE DRILLS. Fire drills are held periodically under the direction of the Self-protection Organization. They familiarize you with evacuation procedures. At the sound of the alarm, leave the building without delay as you have been directed to do in the case of actual fire.

4. FIRE PREVENTION RULES. The following are the ground rules for fire prevention in (*our building*). (*Add any others required to lessen or eliminate the danger of fire from mechanical, chemical, or structural faults or human error.*)

a. Maintain good housekeeping in all areas of the building, as this is one of the most effective means of preventing fires.

b. Bring to the attention of your supervisor any apparent fire or safety hazard existing in the building.

c. Obey "No Smoking" signs.

d. Do not throw matches, cigars, cigarettes, or pipe ashes into waste baskets, or into any type of receptacle containing combustible material.

e. If you are a smoker, provide yourself with noncombustible ashtrays and see that all cigars, cigarettes, pipe ashes, or matches have been completely extinguished. Never empty your ashtrays. Let the building cleaning force do this.

f. Oily rags or similar flammable materials in the building must be placed in approved metal containers provided for the purpose.

g. Hotplates, irons, or similar electric equipment with heating elements, may be used in the building only when the installation, including the stand, is approved.

h. Deposit all trash in receptacles provided for the purpose, or arrange for its storage in regular trash rooms.

i. Clear adequate passageways in corridors, to stairs, between stacks and to firefighting equipment, must be maintained at all times, and storage must be in accordance with safe practices.

5. ATTACK INSTRUCTIONS. A nuclear attack could occur with no warning, or with warning of minutes, hours, or possibly days. Whatever the time of warning, all building personnel must be prepared to take immediate action upon being alerted.

a. THESE ARE THINGS YOU SHOULD KNOW NOW

(1) Location of your shelter area.

(2) The attack warning signals.

(3) What to do when they are sounded.

b. THE ALARM SIGNALS USED IN THIS BUILDING ARE

(1) ALERT SIGNAL. A warning signal consisting of a steady siren wail for a period of 5 minutes, and a series of horn blasts, each 16 seconds long at 16-second intervals from within the building. It means Attack is probable: go to shelter and await instructions. Do not take the elevator unless you are physically handicapped. All personnel move from their normal work stations to shelter areas as indicated below:

Occupants of	*Go to*
Floor No. —	Nearest shelter on — Floor
Floor No. —	Nearest shelter on — Floor
Floor No. —	Nearest shelter on — Floor

(2) TAKE COVER SIGNAL. A warning signal consisting of a rising and falling siren wail for a period of 3 minutes and a series of short horn blasts from within the building. It means Attack is imminent; take cover immediately. Go to central corridors and, if possible, proceed to the shelter area as indicated above. Obey wardens at all times. Do not take the elevator unless you are physically handicapped.

(3) ATTACK WITHOUT WARNING. There will be no alarm signal. A blinding flash means attack has come.

(a) Drop to floor; cover head with hands.

(b) Lie against wall or under desk or other large object to safeguard against flying glass or other debris.

(c.) After the blast, go to nearest shelter area where Organization personnel will arrange for first aid, food and water, sleeping arrangements and registration, and will disseminate information as available.

In the event of attack, the following shall apply: After ALERT signals, further instructions will be broadcast by radio, After TAKE COVER signals, further instructions will be given orally to personnel in shelter areas by the organization. Notification to return to the building after DRILLS will be given orally by wardens.

(4) SHELTER ACTIVITY. At the sound of an ALERT signal, all occupants of the building must go immediately to shelter areas. REMAIN CALM. Your organization is prepared to take care of you and make life in the shelter area as livable as possible.

(a) Food. There is food stored in the building to provide about 700 calories a day for each occupant for 14 days.

(b) Water. A quart of potable water per day per person is available.

(c) Medical. Medical kits containing equipment

and supplies required for emergency treatment are stocked and available.

(d) Sanitation. Sanitation kits to accommodate all occupants of the shelter for a period of 14 days are stocked in the shelter area.

(e) Radiological Monitoring. Radiological Monitors will survey all shelter areas to assure that occupants are being housed in the areas least exposed to fallout.

(f) Communications. A battery-operated radio will be located in our command quarters. As information becomes available, it will be disseminated to those in the shelter.

PUBLICITY

Informing the employees of the plan cannot stop with training. This training must be reinforced with appropriate publicity and promotional activity to keep them appraised of any changes and to keep them constantly aware of the part they will be required to play either to help themselves or to assist in implementing the entire program. This can be handled in a variety of ways, among which are the following:

1. Using disaster emergency and disaster posters throughout the plant on bulletin boards.
2. Making distribution of pamphlets to employees.
3. Setting up special displays in lobbies, cafeterias and other departments.
4. Publishing emergency defense articles in the company newspaper.
5. Discussing emergency defense at employee meetings.
6. Making movies of the exercises which can be shown to employees and to other groups in the community as a means of informing them regarding the overall company plans for emergency defense and enlisting their co-operation, support and assistance.

Chapter 18

DISASTER CONTROL SERVICES AND ORGANIZATION

DISASTER CONTROL SERVICES

T HE PROTECTIVE SERVICES of the disaster organization are not intended to replace the usual personnel and security measures which customarily function under normal conditions. They are formed within the installation to perform the specialized services necessary to safeguard and salvage the property and occupants in the event of a major disaster. At small plants it may be possible to combine two of the services, such as the rescue and the emergency maintenance services. At others, the existing organization may provide sufficient strength to eliminate a service, such as public information. The following services based on civil defense organization and tailored to fit the individual installation are considered mandatory in disaster planning:

1. Police service functions are traffic control, protection against looting, panic prevention, employee identification, plant guard and patrol, enforcement of regulations, un-exploded ordnance reconnaissance, safeguarding records and valuable instruments, holding and reporting all suspicious persons, training policemen, liaison with local police department and civil defense organizations.

2. Firefighting and control services' responsibilities are fire prevention, fire fighting, plant fire inspection, reporting fires to local fire department, assignment of duties, training firemen, liaison with local fire department and civil defense organization. Also, in large plants, a deputy in charge of fire drills has searchers, runners and exit guards. This group assigns and trains employees in fire drill duties, searches assigned areas, transmits fire alarm,

develops exit plans and, in the absence of wardens, directs exit traffic.

3. Engineering services' functions are selection and designation of shelter areas; construction and maintenance of shelters; maintenance of buildings, equipment and supplies; shutdown procedures; demolition of unsafe structures; removal of debris; disinfection of water mains after repair to insure potable water; water supply for drinking, washing and cleaning; restoration of the plant; decomtamination of buildings and their contents; repair and restoration of machinery and equipment; supply power, light, compressed air and heat; rigging and millwright service; provide materials handling and industrial transportation equipment (trucks, tractors, lift trucks, portable cranes, power shovels, etc.); train workers in disaster control duties; liaison with local and national engineering societies, neighboring plants, local utilities service specialists and civil defense organization; maintain internal communications systems; assist in preparation and maintenance of the evacuation site.

Another important responsibility is to minimize damage to or by utilities. This function provides for control of utilities in the facility or building in the event of a disaster. This function is concerned with the control of all electrical and mechanical controls and equipment; water gas and steam valves and conduits; power switches; ventilation and refrigeration devices; and any other similar equipment in the facility or the building.

4. Welfare services' functions are to supply emergency food; organize emergency personnel transportation; provide counsel and information and emergency financial aid; reunite separated families; locate housing; assist in preparation of evacuation site, and provide liaison with local public and private welfare agencies and civil defense authorities.

5. Warden services have two main functions. The primary function is to plan and direct movement of all building occupants in the event of an emergency, whether the

plan requires withdrawal to a designated shelter, the entry of additional persons from outside the building, or the evacuation of building occupants to the street or elsewhere outside the building; report damage; restore order; direct workers to safety; first aid; psychological first aid; maintain morale; take the roll call after the emergency; distribute warden equipment; protect records, blueprints, or valuable devices in case of emergency or attack warning; keep records of building occupants; train workers in warden service, and guide personnel arriving at evacuation area. Their secondary function is to reconnoiter unexploded ordnance or hazardous material; decontaminate material and areas; aid police, fire and rescue units, and provide liaison with local civil defense and neighboring plants.

6. Some personnel should be in charge of the rescue of injured or trapped persons. This function provides for seeking out, rescuing, providing on-the-spot lifesaving aid, and removing injured or trapped persons to a place where they can be cared for in an emergency. Personnel assigned to this function should be selected, if possible, from employees who have had construction, engineering, or other experience that will enable them to evaluate structural or load-bearing factors incidental to rescue work or light demolition work. Previous first-aid training and experience is also useful in connection with this function.

7. Health services' functions are emergency medical treatment; first aid; transport and identify casualties; protect against atomic, biological, and chemical warfare agents; decontamination of personnel; radiological monitoring; advising on environmental health hazards; training of health workers; liaison with local medical groups.

8. Personnel should be detailed to the function of public information. Only by keeping all concerned informed on all matters pertaining to the defense of the installation can complete understanding and cooperation be assured. One individual should be given leadership responsibility for each service. They should be selected for their experi-

ience, maturity, emotional stability and stamina. Provision should also be made for alternate chiefs of services. In particular, this should be done at installations employing more than one shift of workers. Full use should be made of the normal service departments within a facility and in many cases the chief of such a department will be the logical selection as chief of the comparable emergency service.

In augmenting existing services and in the formation of new ones, individuals should be chosen on the basis of their natural talent or past experience. Many positions may be filled by women employees. These include clerical workers, telephone operators, nurses aides, cooks and drivers.

DISASTER CONTROL ORGANIZATION

One of the first steps in the actual preparation for disaster is the appointment of a single individual at both the company level and in each plant—a disaster planning or civil defense coordinator—to provide coordination and direction of the overall corporate disaster plan and the disaster plan in each plant.

Second, at both the company and plant level a disaster or civil defense advisory committee should be designated representing various departments of the company, to assist in the development of the various phases of the disaster planning or civil defense program. The duties and responsibilities of the disaster control officers of the organization should be defined in the facility's plan. Collectively, the duties and responsibilities assigned to all of the principal officers should provide for the performance of every major function specified in the facility's plan.

DISASTER CONTROL DIRECTOR

In large organizations where the problems are complicated the position of disaster control director or defense coordinator may require the services of a full-time employee, while in smaller activities this duty may be assigned as an additional responsibility. His position in the organization must be such that he has a comprehensive knowledge of all influencing factors of management, and his authority in protective measures must be complete. He should be a man of proven executive and administrative ability and should be thoroughly familiar with the installation.

It is the responsibility of the coordinator to set up a system of communication and control which will enable him to operate successfully the disaster-control program until the emergency condition returns to normal. Disaster means dictatorship. To be effective, the coordinator must have full authority and responsibility to speak for management in all matters relating to disaster control. Thus, it is imperative that he either be a top executive or speak for an executive charged with this responsibility.

The director should provide coordination and general direction within his organization. He should immediately prepare appropriate policy and administrative directives for the establishment of the program. In addition, he should establish close liaison with civil defense authorities, local government agencies and the program directors of other plants in the community. These representatives have a storehouse of information in the form of up-to-date emergency planning and guidance information. Such material is invaluable as a help in the saving of time and work. The program coordinator also needs to tie his program in with these representatives so that his plan will be coordinated with the community-wide plan.

The coordinator performs at least the following duties:

1. He establishes the protective plan and insures that the basic provisions of the plan are disseminated to all persons employed in the building.
2. He appoints deputy coordinators, one for each working shift, and other assistants as necessary. He designates a records officer responsible for duplication, dispersal, and general protection of valuable records and documents. He recruits or selects qualified emergency service personnel.
3. He establishes a control point from which the activities of the organization may be directed in an emergency.
4. He maintains liaison with and cooperates with the principal officers of the facility. He provides periodical reports to management on the plant preparedness and the

operational readiness of the disaster-control program.

5. He maintains liaison with the local civil defense director and the local fire department.

6. He insures that appropriate training courses are set up and that arrangements are made for obtaining assistance from the local civil defense organization; the Red Cross, local fire departments, and other sources, as required, to train the organization staff.

7. He coordinates arrangements for and directs civil defense and fire drills.

8. He arranges for posting on appropriate bulletin boards (or otherwise publicizes within the building) a roster of organization personnel who have responsibilities for the shelter or other emergency operations.

9. He takes all necessary actions to insure that the facility's organization operates safely and efficiently in emergencies.

10. He exercises command responsibility for the orderly movement of all personnel in the building (including nontenants) in accordance with the plan.

11. He arranges for and supervises the storage of required supplies and equipment in facility shelters.

12. He conducts periodic inspections of the facility.

13. He directs all shelter management operations when fallout shelters in the facility are occupied.

14. He works closely with and cooperates with the local or state civil defense director in all matters involving the facility's plan, organization, or operations.

15. He develops mutual aid plans with neighboring plants.

16. He plans and secures the installation of necessary alarm systems and arranges with local civil defense and municipal authorities for transmission of warnings directly to him or his deputies.

17. He strengthens the plant's fire protection and accident prevention program.

18. He directs emergency activities from the first warning until normal conditions are restored.

19. He provides and maintains a written record of all

operational and policy proceedings for the duration of the emergency, from the warning until normal operations are resumed.

20. He authorizes reentry of the plant and/or area after a disaster.
21. He officially designates the end of the emergency period.
22. He directs or coordinates postdisaster recovery.
23. He assesses the nature and extent of damage resulting from disaster and reports this to the local civil defense director.

Regardless of the size of the company, all of the foregoing should be done. The owner of a small shop may act as coordinator too. Perhaps some of these functions will require condensation or services will be consolidated. Nevertheless, these elements must be covered. In a large company having many plants at scattered geographic locations, each plant should have the complete system starting with the plant defense coordinator. Then the parent company or home office should provide an executive and staff of experts to coordinate the entire program.

A disaster committee, representing various departments in the organization, can be of invaluable help to the program coordinator by advising him and assisting him with his planning.

The proper selection of key administrators for the implementation of the program is an important step the program coordinator must take to establish an organization to function during an emergency. These administrators should be carefully chosen from among the service heads, employee representatives and key personnel.

DEPUTY DIRECTOR

The organization should generally include a deputy director who serves as the principal assistant to the coordinator and acts for him in his absence. The deputy performs such duties as the coordinator may assign. In small facilities, for example, he may serve in any of the positions described in the paragraphs immediately following, in addition to serving as deputy. In a larger facility, the coordinator may assign to the deputy the responsibility of training the organization personnel, including arranging for instructors, obtaining appropriate training materials, arranging training schedules, etc.

PERSONNEL MOVEMENT OFFICER

The personnel movement officer is responsible for supervising and expediting the planned and controlled movement of all building occupants in an emergency. Under the general direction of the coordinator, the personnel movement officer serves as head of the personnel movement service and performs the following duties:

1. Plans personnel movement routes and establishes movement procedures to give effect to the personnel movement provisions of the plan.
2. Assigns and trains floor, area, wing and other wardens and related personnel, as required by building configuration or layout.
3. Supervises and directs the movement of personnel within, into, or out of the building as required by the building self-protection plan, during drills and actual emergencies.
4. Assures that all building occupants, including members of the organization, comply with procedures indicated by the alarm signals, as specified in the plan.
5. Cooperates with the first aid officer in providing first aid instructions to personnel movement personnel.
6. Supervises and directs the activities of floor, area and wing wardens; room, stairway and elevator monitors; and any messengers assigned to duty with the warden service.

   a. Floor, area and wing wardens, under the general direction of the Personnel Movement Officer, supervise and expedite the movement of personnel within, into, through, or out of their assigned part of the building. Their duties include the following:

      (i) Making certain that personnel movement routes are clearly identified and are made known to the regular occupants of that part of the building.
      (ii) Maintaining a roster of physically handicapped

persons regularly in the areas and making appropriate special provisions for their movement in an emergency.

(iii) Assuring that the procedures to be followed on the receipt of warning signals are known to all regular occupants of the floor, area, or wing, including specifically the procedures to be followed on "Alert," "Take Cover," "Fire," and any other alarm signal provided for in the plan.

(iv) Directing the brisk and orderly flow of personnel, during drills or actual emergencies, along the prescribed personnel movement routes.

(v) Assuring that all persons have vacated the assigned area when this is required by the plan.

(vi) Coordinating the activities of room, stairway, and elevator monitors in the assigned areas.

b. Room, stairway, and elevator monitors assure that rooms are vacated, that windows and doors are closed, and that electrical appliances are turned off in their assigned area of the building. Stairway monitors are posted at assigned places or stairways; they keep doors to the stairwell open and control the flow of personnel into the stairwell. Stairway monitors may also be assigned to positions on landings to regulate and expedite the safe flow of personnel. Elevator monitors take their positions at assigned elevators to assist the movement of the physically handicapped, and to restrict elevator use by others.

MEDICAL OFFICER OR FIRST AID OFFICER

The medical or first aid officer serves as head of the health and first aid service. He is responsible for training and equipping all personnel assigned to perform medical or first aid services in an emergency, and for supervising emergency first aid or medical selp-help operations within the building during an emergency. His duties include the following:

1. Arranging with the American Red Cross or other sources

for first aid and medical selp-help training for all organization personnel who need it.

2. Supervising the selection of first aid or medical treatment areas in shelters and elsewhere, as required.
3. Directing first aid or medical self-help operations and controlling access to medical supplies, as required, to assure their proper use, conservation and availability for emergency use.
4. Establishing policies and rules governing the emergency treatment of ill and injured persons, the maintenance of adequate sanitation and hygenic standards, and other matters relating to emergency health, hygiene and medical activities within the building during an emergency, or during the occupation of shelter within the building.
5. Inspecting the storage and handling of food and drinking water in shelter areas within the building.

A plan should be worked out by the medical officer of the facility with the local chief of emergency medical service. It should be broad enough to provide for efficient service in the event of an emergency to the plant alone or to the entire community.

Details of the technical and professional aspects of the plan are not discussed here, but generally it should include the following:

1. First aid at the site of the incident.
2. Adequate ambulance service.
3. Hospitalization of casualties.
4. Transportation of casualties.
5. Identification of casualties.
6. Casualty stations
7. Sanitation service.
8. Medical auxiliaries.
9. Emergency supplies.

RESCUE OFFICER

The rescue officer is responsible for locating injured or trapped persons and removing them to a place where they can be cared for safely in an emergency. Under the general direction of the coordinator, he performs the following duties:

1. Organizes and trains the rescue team or teams provided for in the facility plan.
2. In cooperation with the first aid officer, insures that the members of the rescue team are proficient in on-the-spot first aid techniques.
3. In cooperation with the fire marshal, obtains appropriate equipment (e.g., handtools, ropes, etc.) for rescue operations.

FIRE MARSHAL

The fire marshal serves as head of the fire service. He organizes fire fighting teams (or brigades) for initial fire-fighting operations. His duties include the following:
1. Providing fire-fighting instruction through available sources, such as fire-fighter schools, local fire departments, etc.
2. Assuring that fire-fighters know their stations, locations of fire-fighting equipment in their area of responsibility, and the alarm signals which direct them to their stations.
2. Deploying fire-fighting personnel to fire areas to extinguish or contain fire pending the arrival of other fire-fighters forces, and cooperating with such forces upon their arrival; coordinating with the first aid officer and arranging for the first aid training for fire-fighters. In some buildings, it may be feasible for the fire marshall to organize and train a fire brigade or brigades, which respond wholly or by sections to the fire alarm. Under this system, fire-fighters should be instructed that each should go into action at the point of fire in his area of assignment. In the absence of fire in his area, he should report to a prearranged shelter area for instructions.

UTILITIES OFFICER

The utilities officer is responsible for controlling utilities in the facility or building during an emergency. He serves as head of the utilities service or utilities control team.

The functions of the utilities officer include the following:
1. Establishing a plan to attend mechanical devices, ventilation, water, gas and steam valves, power switches, etc.

2. Dispatching individuals or teams, at the sound of emergency alarms, to prearranged control points for preplanned or directed action.
3. Deploying personnel, after the fire, to reconnoiter and correct damage to utilities or to report conditions which require other assistance.

RADIOLOGICAL OFFICER

The radiological officer serves as the head of the radiological monitoring service. He is responsible for organizing, training and equipping personnel assigned to radiological monitoring functions and for supervising and directing all radiological monitoring activities within the building after a nuclear attack. His duties include, but are not limited to the following:

1. Training or arranging for the training of all personnel assigned to the radiological monitoring service.
2. Obtaining and providing for the safe storage of radiological monitoring instruments, personnel dosimeters and related equipment.
3. Insuring that all radiological monitoring instruments assigned to or in the custody of the organization are maintained in effective operating condition.
4. After occupation of any shelter area in the building, directing and supervising radiological monitoring activities and serving as technical advisor on radiological defense to the shelter manager and coordinator.

SUPPLY OFFICER

If the facility or building includes a fallout shelter, the organization should include a supply officer, who will be responsible for the care of and issuing to appropriate officials all supplies, especially essential stocks of food, water, medical supplies, sanitation equipment and radiological monitoring instruments. He serves as head of the supply service and performs the following functions associated with shelter management:

1. Plans and scheludes distribution of supplies to appropriate leaders, and regulates consumption in conformance with established shelter management police.

2. Procures, as available, and to the extent storage space permits, desirable items over and above basic stocking through the local government, or other possible sources.

3. Establishes a program covering the acceptance and storage of issued supplies and personal supplies; inventory of all supplies and equipment; provision for, and maintenance of, security storage of such materials; and issuance of daily rations, medicines and other items to appropriate organization officials, based upon the quantity of supplies, the minimal health requirements and the estimated length of stay.

4. Trains supply assistants in sufficient numbers to carry out the supply functions of the plan.

WELFARE OFFICER

If the facility or building includes a fallout shelter, the organization should include a welfare officer, responsible for planning and supervising emergency welfare activities involved in shelter management. Such services include registration of shelter occupants, distribution of food, sleeping arrangements, recreation and emergency information. The plan shall provide for an appropriate number of assistants to insure the adequate performance of these functions in accordance with the requirements of that particular facility.

Registry

Establish a registration section to record statistics on each person in the shelter area (such information should include, but not be limited to, name, address, marital status, occupation, special skills and health perculiarities such as diabetes, contagious diseases, epilepsy, etc.); develop and prescribe the format for a registration form; from review of the information received, advise the shelter manager or coordinator regarding persons available for duty in the shelter, those who require special care, and those who, because of special skills (or lack of them), should be given special assignments; and maintain an official record of incidents, problems and activities, such a record to include, but not be limited to, vital statistics, health problems, critical conditions, operation problems and any information which should be preserved for historical or legal purposes.

Feeding

Plan and supervise the distribution of food in the shelter. Arrange to obtain the necessary equipment and staff assistance to provide for the planned, orderly, sanitary preparation, distribution and consumption of food in the shelter and the disposal of related waste materials. Feeding responsibilities include, but are not limited to, maintaining an inventory of the quantities of food and water on hand; issuing food and water rations to occupants; determining number, types and schedules of feedings in accordance with the available supplies, shelter population and physical arrangements for feeding, preparing for and arranging any special feedings that may be required (e.g., for infants, the injured, the ill, etc.); monitoring water drums to minimize leaking and contamination, and providing for the cleaning of utensils, equipment and areas used for feeding occupants.

Billeting

Plan and supervise sleeping arrangements based upon available space, floor layout, size and nature of shelter population, etc. Arrange to obtain the necessary supplies and equipment and sufficient staff assistance to set up any necessary partitions, screens, bunks, or other structural adaptation required to prepare the sleeping area for use. Billeting duties include, but are not limited to procurement of sleeping supplies and any partitioning material, and other furnishings required to provide sleeping accommodations; sleeping area allocation (based upon sex, age, family status and other like considerations); designation of sleeping hours by shifts, groups, etc., and preservation of orderly conditions conducive to sleep in the sleeping area; and maintenance of an inventory of additional supplies, such as blankets and extra clothing brought into shelter area by occupants, with a view toward accepting voluntary donation for urgent needs, or confiscation for such needs, if necessary.

Emergency Information

Keep shelter occupants informed about the attack and postattack situation; indoctrinate them in the basic elements of shelter living; and, to the extent possible, instruct them on survival measures they will have to take upon emerging from shelter. Emergency information duties include developing a program aimed at maintaining morale, health and adherence to shelter rules, and may involve organizing an information service and training schedule for shelter occupants. Emergency information objectives may include the following, at the discretion of the shelter manager: description of the attack situation and how it will affect the shelter occupants; the shelter situation, including the organizational structure of the shelter; the supply levels and the supply discipline required; and the need for tolerance, cooperation and adherence to regulations required for orderly shelter administration.

Miscellaneous

Insure that plans are developed and appropriate arrangements made for a diversified recreational program adapted to the requirements of shelter living; care for the religious or spiritual needs of shelter occupants; and provide for any specialized training that may be required for effective shelter administration or the survival of shelter occupants.

PUBLIC RELATIONS

The scope and importance of public relations during and after a disaster are too often underrated if not completely ignored. In such a situation public relations are of the utmost importance. At such times it embraces relations with the community and area, with fellow manufacturers, with vendors and suppliers, and with customers. They are all interested in what you are doing, what has happened to you, and what is ahead. To meet the demand that will be made, a public relations director should be designated.

As soon as possible after a disaster and after having made an evaluation of the damage and prospects for reestablishment of production, notify your suppliers of your position, plans and needs. Let him know you want him to keep on filling your orders on schedule. Then tell him where to ship. The same policy holds true in handling customer relations. If you do not quickly see that customers are given the full facts, you may find they are moving over to your competition for guaranteed deliveries.

Insofar as community public relations are concerned, there are numerous steps that can be taken. First and foremost is to supply the news media with full and complete facts. Naturally, full cooperation should be given to the news media.

It is possible to go further and sponsor radio and television announcements thanking those who assisted during the emergency. Personal letters of thanks are also appropriate and appreciated. Reports of future plans may be made on radio or television or through letters and booklets. But whatever is done, it should be done quickly, with good taste, and be scrupulously accurate.

CONTROL SYSTEM

Every plant protection organization, to operate effectively, must be centrally controlled and directed to take needed action by a single person who is responsible for the organization as a whole. In small organizations, direction can be exercised by the issuance of orders directly from the plant security officer to the individuals who are to carry out the instructions. In large organizations, composed of many units operating at widely scattered points, some means must be devised to enable the responsible individual to know all that is happening in the area under his jurisdiction and to send orders quickly and accurately to all units under his direction. This system is known as the "control system." It is directed from one point, known as the "control center."

The control system should provide the following:

1. A suitably protected headquarters or control center.
2. A reporting system to keep the security superintendent constantly informed during an emergency or disaster.
3. Communications for the transmittal of orders without error and secure against enemy use to create confusion.

The staff of a control system is usually composed of the chief of plant protection and those responsible to him—the fire, guard, reserve, medical and other chiefs.

The control center is the most vital part of the protection system. It is the nerve center when an emergency arises and the whole plant protection organization must go into action. The control center should be carefully selected and should be located in a strategic place so that it can operate with maximum efficiency even under adverse conditions.

The primary purpose of the center is to coordinate all disaster operations, i.e., fire fighting, rescue, repair efforts, etc. Thus sufficient office space should be provided adjacent to the control center for use of members of the disaster control organization or members of top management of the company during an emergency. Telephones, office equipment and supplies as well as other necessary items such as maps, should be stockpiled so that this area can serve as an operational headquarters. It should be equipped with the necessary maps, charts, control panels and communication systems (telephone, radio, messengers). There should be an alternate location for setting up an emergency control center to provide for damage to the regular control center.

The following steps should be taken to make the center effective:

1. Provide for at least two systems of exterior communication, one of which should be radio, operated by either independent or an emergency source of power.
2. Provide for at least two systems of interior communications covering all important fixed areas, of which one system should have an independent power source.
3. Provide for a system of radio communication connecting all necessary plant organizational services in which each fixed and portable ground station is provided with either an independent or an emergency source of power.
4. Provide for a separate warning and alarm system to alert personnel in the event of enemy attack.
5. The disaster and/or damage control center should provide for the following:
   a. Telephone control boards.
   b. Auto-call and alarm signal system contact makers.
   c. Radio for newscasts.
   d. Voice radio set (walkie-talkie) for intraplant emergency communication.

6. Formulate a plan whereby necessary personnel would report to the control center immediately after alarm has been sounded.

7. Provide for direct contact with the office of the civilian defense communication center of the area or other officially designated "alerting" headquarters when such is established.

8. Provide a complete scale map of the plant or establishment showing the location of all buildings, equipment, structures, emergency first-aid stations, casualty stations, reserve squads, etc.

9. Maintain prints showing the location of all shutoff valves for water, steam, gas and air service lines, sprinkler-system valves, hydrants, hose houses and gas masks.

10. Maintain a suitable area (city) map showing hospitals, aid stations, fire stations, civil defense headquarters, medical supply depots as part of the equipment in the control center.

11. Select a number of men who have previously volunteered to report to the control center to serve as messengers or runners in the event methods of communication are disrupted.

12. Make provisions to plot incoming information regarding the progress and activities of the various emergency crews. As these crews complete a job, they may be dispatched by telephone or radio to the next most urgent job.

COMMUNICATIONS

Effective and invulnerable lines of communication are a key factor in the entire disaster control program, necessary during preparation, at the time of disaster, and for reorganization. Communication systems must be established and alternate means of communication provided in the event a failure of the primary means occurs. Unless communications are established and maintained under all conditions, top management may find itself commanding only the desk in front of it and the people within hearing of its voice. Lines must be established, by one means or another, to everyone in the facility, bar none.

Mediums of Communication

Important mediums of communication are person-to-person conversation, a public address system, radio, telephone, bulletin board, plant newspaper and personal letter. The personal touch of individual contact is probably most effective. However, it is time-consuming, hence can be accomplished to only a limited degree. Do not neglect this method altogether. Top management, especially, should make it a point to see every person in the plant as often as available time permits. Other than this, supervisors and foremen should keep employees personally aware of their importance to the plant and of the conditions which may affect operation. All of these will increase the employees' sense of security.

Personal letters sent at regular intervals, by the company president or general manager to each employee, are also effective. Through these management can explain its position on important matters or subjects of special concern. This genuine "straight from the horse's mouth" approach has considerable appeal to most employees. It should never be scolding or threatening, but rather guiding and informative. Disciplinary messages should be conveyed by the receiver's immediate superior only. The top-management letters should be sent to each employee's home, rather than delivered at the plant. Thus it is read at home and shared with the family. Each member realizes that he, or she, is recognized as an individual and that his welfare is important to the plant management.

Bulletin board, public address system, and house organ are means for general message distribution. Wisely used, they are extremely effective for maintaining good morale, for psychological control, and for keeping rumor to a minimum. Everything said must be honest and straightforward in presentation. Plant newspapers, particularly, must be kept free of company propaganda. No announcement or bulletin should be released until all of management including supervisors and foremen are fully aware of the message. Never leave them in the dark before the people they supervise. This, of course, does not apply to emergency messages, although even with these, if time permits, it is better to forewarn leaders first. This may be done by telephone, using a prearranged signal code to maintain confidence and order.

Public address systems have multiple value to psychological control. In addition to delivering general messages, they may also be used to assemble special groups, provide warning, direct evacuation or rescue, and deliver carefully selected, soothing or invigorating music, as may be required. Certain types of training are also possible through this medium. Sirens or whistles used as a signal for starting, stopping, announcing lunch and rest periods, and as a protective warning are a special means of communication.

Forms of Messages

Three particular forms of message are important: protective, directive and definitive. Fear, you will remember, results either from a lack of knowledge, distrust of leaders, or overwhelming danger. Through protective messages you provide the assurance of safety and security. Directive information is the guide and go-ahead that gets things done; here is where true leadership shines brightest. Definitive messages will give your people a true picture of events in process, clarify otherwise puzzling acts and dispel rumors; hence make plant personnel understand the problems at hand. No condition should exist without explanation simply because you feel that the answer is too elementary to bother with. Take the time to have each employee get as much information important to him as security regulations permit. Try to make every message positive.

Noise

Noise level in the plant should be reduced to the smallest practicable figure, each individual and unnecessary contribution to clamor weeded out. People, as a rule, become accustomed to the routine noises associated with their work. Strange or high-pitched sounds are distracting and wear heavily on the nervous system. Sometimes unrecognized, they do contribute to morale breakdown. Slapping belts, hissing safety valves, overdone lift-truck horns, cutting tools screaming without real need, slamming overhead conveyors, and constantly clanging call signals are just a few of the sources you may find in a real noise reduction campaign.

Proper planning will provide for certain actions to take place automatically in the event of emergency, bypassing the usual necessity for communications in order to set some action in motion. The concept of fail-safe provisions should be used so that in the event of loss of contact between units, certain precautionary measures may be taken without specific instructions.

WARNING SIGNALS

Arrangements should be made to receive a warning and to disseminate it quickly to employees throughout the plants. Where buildings are spread over a wide area or located beyond hearing distance of a community warning, a separate warning system may be necessary. In many instances the existing public address system can be used. However, the warning system must be adequate to reach all office buildings, plants, laboratories and all other places where employees are located. When time permits, such as a slowly rising flood, a warning by normal means of communication is preferable and will result in the smoother operation of emergency plans than if the signal is given by a sudden and exciting alarm. However, sudden contingencies must be provided for; a flash fire in an explosive area might call for immediate evacuation. Inform employees of how they will be warned of impending disaster, what the warning signals are, what they mean, and what action should be taken upon receipts of the warning signal.

Authority to Give

It is of importance that authority to give an emergency signal be clearly delineated. The action resulting from a mistaken or accidental alarm can be very costly in manpower and material. The means of activating an emergency signal should be so positioned or protected that accidental initiation is impossible. Authority to activate an emergency signal should be based on the chain of command, culminating in an area or process supervisor. In certain highly sensitive process or storage areas, it is advisable to train all assigned personnel in the recognition of a potential disaster and to delegate alarm authority to all.

Types of Signal

The type of signal used is of little importance. What is important is that there be quick recognition of the signal and complete understanding of the action to be taken. Thus the warning signal must be kept distinct. There must never be any question of the meaning. It is especially important to coordinate with other plants in the vicinity having similar devices. Two sirens with a similar sound may cause considerable confusion—particularly when nerves are taut. Be certain that the device can be heard clearly in every part of the plant, but is not overdone.

Civil Defense Warning Signals

OUTDOOR WARNING DEVICES. The local government is responsible for disseminating attack warning to the public. This usually is achieved through a system of centrally controlled outdoor sirens. There are two civil defense warning signals.

The "ALERT" Signal is a three- to five-minute steady blast or tone and means that attack is probable. Upon receipt of this warning, the public is expected to take action as directed by the local government. Building occupants take action as prescribed by the building attack warning instructions; these instructions will be based upon local government plans.

The "TAKE COVER" Signal is a three-minute warbling tone or a series of short blasts and means that attack is imminent. Upon receipt of this warning, the public is expected to take cover immediately.

INDOOR CIVIL DEFENSE WARNING SIGNALS. Warning of imminent danger for all offices, rooms, laboratories, shops, or other places where people are located in the facility is a basic requirement in the plan. This may require the installation of a system of indoor warning devices. When internal alarms are installed, there must be a readily distinguishable difference between the sounds of the civil defense alarms and the sounds of the fire and other alarms.

Warning Signals Not Associated with Civil Defense

THE FIRE ALARM. This may be a series of coded gong strikes, each code designating a location in the building; a continuing striking of gongs; or any system of recognizable signals which is identified as meaning "Fire," and are readily distinguishable from "Attack Signals." It means there is a fire in the facility and building occupants should vacate the building (or only parts thereof, if the plan so dictates), clear the immediate area, and wait for oral instructions from the organization officials.

OTHER ALARMS. Local plans and preparations should also include warnings for two other disaster categories: (1) natural disasters such as hurricanes, floods, earthquakes and storms (including snow, dust, tornado and rain), and (2) other disasters such as explosions, escape of lethal gases, etc. Warning signals for these categories of disasters must be by a different sound medium than that of civil defense warnings. In cases of some natural disasters, the weather bureau will, through radio and television, disseminate warnings, and instructions which may call for evacuation of the danger area. Building evacuation will be accomplished by the use of fire alarms (or other established signals) and oral instructions from the personnel movement service of the organization.

Prerecorded messages broadcast over the facility intercommunication network can also be used to warn the employees of danger. It will be reassuring to the employees to hear a calm, well-modulated voice telling them to evacuate the buildings, particularly if they have grown accustomed to hearing that same voice broadcasting "test" warnings to which they had responded.

Tests

Warning signals should be tested on a regular schedule to insure that the equipment is working properly. Drills should be scheduled often enough so that employees will become familiar with the alarm and thus instantly recognize the sound and meaning of the specific warning conveyed and automatically take the proper action.

Procedure

It is essential that professional personnel such as guards, firemen, medical, maintenance and transportation be alerted ahead of the general plant population so they can be placed at predetermined posts to prepare for mass movement of employees. Gates must be opened, fire trucks dispersed, traffic obstacles removed and guards stationed at bottleneck locations to aid in an orderly movement. All this must be done quickly. Therefore the first step is to disseminate the warning by direct emergency lines to professional personnel who then start their plans in action. Guards open gates; electricians and maintenance personnel take their positions and cut predesignated utilities, steam lines, oxygen lines and boilers; the fire and transportation departments disperse equipment. After the professional personnel have been notified, the general notification to employees is made over the public address system.

Chapter 19

MINIMIZING THE EFFECT OF
A DISASTER

SURVEY

T HE FIRST STEP in minimizing the effect of disaster is to locate potential trouble spots. A survey of the plant should be made to ascertain possible trouble sources—weak construction, poor wiring, inadequate sanitary and drain facilities, poorly stored material, dangerous overhead structures, broken windows, decrepit tools and improperly functioning or unused machinery—everything that reduces strength. Also, inspect drawings, operation sheets, production schedules and work orders; scrap, inventory, and output records; personnel records, pay records and procurement activity. See all of these are posted to date, that shop-floor changes have been noted on office copies, that inventoried items are all well above minimum balance. All discrepancies should be corrected. Repair or remove all damaged or faulty mechanical equipment. Reorder inventory items in short supply. These things should be done at once, even if overtime effort is required. Everything affecting production must be dependable.

Next, have each foreman fill out a report for his section which answers these questions:

1. Can you suggest alternate tools, materials, or supplies in case present ones become unobtainable?

2. Which of the essential equipment, materials, or supplies do you consider vulnerable or likely to be damaged in event of a fire, a flood, or an explosion?

3. List the names of all employees in your section, followed by a brief description of his principal occupation or specialty. Then, for each person, describe all other jobs or function he is capable of performing—including personal mechanical hobbies.

4. List the names of all indispensable employees.

291

5. List all special jigs and fixtures.

6. Which parts wear rapidly; are damaged easily?

With the answers to these questions compiled and analyzed, you can determine many needs:

1. Sections which are vital, also the weakest link which will require the greatest protective effort.

2. Which utilities, plus how much, each section needs to operate.

3. What tools are likely to be damaged, hence need special protection. Also those jobs requiring alternate operation planning.

4. Similar equipment, and location, to plan machine-part interchangeability for repair.

5. A list of vital parts to be stocked for emergencies.

6. A guide to seeking tools and parts under debris.

7. Alternate occupations among available employees to provide replacements for key men.

8. Where training is needed to provide replacements for key personnel.

9. A guide to splitting and separating production lines to offset total destruction of a key operation.

10. A guide to stocking materials and supplies, also doubling-up on delicate instruments.

PROTECTIVE MEASURES

When fighting a disaster it is extremely important to protect the equipment that will be needed for rescue, recovery and emergency repair after the disaster. These things take on much more value than they might have under normal circumstances. Saving several lathes instead of a gasoline winch, or a crane, may add days to your recovery time. A list of key recovery items should be prepared and used as a guide in time of emergency.

One more principle should be remembered: Do not waste time on areas or equipment that cannot be saved. It is better to concentrate the effort and save something than to spread the forces too thin and lose everything. Sometimes this is a painful decision, but it must be made.

SHUTDOWN PROCEDURE

Orderly and speedy shutdown in industrial plants is vital

in time of emergency. Whether simply pulling a switch, closing a valve, or cooling a large furnace, all must be planned for in advance. This is true also in office buildings and institutions. In large buildings explosive gas fumes, high volage lines, fire and similar hazards can be almost as deadly as natural forces or military weapons. In some instances, due to the manufacturing processes, disorderly shutdown could result in self-destruction of the plant—destruction almost as great as that resulting from an attack. Procedures must be planned and tested. Mock shutdown training exercises are conducted regularly in many plants.

In some areas, pulling a switch or closing a valve may be all that is required to shut down equipment safely. Other areas may require the concerted effort of several people and take several hours to accomplish. In other cases, the safeguarding of certain machinery and equipment may require that certain auxiliary protective equipment be kept in operation, such as fan motors for furnaces; or it may be necessary to cover or remove small tools and delicate instruments to safeguard them against flying debris and dust.

Types of Shutdown

Because of these factors, some areas may require plans for two types of shutdown, the choice hinging on the circumstances at the time of the disaster.

STANDARD (OR ROUTINE) SHUTDOWN. When advance warning time is sufficient, an orderly shutdown may be accomplished without loss of equipment or product.

CRASH SHUTDOWN. An alternate "crash" shutdown procedure should be planned where immediate shutdown is required and where safety of personnel assigned this duty is involved. This type of shutdown would likely result in loss of product and possibly some equipment, but would remove the hazards of additional damage to personnel, property or equipment.

Unit Emergency Plan

Because of the dissimilarities between various units that exist within a facility, each should prepare its own shutdown plan tailored to its own special requirements. These unit emergency plans developed by each separate shop, laboratory, production unit, or service department are the key to the effectiveness of the entire emergency plan organization.

Basic Steps

Five basic steps should be considered in establishing the procedure for shutting down operations for each unit:
1. Who is responsible for establishing shutdown procedure.
2. Who will activate the shutdown.
3. Methods to be used
   a. Routine shutdown.
   b. "Crash" shutdown.
4. Assignment of individual wardens or groups to carry out definite assignments.
5. Color coding and special marking of utilities should be considered.

Responsibility

Responsibility for an emergency shutdown should normally rest with the engineering or maintenance services, but they, in turn, should assign the tasks to people familiar with the various processes and equipment. Crews should be kept as small as possible and frequently drilled in fast shutdown procedure. Simulating shutdown and tagging controls will develop speed in training. Security personnel should be trained to accomplish shutdown procedures in the absence or disability of personnel primarily assigned this responsibility.

Testing

The training of personnel in shutdown procedures is a necessity. It is equally as important to conduct periodic "dry runs" to familiarize employees with the procedure and to keep the program alive. With contant training, it will become an involuntary action in time of disaster for employees to perform in the manner to which they have become accustomed.

The test may also consist of special written problems. The reaction of the unit to the practical problem is observed and a report is made indicating how the unit plan worked together, with any suggested revisions in that plan. These reports should be studied by the disaster director and appropriate action taken. The overall result is the education of all plant personnel in spotting and correcting weak spots in shutdown procedures, evacuation and the associated problems they would encounter in an actual disaster.

Utilities

The protection of utility installations, because of their importance to production, should receive particular attention in all facilities. Considering those measures which will assure continuous functioning of these installations, attention should be given to the following:

1. Dual supply of essential utilities.
2. Provision of diesel-powered or gas turbine generators.
3. Definite plans for interconnection of transformers.
4. Plans with neighboring plants to obtain power if they have generating equipment.
5. Plan for acquisition of trailer-mounted air compressors, motor-generator sets, "package" boilers, etc. This equipment, being transportable, can be moved about within the plant and to other stricken areas or plants.
6. Auxiliary gas supply through use of butane.
7. Alternate connections for electrical power, gas and water lines to bypass any damaged section.
8. Provision for continuous adequate water supply for fire fighting and processing through use of wells and storage tanks in addition to normal supply. Necessary pumping equipment should be independently powered.
9. Connections of appropriate size to receive the portable or standby equipment available.

The closing down of a plant's utilities should be on a very selective basis because of their vital contribution to the operation of the plant under emergency conditions. Water may be required in some mains for fire fighting. Electricity may be needed to run emergency lighting and ventilating equipment. Normally, a plant's water and electricity should be shut off at individual meters and valves rather than at the incoming mains. After properly closing down or safeguarding machinery, equipment and utilities, plant personnel should report to their assigned emergency duty stations with the plant protection organization or proceed to the shelter areas.

POWER

Emergency Electric Power

Emergency electric power, standby equipment and electric supplies should be available for immediate use in event the main source of electric power is damaged, cut off or severely curtailed.

Repair of Transmission Lines

Frequently, the plant's electric power supply is dependent upon overhead lines: In cases where such lines enter the plant at a single point, the problem of speedy repair of broken lines should be discussed with the utility company. A plan and bill of materials should be made which will expedite the restoration of service.

Temporary Generating Equipment

In case the main power station or substation is destroyed, plans should be made to obtain electric power from an alternate source. Frequently, plans may be made to borrow a portable electric generating set for use until service may be obtained from other points on the electric system.

Minimum Standby Generating Equipment

Plants which purchase electric current for lighting and manufacturing needs should provide standby generating equipment at least sufficient for maintenance and protective services and for emergency lighting and communication.

Standby Battery Equipment

Perimeter boundary lights, lights at entrances and exits and at sensitive points within the plant should be provided auxiliary automatic battery lighting sufficient to maintain lighting and alarm systems until an auxiliary generating plant is set in operation or power is obtained from other sources.

Chapter 20

MEASURES FOR PERSONNEL CONTINUITY

INTRODUCTION

THE MOST IMPORTANT REQUIREMENT in rebuilding a plant that has been damaged by explosion or some other disaster is people. New equipment can be purchased, but personnel "know how" is the key to getting back into production.

EVACUATION

The first step to take in assuring a continuity of personnel is to protect them. Thus evacuation plans should be developed for the safe evacuation of all employees in case of disaster, and should include the safe shutting down of equipment.

Employees should receive complete instructions on procedures to be followed in evacuating their respective work areas and in the evacuating of the entire plant area in event of a disaster. Periodic evacuation drills should be held to thoroughly familiarize and train employees in the rapid shutting down of their equipment and the evacuation of their work areas and the grounds of the plant proper. One person should be designated to be responsible for evacuation planning.

A general traffic plan both within and outside the plant must be made, directional signs and instructions posted, and adequate transportation provided. The plans should include a priority of departure for operating and maintenance personnel so that plant shutdown can be orderly and production resumed as quickly as possible.

It should be kept in mind that there never will be definite assurance of a specific amount of warning time. Therefore, plant personnel must be ready to move fast if local authorities decide there is enough time for evacuation.

INVENTORY OF JOBS

All current functions of the company should be analyzed from the standpoint of their value during emergency conditions. Some jobs could be eliminated, if necessary. Many jobs, although important, could be postponed for the duration of the emergency. In others, the work could be doubled up. A study should be made throughout the company to distinguish the vital functions from the nonvital. The study should be initiated and coordinated by the disaster director.

INVENTORY OF SKILLS

A similarly coordinated project to inventory all personnel skills is advisable. Such an inventory will open up the untapped reservoir of employee background abilities and training, for numerous employees have a variety of special skills and abilities that have no relation to the particular requirements of their job classification. Obviously, from a cost standpoint, a saving is effected for employees already possessing the training needed, thereby either eliminating any need for establishing a large-scale training program or minimizing the number that have to be trained.

A simple method for obtaining the information is through the use of a questionnaire which each employee is asked to fill out. Questions should be designed to obtained information regarding the nature and extent of the background, training, skills, hobbies and interests of each employee that can be effectively utilized by the company in the event of a disaster. It should be made clear to each employee when requesting the information that it is to be furnished on a purely voluntary basis and the purpose for which it is being collected should be clearly delineated. In those few isolated cases where an employee declines to submit a questionnaire, the matter should be quickly dropped without any attempt to persuade him to change his mind.

Of course, a great deal of information requested on the questionnaire will be depulicative of the information previously requested on personnel forms and already contained in personnel records. However, in many instances much of the information in the personnel files is no longer current or complete, and an up-to-date inventory of skills currently possessed is desirable.

If a company has a bargaining unit, the shop steward or other union representative should not be overlooked. Some unions have taken full responsibility for obtaining appropriate background information and will make it available for industrial defense purposes. Employee clubs and intracompany social organizations may be able to provide a source of information regarding employees' skills and interests.

Suggested Questionnaire

The following is a suggested questionnaire.

Inventory of Employee Skills

1. *Personal Data*

Name _____   Sex _____
Date of Birth _____   Height _____
Job Classification _____   Weight _____
                                  Department Number _____

2. *Military Status.* Indicate Reserve status (Active or Inactive National Guard, Branch of Service, highest rank attained and total number of years of military service, including active duty and reserve time).

Reserve status _____
Branch of Service _____
Highest rank attained _____
Years of service _____

3. *Work History.* Indicate the "type" of work (not classificaitons) in the order in which they occurred and indicate the duration of each, up to and including your present type of work for the past ten years.

        Type of Work   *Duration*

4. *Formal Education.* Indicate the highest level of formal education you attained, the kind of school it was attained in, whether you were a day or night student, whether or not you graduated, the kind of certificate or degree you were awarded, and your field of study. (For example: 4 years technical high school, days, graduated with technical diploma, mechanical drafting.)

5. *Special Courses.* Indicate any special courses you may have taken either at day school, night school, adult education classes, correspondence schools, military schools, company out-of-hour classes, etc., and so indicate which kind of courses they were. (For

example: Electricity—Adult Education, Basic Math—Correspondence.)

6. *Languages.* List any language other than English that you can speak, read, write and indicate your proficiency in each such as slightly, moderately, fluently.

Language	Speak	Ability to Read	Write

7. *Hobbies.* List all your hobbies, such as photography, radio construction and repair, television repair, woodworking, etc.

8. *Military Experience.* If you have had military service, please indicate the branch of service and kind of experience, such as Navy-radar operator, Army-medical corps., etc.

9. *Civil Defense Experience.* If you have participated in civil defense in your community, indicate the activities you were engaged in, such as first aid, demolition squad, warden, etc.

10. *Community Interest.* Indicate any position or membership held in your home community, such as mayor, freeholder, fire marshal, board of education, etc.

11. *Company Clubs and Associations.* Indicate the activities you have participated in and any offices you have held in intra-company clubs and associations.

12. *Other Organizations.* Indicate the activities in which you have participated and the organization titles you have held in fraternal, professional or civic organizations outside of the company.

13. *Other Personal Background.* Please indicate any information about yourself that you feel should be included in your background history that is not covered elsewhere in this questionnaire or any special interests you have.

14. *Comments and Recommendations.* Include in this section, if you wish, comments on the merit of this project, the questionnaire format, additions or deletions you believe would improve the questionnaire or anything you can think of that would increase the usefulness of the plan.

Processing

Once the inventory of employees' skills has been accomplished a systematized cataloging of the various skills applicable to disaster control should be established for ready reference. Again, the size and organization structure of the particular company involved will determine the system to be adopted. In small companies a simple three-by-five card index file of employees, filed alphabetically under the particular industrial defense skill category will suffice. For example, under fire-fighting, rescue squad, first aid, auxiliary police, etc. would be placed the cards of those employees who have the requisite skill listing their names, duties, industrial defense assignment, etc. If an employee is trained in more than one skill he would have a separate card under each category. In large companies a more refined system using a mechanical or electronic data processing system of coded punched cards similar to that in use in large law enforcement agencies would be more appropriate. Whatever system is used, it must be kept relatively current for maximum usefulness. Therefore, it is suggested that the system adopted should be susceptible to additions and deletions as employees are transferred, leave, or acquire additional skills or training.

REMOTE REPORTING CENTER

The alternate company headquarters is intended only for key top executive personnel. It is neither large enough nor convenient enough to serve as office and home for the entire administrative staff. Also, many people prefer to stay near their own homes during emergency situations. Thus, additional emergency reporting points, called remote reporting centers, provide rendezvous locations for other levels of designated key personnel so as to maintain operational continuity. These centers are established in the homes of dependable employees; their location is determined by geographic distribution plus the concentration of employees living in any given sector of the area surrounding the plant and within a fifty mile radius.

Start the system with a large-scale map of the area in which the company or office is at the approximate center and with a card index of all employees. Each card should describe one employee by name, address, telephone number, job description, additional or secondary skills, job code number, age, marital status, number of dependents, house or apartment, own or rent, own an automobile, number of years with firm, number of rooms in house and blood type.

With the plant as the center, on the map lay out three concentric circles at radiuses of fifteen miles, thirty miles and fifty miles. Then divide these circles into sixteen segments or sectors (N, NNE, NE, ENE, E, etc.) and number them. Next use pins to indicate the home of each employee, to develop the pattern of concentration, and put the sector number on each employee's card.

Select in each sector, between the fifteen- and thirty-mile circles and preferably in the center of any concentration of employees, a home that seems practical for tactical operations. It should be big enough to house a few additional people in an emergency and owned by an employee, preferably in supervision, who expects to stay in the firm and the house for some time. Ask the employee for permission to use the house as a reporting center, If permission is given, the center is equipped with one telephone, one typewriter, one file cabinet, one complete set of cards covering all employees, indexed alphabetically, and one complete set of cards covering all employees, indexed by job title.

If the concentration of employee residences in a given sector is high, or the number of employees is very large, reporting centers can be established at the fifteen-mile and the thirty-mile circles, or even at the fifty-mile circle. The plan should be flexible.

Every employee is given an identification card that shows exactly where he is to report if the plant is out of action. This card also includes a list of the other reporting centers. The employee is free to report to any of them if he is out of his own sector when the trouble arises. In addition, each card describes the employee's next-of-kin so that either the employee or his next-of-kin can report to the center for information, help, or even salary, which is available from funds kept at each center in check form. It is easy to see that if a disaster occurs while the employees are away from the plant, or if they are evacuated when trouble starts, any member of management need only report to one of the centers (it might be his own home) and he can quickly form a group for any required task.

By making previous arrangements with the telephone company, telephones at the reporting centers can be kept open during an emergency. Thus, liaison can be maintained with all centers. Much routine office work can be done at each center without the necessity of employees traveling too far from home. Most of all, the organization is integrated and operational continuity is maintained.

A severe disaster may in a short time convert an organization into quite a different type of organization structure with altered or expanded functions. Such drastic changes inevitably mean that some groups of employees will become surplus at the same time that other units urgently need more workers. A systematic placement and transfer plan will prove of great value under such conditions. Also, plans should be made for quick emergency employee training, revised salary and wage administration, hours to work, protection of the retirement system and union relations.

Chapter 21

REPAIR AND RESTORATION

INTRODUCTION

T HE RECOVERY OF PRODUCTION in a reasonable time after a disaster is not due to luck but results from minute planning to prepare the plant, its equipment and personnel to meet a disaster and to deal with it effectively. Success of any plan is dependent on good organization and thorough training.

Only in extraordinary circumstances will an industrial plant be totally demolished by a disaster. At first look a plant may appear hopeless, but closer investigation usually is more encouraging. Initial estimates of damage are frequently higher than a detailed survey indicates is the actual condition. After a disaster dirt, carbon, silt, mud, slime and rust cover everything. Wires are twisted or burned. Structures are distorted. Debris fillls the area. After a little clearing, cleaning and straightening, hope for recovery improves. Even a nuclear attack will only demolish plants in the target area, while most of the plants in the outlying zones will be recoverable.

Machine tools and other heavy machinery are seldom damaged beyond repair by flood, fire, wind, or blast. However, the repair necessary to make them usable may be extensive because motors, controls and bearings are subject to damage from heat and water. It is also unfortunate that the best alloy steels for tools, wear surfaces and instruments are neither rust resistant nor stainless. In addition, some forms of damage may be hidden, may not show up until the machine has been back in operation for several weeks. Thus, while the basic machines may not be seriously damaged, one of the first jobs will be to minimize further damage to the auxiliary equipment and drives. Even a little "first aid" will greatly simplify repairs later.

PLANNING

In planning for restoration of production, management faces two basic problems: first, to determine where, within the plant, post disaster restoration of production would produce the greatest benefit; and second, to decide on a method of approach which will yield meaningful results.

The basic unit of planning for restoration of production is the functional area. A functional area is composed of a group of machines or equipment performing related functions or operations. Obviously all such areas are not of equal importance. Thus the first step is to evaluate these areas and determine which should be given top priority for restoration. At the top of the scale would be a plant area which is highly important to the overall production and susceptible to damage. At the other extreme is a relatively unimportant area not susceptible to damage.

After identifying the functional areas in which planning for restoration of production is most needed, management must next plan on the steps to be taken to restore key areas.

There are four basic ways to restore production:

1. Repair damaged equipment.
2. Procure new machines and equipment.
3. Have subcontractors perform certain operations formerly handled in the damaged plant.
4. Develop alternate methods on substitute machines.

Management's approach to planning for restoration of production should be based on exploring these factors in an attempt to identify the best combination which will minimize production recovery time. Each area should be continually studied.

SURVEY OF DAMAGE

Preliminary to repair, make a complete survey of the damage. The first thing to look for is structural weakness—sagging roof, floor and walls, warped beams, holes in floors and collapsed stairs. Simultaneously, check for exposed wiring, broken gas and water lines, spilled chemicals and similar hazards.

These conditions may be corrected and dangerous areas barricaded or roped off before any personnel other than necessary rescue and repair crews enter the plant. The plant engineering crew should survey the building and grounds, particularly the building structure. At least one person must be technically competent to recognize structural weaknesses.

If fire was the principal trouble, the things to look for are warping and buckling plus weakened supports and loose masonry. Following a flood, particularly a flash flood rather than rising water, the great danger is from undermining of walls, floors, foundations and roadways. Use a heavy hammer or bar and really wallop critical or suspicious areas. A hollow sound indicates trouble. Do not hesitate to tear out plaster walls or to open flooring if you suspect danger. Most decorative construction will have to be replaced anyway. Be cautious about starting heavy machinery.

Regardless of the disaster's nature, examine the building and the grounds for unsupported walls, equipment ready to topple, live and exposed electric wiring, escaping gas, dangerous chemicals, damaged bottles or tanks of oxygen, acetylene, hydrogen and other hazards. Immediately, if you can, shut off gas at the main valve and power where it enters the plant. Put warning tags on hazardous valves and switches. They can be turned on later, following a thorough investigation. Even if community utilities are off, they may be started unexpectedly.

CLEARING

As the plant survey crew proceeds, provide a crew of plant maintenance people to handle these critical tasks. If possible, start the survey and the safeguarding at one end of a main aisle so that, as soon as it is safe, the cleanup crew can clear one broad aisle right through the plant. This will simplify moving people and equipment in and out, and will also provide a central escape route if further trouble develops. As soon as this aisle is complete, with an exit at each end, start some of the crew at making a second aisle centrally along the first and at right angles to it. Don't worry about cleaning dirt and debris out of the entire plant. That can be done in spare time and after more critical tasks are finished.

Almost simultaneously, a number of jobs must be started.

However, never assign so many people to one area that they interfere with safety and efficiency. Assign group leaders, one for every six or eight people, and keep a list showing assignments of all personnel. Use discretion in assigning people unused to exertion or to manual jobs. Try to limit working time to eight, at most twelve, hours with frequent breaks. Watch for exhaustion, irritation, temper, illness. Your people will tend to overwork, even without realizing it. As rapidly as the survey team shows an area safe, start the removal of debris. As each section is cleared, it is resurveyed for damage to tools, equipment, materials, parts, supplies and paperwork.

Start by repairing equipment that can be used to further the recovery program, including lift trucks, winches, cranes, engines, pumps and general purpose machine tools. Then proceed to the critical production equipment, repairing the simplest damage first to make room in the shop and get more operations going. Use replacement parts not otherwise available.

When a part is taken from another machine, tag the machine to indicate what was removed and where it was located. Then order the part purchased or made and indicate this and the date on the tag.

TOP PRIORITY TASKS

Get as much of the equipment workable as quickly as possible. Here are some top priority tasks.

1. Set up a "command post," a centrally located headquarters for control. Use a large room in the building or in a building adjacent to it. If no sheltered place is available build a shack or use a tent.

2. Provide toilet facilities. Nervousness, working in water or outdoors, unusual exertion and fatigue will make these facilities an urgent requirement. Use the sewers and established facilities if possible; if not, dig a slit trench and build "country accommodations." Canvas sidings will supply privacy. Select a location that will not contaminate drinking water. Use plenty of chloride of lime to avoid disease. Beware of flood-soaked toilet tissue, it may be contaminated.

3. Set up a medical station that includes washing facilities. Insist that all employees wash hands and arms, to the elbows, in antiseptic solution at least once every two hours. Also insist that every scratch, cut, or abrasion be treated at once. Infection spreads rapidly under these conditions. Provide litters and an emergency ambulance.

4. Develop a communications systems, external and internal, to reach as many as possible of the working force as quickly as possible. If other systems are out, messengers or runners can provide emergency service. As soon as possible get telephones, intercoms, public address systems, call bells, and/or other devices back into operation.

5. Locate usable power sources such as gasoline-powered generators and other portable power packs. Salvage any available wire and string emergency lighting, with floodlights at critical places.

6. Obtain usable water. After fires, and especially after floods, "safe" usable water may be hard to find. Initially, the city sanitation department, local farmers, or the park department may be able to provide tank trucks. Test all water used for drinking, then use purifying chemicals anyway. Store it in milk cans if possible. For hosing, get engine-driven pumps. If they are not available, sump pumps, coolant pumps, etc. from the plant might be pulled out to provide pressure. Portable aluminum irrigation pipe might be available from farmers to replace the damaged supply system.

7. Locate and collect essential records. Clear offices enough to permit skeleton operation, but avoid complete overhaul later. Itemize and check surviving records and start reconstruction work.

8. Locate and order large quantities of alkaline cleaner, soap and chloride of lime. The cleaner will remove rust from tools and instruments as well as prevent further rusting and will speed the cleaning of other equipment. The lime will disinfect and minimize the sour "disaster smell." Similarly, order strong work gloves and dun-

garees or coveralls. This work will be hard on gloves—and they should be used. Clothing should be washed daily, and it, too, will wear out rapidly.

9. Start the salvage of tools and instruments as quickly as possible after areas are declared safe. You will not have time to clean and dry everything before rust makes it unusuable. Do not waste effort trying. Fill fifty-five-gallon drums (tops cut out), ash cans, and large tanks with alkaline-cleaner solution and place everything badly rusted or wet in them. This will loosen rust and prevent further damage until your people have time for thorough cleaning and repair. Do not put aluminum parts in alkali.

Be especially careful about handling bottles and carboys after a fire. The heat may have cracked them, or set up severe strains, and left them intact. Even a slight disturbance may cause breakage. Similarly, be careful about opening sealed metal containers. Heat may have caused a pressure buildup that will eject the contents when the cap is loosened. Approach each recovery task remembering the conditions to which the materials and equipment have been exposed. Until thorough tests have been made, double the factor of safety on all lifting, pulling and supporting devices. Watersoaked or burned ropes will be weaker, elevator and crane cables may have been affected, pulleys may be distorted, and wet clutches or brakes may slip or grab.

10. Pay early attention to bearings, journals, ways and gibs on machinery and equipment. Delay will greatly reduce the chance of salvage. Partially disassemble them, to get at each element, then wash them out if flooded and rinse with alkaline solution. Flush out coolant tanks and chip boxes to provide individual containers for segregating parts for individual machines. Soak rusted parts in alkaline solution until you can do a thorough job of cleaning, drying and oiling.

Following a fire, drain oil that may be contaminated

by water or broken down into an acid condition by heat. Then dismantle as above, rinse with an alkaline solution and reoil. Steel parts that are badly discolored or "varnished" by burn oil can be soaked in alkaline solution until you are able to clean them. Cutting tools, drill bushings, gauges and hardened wear surfaces will have to be tested for hardness. Intense heat may draw the temper on these items.

Tag all machinery and equipment, plus each part as it is removed, with cloth tags, or tracing cloth, and use India ink for marking. When parts are immersed in solutions, allow enough string to permit the tag to hang over the edge of the container. Thus you can locate a part more easily and tags are protected. Basic information should include, for machines: name, serial number, manufacturer, location in shop, special tooling or product if machine is special and a complete description of the motor or drive. For parts, list name, serial number if any, name of assembly it comes from, plus name and serial number of machine and machine's location. Motor tags should contain all of the nameplate information and the name, serial number and location of the machine, the motor powers; also a description of any drive used with the motor.

11. Get some kind of heat to stop further damage and to dry the atmosphere, particularly in cold weather or after a flood.

12. Clear a large area, perhaps a storage section, of dirt, debris, machinery, equipment, crates, etc. Most practical would be an area, well sheltered, close to a good road and a railroad siding. Enclose the area with a fence but provide several large gates. This space is the "marshalling" area for damaged machinery, motors, tools and equipment prior to shipping them to other sections or out of the plant for repairs.

Keep a list of everything brought into the area and everything sent out, plus its destination. Do not allow the "emergency" to destroy systematic control. Keep the system simple, but keep it. Once you lose control of people and/or equipment under these conditions, chaos takes over. Regaining control may cost you days of delay.

13. Disconnect and remove all damaged or water soaked motors, generators, transformers, rectifiers, controls and other electrical equipment. Be sure to tag them prior to removal. Move them into one large area cleared as an electrical overhaul section. There each item can be cleaned, inspected, dried if necessary and repaired, or sent out for repair or drying. Equipment beyond repair can be stripped for usable parts and replacement units can be requisitioned.

Set up an electrical equipment pool with usable units from unused machines, machines with major damage and requiring lengthy repair. This will supply an immediate source of electrical units for machines otherwise satisfactory, also will further recovery tasks.

14. Get your purchasing agent and his staff into action as quickly as you can arrange office space and a telephone. If necessary, have him go outside the geographic area to locate needed supplies. Notify equipment manufacturers and suppliers of the problem. Order needed supplies and spare parts. Request technical assistance in disassembling, overhauling and repairing. Make arrangements with manufacturers for returning machinery and equipment to them for rebuilding. Call in local building contractors for help in cleaning up; use of power equipment such as scoops, shovels, dump trucks, cranes, air compressors, generators and pumps. Make arrangements for structural repairs and rebuilding.

One point to remember: machinery builders and suppliers will be more than willing to cooperate in every way. Many of them have had experience with damaged equipment. Do not be afraid to ask their help. However, do not expect miracles, and give them all of the time you can.

15. Do not neglect your personnel relations. Send letters

to all employees describing the extent of damage and telling them of the progress made in recovery. Give some tentative date when they can expect to be called back to work. Be fair in calling in help for the recovery jobs. Skill and seniority will be the principal criteria, but physical strength may also be a factor. Strength and skill will be needed. However, do not call in more people than you can use. They will only impede progress and increase the safety problems.

16. Round up all of the transportation equipment you can find. Ask employees to bring in small trucks, tractors, jeeps, etc. Pay them for the use of these vehicles. Make one or two people responsible for assigning, and assuming proper maintenance of, all vehicles.

17. Provide food and beverages for clean-up crews. Ask employees' wives, local community groups, your own cafeteria personnel, or a caterer to supply simple fare. Hot food will improve spirit and energy. Sugared doughnuts, for dessert, give quick energy and boost morale.

18. If you belong to a mutual-aid group, call a meeting. In a community-wide disaster, discuss how you can pool your resources and talents, to expedite recovery. Exchange ideas. If you are individually affected, tell your fellow members what you need and how they can help. Their purchasing, medical, transportation, engineering, safety and other departments can greatly expedite your recovery.

19. Immediately advise your local authorities, services and utilities of the extent of damage and your situation. These include police, fire and medical departments, civil defense and Red Cross directors, electric, gas, water and telephone companies. Can the police provide extra protection or traffic control? How about fire protection? Have you discussed emergency service with the telephone company? Will Red Cross assistance be needed for employees without an income? Can local civil defense provide some of the answers for faster recovery? If this disaster results from an enemy attack, your civil

defense director will have tremendously broad authority, will probably control the area.

20. Keep a detailed record, complete with photographs, of the extent of damage, recovery measures and all pertinent factors. You may need this information for legal or insurance purposes at a later date. Make it chronological and include survey findings, details of work performed, all expenses, delivery promises and a progress report.

SALVAGE OPERATIONS

Machinery totally damaged beyond repair necessarily must be scrapped. However, machinery is rarely totally destroyed, and normally if it cannot be repaired, undamaged parts can be and should be salvaged from such machines and composites assembled therefrom. Set up machine repair on a production line basis:

1. Provide facilities to clean every machine thoroughly as it enters the depot.
2. Inspect each machine and prepare a detailed list of items to be corrected.
3. Assign a priority number to the job and specify which line will handle it.
4. Remove motors, pumps, controls and other accessories; tag them and the machine.
5. Provide necessary repair or temporary storage until repaired.
6. Inspect.
7. Replace motor and accessories.
8. Make the final inspection.
9. Apply protective treatment.
10. Ship.
11. Install the machine at the plant.
12. Give it trial run.

A mobile repair shop crew can remove damaged components from machines too large to move from plant. Have them repaired at the depot and replaced.

Motors, drives and accessories should be sent to a separate

area adjacent to the main depot shop where facilities are provided for special cleaning, drying, repairing and testing. Drying of electrical equipment may take longer than repair of the machine it controls or powers, hence a reserve of this equipment and top priority for its treatment are essential. Machines might be made productive more quickly by supplying electrical equipment from a common stock, where this is possible.

STOCKPILING OF MAINTENANCE AND REPAIR PARTS

Stockpiling of maintenance and repair parts is another important measure which, if properly carried out, will aid in reducing loss of production to a minimum. A review of presently stockpiled items should be made in conjunction with the development of a mutural aid program. Stockpiling should include but not be confined to the following:

1. Compressors, electric cable, piping, valves and regulators.
2. Special alloy piping and valves in those industries where other materials would adversely affect the process.
3. Spare parts for overhead cranes and other material handling equipment.
4. Pipe, which can be filled with sodium, to be used as temporary buss bars.
5. Standby electrical transformers in the most commonly used sizes.
6. Flexible hose for emergency repair of gas and water mains.
7. Tarpaulins.

Standby Equipment

When a plant's entire production or a major portion of its production depends on the continued operation of certain sensitive points, duplicate reserve units, substitute standby equipment or stocks or critical parts should be obtained and maintained or safely stored.

Standby Equipment Pools

When it is impractical for each facility within an area to individually stock standby equipment or stocks of critical parts, facilities engaged in similar operations or requiring similar standby equipment should join together to obtain such material on a pool basis and all participating members should be permitted to draw therefrom as required by an emergency.

EMERGENCY REPAIR CREWS

A well-equipped emergency repair crew should be organized, and plans made to expedite restoration of each specific service or protect critical equipment from further damage after a disaster. Emergency repair crews should be recruited in part from the normal maintenance crew and supplemented by other workers having appropriate skills and capable of working rapidly and efficiently.

Training

Emergency repair crews should be trained in methods of rapid and emergency repair and should plan the steps and means by which utility services, communication services and production equipment may be cleared of debris and restored to operating conditions. Test runs should be made simulating disaster conditions and the crews dispersed over several alternates routes to perform their prescribed tasks.

Equipment

To be held in reserve for emergencies, one or more vehicles or small trucks patterned after public utility company trucks should be adequately provided with tools and special repair materials such as a power generator, an air compressor, portable power tools, mechanics tools, welding apparatus and some small machine-tools such as toolmaker's lathe, miller, drillpress, and metal-cutting saw. It should also have a stock of standard parts, V-belts and roller chain.

When the plant has power equipment such as bulldozers or portable cranes capable of clearing debris, definite plans should be made for their use during an emergency. Operators of such equipment should be assigned to the emergency repair crews.

Headquarters

A headquarters should be established for the emergency

repair crew. This headquarters should have direct communication with the plant control center and the plant guard headquarters and should be provided with utility service maps (indicating location of power switches, gas regulators, gas lines, steam lines, oil lines, valves and fuse boxes), crash helmets, electric lanterns, spare bulbs and batteries and necessary supplies required for use in an emergency.

PROVIDING OUTSIDE SPECIALIZED TECHNICAL SERVICES

In many instances management will need technical advice and assistance beyond its own personnel. Prior arrangements should be made with the following:

1. Builders and contractors for cleanup and building reconstruction.
2. Architectural engineers for assessing the extent of damage to buildings.
3. Machine and equipment rebuilders.
4. Electric motor repair.
5. Public utility companies.

EMERGENCY PRODUCTION SITE

Despite measures to prevent damage, some buildings are likely to be destroyed. Production recovery may be shortened by the following:

1. Planning for the use of temporary production shelters, such as tents, quonset huts, etc.
2. Surveying buildings in area to ascertain their adaptability to production in case of emergency.
3. Developing standardized construction procedures which will reduce building reconstruction time. Plans could be developed for a standard bay.
4. Checking present building drawings for accuracy.
5. Considering the use of substitute construction materials and standard shapes.

Under conditions of major destruction, it might be practical to lease another building in an area and organize production there until the damaged plant is repaired, moving machinery into the new quarters as it is overhauled. Another way is to lease only enough space to provide offices, assembly area, critical production and inspection operations; then subcontract the remainder of the manufactured parts. By doing this, you can gradually pick up production operations again as areas of the plant under repair become usable.

DEVELOP SUBCONTRACTS

The use of subcontractors will contribute materially to maintaining production as well as starting quickly after an emergency. Recommended actions are as follows:

1. Develop additional production sources.
2. Eliminate dependence on one source. Establish duplicate sources of supply.
3. Compile and maintain up-to-date records on potential suppliers for building materials, replacement equipment and spare parts.
4. Maintain information on facilities having similar equipment and machine tools.
5. Establish machine loading records and amount of available production time in subcontractors' plants.
6. Plan for the acquisition of essential production tooling, such as jigs, fixtures, etc.
7. Assure that contractor's continued production load does not exceed his ability to deliver.
8. Consummate planning in a formal agreement.

RULES TO REMEMBER

Things to Do

1. Set up emergency headquarters as quickly as possible after disaster.
2. Provide briefing for employees returning to work. Advise them of danger zones, special safety requirements, compensation and provisions for eating, personal comfort and first aid.
3. Make an effort to return from emergency operation to normal procedures and practices. Thus you minimize extraordinary responsibility on personnel and simplify

the supply system. Also, you limit the establishment of precedents that are difficult to overcome.

4. Give some thought to emergency work pay plan for employees. Decide on basis of company policy, company resources and prevailing practice in your area. During a flood, some plants paid the regular rate, some paid a straight hourly rate for all, others paid time-and-one-half, still others paid the regular rate plus a bonus for completion of recovery ahead of schedule. Take your choice.

5. Accelerate efforts to make locker rooms, service rooms and toilets usable.

6. Keep employees informed of conditions and the extent of recovery. Tell them when you expect to call them back and on what basis. When recalling them, advise them of shift hours, pay rate, whether to bring lunch, type of clothes.

7. Guard remote areas to prevent hazards, pilferage, looting and the natural tendency for crowds to gather at a colorful operation such as bulldozing or blasting. Use a pass system to determine who will enter the area.

8. Make one person responsible for health and sanitary conditions. Have him tag drinking fountains, toilets and washbasins, approved for use.

9. Have another person, perhaps your safety engineer, supervising safety practices. Make this a full-time job. Unusual tasks, plus excitement and fatigue, create unusual hazards.

10. Maintain routine records. Later on you will need them to settle insurance claims, tax deductions, legal questions and payroll arguments. Assign people to gather required data, take photos.

11. Judge people by what they accomplish, not by display of activity alone.

12. Encourage imagination. Only real ingenuity will solve many of the problems. Think of similarities, differences, substitutions, of objects that might serve some emergency purpose.

13. Check all grinding wheels and other high-speed, rotating, equipment prior to using it after exposure to heat or water. If in doubt, return the equipment to maker.

14. Keep close control on internal and external temperature when drying electrical equipment. Insulation may break down above 190° to 200° F.

15. Weigh the cost of unwrapping each package, cleaning, drying, testing and inspection for confidence against a superficial estimate that rewrap only is needed. How will this effect operations and customers later on?

16. Send as much damaged equipment as possible back to the manufacturer for overhaul or, at least, send it to a service shop. You will need all of your available talent and space for things that cannot be sent out.

17. Tell employees the true extent of your insurance coverage. Minimize strange notions that insurance covers everything, that nobody loses.

18. Assign long-range responsibility, beyond the immediate emergency. Make sure that temporary repairs are redone to permanent specifications as soon as possible. Audit your condition as soon as you are in full operation, then again in ninety days. A heating system will not be missed all summer—until the first freezing day in the Fall.

19. Take time out after the first crisis to reevaluate your situation—to plan briefly next phase. List the jobs you want done. Chart your temporary emergency organization. Compare the two. Thus you will not overlook anything or assume that "someone is taking care of it."

20. Put someone in authority in charge of night activities. Important decisions must be made then too. Use more than enough guards and be specific about what you want watched.

21. Try to help the employees at home.

Do Not

1. Jump to conclusions based on hearsay. If time doesn't permit thorough checking, at least get some facts before making important decisions.

2. Be overcritical of people and work. All of you will be anxious, tired and tense. Exercise self-control. Show appreciation for effort. Give clear instructions. Many tasks will be new and unusual to the workers.

3. Waste time trying to reclaim unsalvageable items or items cheaper to replace than reclaim. In this category are wet cardboard, stationery, contaminated oil, some softened cutting tools and much instrumentation.

4. Forget the danger of spontaneous combustion. Wet rags, paper, etc. can start to burn within twenty-four hours if ignored and conditions are right. Clear out dark, damp, corners; paper stores; rag bins; also adjacent combustibles.

5. Wait too long for local services and supplies. Place orders outside of your geographical area. Ask your supplier to help you locate sources.

6. Believe lightning cannot strike twice. Protect against a recurrence—tomorrow, perhaps. For instance, flood silt raises river bottoms, makes the danger of recurring floods even greater.

7. Spend manhours on routine clean-up until key jobs and critical situations are under control. There will be plenty of time for scrubbing floors later.

8. Lift emergency precautions too soon. Recovery to a 100 percent safe condition takes longer than you think.

9. Quibble with the union over small change during initial stages, but get policy back to normal quickly. Most unions will cooperate in recovery.

10. Be overcautious. Use good judgment. Weigh the risk of shutdown or slight damage to a questionable system against its value if it works and/or the cost of teardown to be certain.

11. Work crews over twelve hours to regain production loss. Use two or three shifts instead. Provide time off and vacations as soon as people can be spared, to allow them to recover after the ordeal and the recovery effort.

12. Ignore your own health. The road ahead will need strength, courage and vitality.

SECTION III

SPECIFIC PROBLEMS AND PROCEDURES

Chapter 22

NATURAL HAZARDS

INTRODUCTION

IN THE AFTERMATH of an earthquake or storm, a vital installation can become the target for portions of the population who have been seriously affected. The loss of protective lighting, perimeter barriers, alarm systems, patrol vehicles and communications creates serious breaches in the security status of an installation. Complete and detailed plans should be formulated in advance to assure that the installation will not remain vulnerable for an extended period of time following the occurrence of a natural disaster.

HURRICANES, CYCLONES AND TORNADOES

Tornado, cyclone and hurricane are all the same thing, differing only in degree. Cyclone is the basic name for a storm in which the wind whirls about an area of low atmospheric pressure at the center. Tropical cyclones are of small average diameter, hence tremendous force, but the area they cover is much greater than that of a tornado. A hurricane is much larger than either a tornado or cyclone. Wherever they are associated with water, the wind danger is usually accompanied by tidal waves. A tornado is a furious conical wind traveling a generally straight path averaging 1000 feet wide, and occurs usually in the central plains area.

The United States Weather Bureau has done much to cut losses from windstorms in general by forecasting and warning, while men have done much themselves by building structures above the reach of storm tides and strong enough to withstand the wind. In badly exposed locations, buildings should be equipped with storm shutters or battens which can be promptly applied. If this is not done to windows, at least on the storm side, failure of windows may lead to lifting of the roof and destruction of the building. In addition, roofs must be securely anchored. Tall structures such as flagpoles or water towers must be designed and constructed to withstand high wind velocities. Walls and window embrasures must be tight to prevent driving rain coming in and ruining equipment or weakening foundations. Other preventive measures include storage tanks for water, staple food to last several days, oil lanterns or their equivalent.

The greatest danger from windstorm is the tendency to believe an area is immune because it has not been hit recently. When the Weather Bureau warns of high winds, begin to get ready for them—even if you are certain they will miss you again.

EARTHQUAKES

Earthquakes are Nature's way of releasing dangerous strains that accumulate in the spinning Earth. They must be expected to occur again and again along existing fault lines. In addition to their destructive impact, there is also the danger of fire from broken gas mains and lack of water to fight it.

The primary preventive action is not to build over an old fault. In an earthquake region, the best preventive measure is to construct buildings so that they are able to prevail against an earthquake. Instead of anchoring to bedrock, float the building above it, ballasting it as a ship is ballasted by making lower stories heavy, upper stories light. Support and balance floors like a waiter's tray on his fingers—by cantilever girders. Keep conduits, gas and water mains flexible—lead, for example—laid in a trench free of the building, rising in open shafts and connected to fixtures flexibly. A "fire pool" will provide water in case city-supply mains are broken and fire starts.

FLOODS

Floods are of two basic kinds, those short and sudden "flash" floods, and those which extend over a considerable period. The first is caused by a torrential rainstorm, bursting of a storage tank, dam, or water main and usually occurs without warning. The second is slow enough in development to provide opportunity for considerable preventive work.

Preparation

The first step is to learn the topography of the facility and the surrounding area. It is essential to know the water level at which your plant will be flooded, and when, in relation to adjacent plants and area. If the plant is on low ground near a river, large storage basin, or the sea, certain long-term precautions can be taken. The simplest of these is an earth dam, a low ridge of earth blocking off the lowest areas and providing a base for sandbagging if the water rise is slow enough to permit it. Sandbagging will also be required at openings in the dam such as roadways, track entries and the like. If the earth-fill dam is inadvisable for esthetic or other reasons, a low concrete or brick wall will serve the same purpose if it is built on a substantial foundation. Openings can be closed with removable flashboards set into slots and calked. If a concrete wall is used to protect a key building, water seepage under a building may exert dangerous uplift pressures if the wall is high—tending to float the whole building or simply to buckle in the floor.

Important records should be kept on upper floors of substantial buildings. Fine-tool cribs, inspection units and other expensive-tool departments should be located above probable high-water level. If flash flood is unlikely, either from natural causes or large water storage in an adjacent plant or other area, it is enough to plan what equipment and records should be moved upward—and where—if the need arises.

A survey should be made to ascertain what pieces of equipment will be of most help in getting back to work after a flood. Special care should be taken to protect those units from harm and to maintain them in operating condition.

Advance planning also requires the stockpiling of emergency equipment such as pieces of tarpaulin, burlap bags, sand, infrared bulbs, shovels, bat insulation, chicken wire, lumber, coal, rope, boats, boots and sand. A gasoline-engined pump is a sound investment, for normal use and emergency. Floor drains and the like may well be provided with flap valves at the discharge end, fittings for pressure plugs at the inlet. Battens should be available for basement and first-floor windows and doors.

During the Flood

Everything possible should be done to keep water out as long as possible. Doors, ventilators, pipe trenches, drains and windows should be closed and covered with heavy bulkheads or battens. If anticipated flood level is low, auxiliary dikes of sandbags or dirt can be built around key areas. All electrical motors and equipment that can be should be removed and stored in a safe place that is high and dry.

Anchor storage tanks for flammable liquids so they cannot float away. Vents should be extended above expected water level and discharge valves closed. Portable tanks or containers should be moved to a high safe place and anchored. Gas should be shut off at the main valve and electricity at the main switch if possible. Tank cars should be moved to high ground. Open flames should be eliminated. Lime or other water-reacting chemicals should be removed. Buoyant materials should be removed from lower floors because they have been known to lift upper floors or float small structures away, and will damage machines and break sprinkler pipes. Underground storage tanks may lift when the ground becomes soft. To prevent this they should be filled or weight placed upon them to hold them in place. Surface condensers should be blocked up on the footing springs and filled with water on the condensate side.

If electrical equipment cannot be removed it must be shut down long enough before the water hits it to allow it to cool off, because hot insulation will absorb more water. Machined surfaces on all units should be coated with grease or slushing compound and barriers set to protect equipment from floating objects. Joints should be sealed with ordinary roofing compound. Breathers should be removed and replaced with pipe extensions or plugs.

Grease should be applied liberally around spindles and other openings to bearings and, if time permits, oil should be drained from sumps and tanks which cannot be sealed off.

Seal conduits and control wiring with muslin and waterproofing compound, oakum and marine glue, or even plaster of paris or roofing compound. Boilers and furnaces should be allowed to cool down so the refractory sustains less damage.

If the plant is very small, or of single-story design and on low ground certain to be flooded, there is little that can be done until the water goes down. If it is a multi-story building or there is higher ground in which equipment can be stored, and boats are available if needed for evacuation, a maintenance crew should be kept on duty. They can prevent a surprising amount of damage by staying ahead of the water. Critical water heights should be established in advance and when the rising water reaches these points the crews should move up to the next highest level.

In contrast to other forms of disaster, a "normal" flood gives warning of its approach and last much longer. In that time, a crew of alert and substantial employees can do much to minimize loss. They can prevent debris pile-ups that may bring down a wall or smash in a door. They can occasionally divert the force of the current, can protect or rescue equipment. They must be carefully picked, for such a group may be cut off from normal communication for days at a time.

If a protracted flood is contemplated, the supplies should be stockpiled before the water gets to the plant. Do not forget books and magazines, cards, games and a portable radio or television set. This will enable the men to keep up with the news, and will give them help on estimating height and duration of the flood, plus warnings of any special danger, as well as providing some entertainment. A gasoline-engine generator set will provide light and radio power, or power to operate a charger. Also acetylene or oil lamps and flashlights should be provided. Cots and blankets are also essential.

Provide a good supply of hand cleaners, soap and towels. A well-filled first aid kit is a must. Regular sanitary facilities will not be operative, so it is advisable to install chemical toilets and a supply of chemical.

The crew will lash or weight down tanks that start to float, rescue electrical units and instruments, grease or slush up machined surfaces ahead of the water and plug openings before water gets to them. The crew should check each part of the facility daily. If a room or building is sealed against water, it should be watched for buckling floors from seepage. The crew can frequently block up or otherwise lift electrical units, precision machines and tool cabinets to cut the work afterward.

When the water begins to go down, the crew can protect equipment against corrosion, can drain and clean gear cases and other containers in which standing water would multiply trouble, remove debris dams that hold water and floating objects that will settle on and smash machines, pump out sumps or low spots and recondition boilers so heat can be gotten up as soon as the grates are clear. Rust, except in sea water, occurs fastest when a wet surface is exposed to air and is subject to temperature variations. Further flood slime may include acid-forming bacteria and almost everything else. Thus, the sooner surfaces are cleaned and oiled, the better. Electrical equipment is slow to dry out, so the sooner this job is started the better.

After the Flood

When flood waters recede, oil, grease, mud and slime cover everything. The first step is to get some kind of heat to stop further damage and to dry the atmosphere.

Next, attack the filth. Sprinkle chloride of lime on floors. Wash down walls, pipelines and machinery. Water alone will remove mud and silt, but will not touch oil and grease. That takes cleaning solutions, which must be chosen carefully. Highly alkaline cleaners may destroy paint and varnish or prevent successful repainting if they are not thoroughly rinsed off. Solvents introduce a fire hazard; carbon tetrachloride avoids fire danger, but attacks many kinds of insulation and the vapor is heavier than air and toxic, particularly for the workman who has been drinking.

Soap and water are advisable on some equipment, detergents safe on some areas, solvents where ventilation is not involved. Start out by washing down with a hose. As any machined surface is cleaned, cover it immediately with a penetrating oil if rusted, with light oil otherwise. Heavier rust may be removed with very fine abrasive cloth or a buffer. Gear boxes and mechanisms in general, unless they were sealed against the water, will have to be cleaned out thoroughly. Sumps will not only have to be cleaned, but also sterilized. Clean speed reducers, drive chains, gears and the like with solvent after the housings have been washed out.

On all equipment, pay particular attention to bearings and journals. Get at them as quickly as possible. Open and clean them, dry the parts, and cover them with water-resistant oil. Clean oil piping, tanks, pumps and sumps, being very careful to get rid of silt. When river water stands for several days over machines, it will penetrate almost any bearing or oil line, carrying silt with it. Unless systems are dismantled, then blown out with compressed air or steam and washed thoroughly, they will carry grit to bearings later. Ball and roller bearings should be cleaned in a good solvent, then relubricated; they will probably be full of silt, too.

Much small electrical equipment, instrument cases and the like, can be dried after cleaning by putting them in a small, heated room from which moisture-laden air can be exhausted.

Larger equipment, particularly electrical equipment that is difficult to move, should be thoroughly cleaned, then it can be dried in place, either by enclosing it and providing external heat or by shorting the windings.

Extremes in Temperatures

Extremes in temperature are classified as natural hazards to security. Extremes of heat or cold can have adverse effects on the human body. Individuals become reluctant to take aggressive action, perform patrol duties, or otherwise expose themselves to such weather extremes. Vehicles, communications and alarm devices may not function properly. Special advance precautions must be taken to insure that acceptable standards of security are maintained.

Reduced Visibility

Any natural phenomena which reduce and restrict visibility are considered natural hazards. Such conditions may arise as the result of darkness, snow, rain, fog, or sandstorm. Individuals attempting to penetrate a vital installation will be afforded an increased degree of concealment under any of these conditions. Planning for adequate security should include provisions for these occurrences.

Chapter 23

FIRE PREVENTION AND CONTROL
INTRODUCTION

T HE POSSIBILITY OF FIRE as the result of a natural phenomenon is ever present. Security personnel must be alert for the early detection of fire to allow for immediate effective action. Fires exceeding the control capabilities of the fire protection unit seriously jeopardize the security of the installation. Fires create excellent diversionary opportunities for pilferers, saboteurs, espionage agents and professional agitators to attempt to breach security measures.

An accidental large fire usually develops over a period of hours from one small start. Under adverse conditions, it may spread to adjacent buildings, but it is eventually halted by effective fire fighters. Deliberate fires caused by enemy action are the product of rapid or simultaneous ignition of a large area by incendiary or other attack. The result is a great mass fire entirely out of control. The type of fire is dictated by the combustibility of the area and by wind conditions.

1. A conflagration is a wind-driven fire that moves to leeward as long as it can reach combustible material. Evacuation of the downwind and the construction of emergency firebreaks are the only defense.

2. A fire storm is intense fire in quiet atmosperic conditions where the aggregate blazes merge into one inferno. The thermal intensity exhausts the air above, causing an influx of new air at the base of the column. The suction effect of the column raises burning particles high into the air and may spew them over a wide area. If these new fires are successfully extinguished, spread of the fire storm will be prevented.

The residential, commercial and business sections of cities

in the United States are made up predominantly of load-bearing, masonry walls and woodframe buildings. Normally, the wooden buildings are highly susceptible to fire. The location of each installation and its surrounding area, the layout, building density, combustibility, terrain and other factors must be appraised individually in order to organize effective emergency measures.

First and foremost, management should strive to prevent fires. When, in spite of all precautionary measures, fires do break out, it is equally important to minimize the damage by promptly extinguishing the fire.

Chemistry of Fire

Fire is the result of a chemical triangle. Its three sides consist of fuel, heat and oxygen. Control and extinguishment of fires in general are brought about by eliminating any side of this triangle. If oxygen (air) can be diluted or smothered out, the flame will go out. If heat can be taken away by cooling the fuel below the temperature at which it will take fire and continue burning, then the fire can no longer exist. When the fuel can be removed or dispersed, obviously there is nothing left to burn.

Combustible solids and liquids do not burn. They give off vapors which burn. The higher the temperature to which material is heated, the faster the vapors are given off, and the rate of burning increases. Fuel is a combustible substance which will burn when it is heated to its ignition temperature or made hot enough to burn in the presence of air or oxygen. Ordinarily wood will not burn, nor will it give off vapors which will burn until it is heated enough to cause gases to be given off which will burn in the presence of other flames. Flames are caused by the hot chemical reaction of combustible gases with air or oxygen. The reaction causes additional heat.

Classes of Fire

Fire may be classified into three groups.

CLASS A. Fires in ordinary combustible materials, such as mattresses, dunnage, piles of wood and shavings, canvas and other materials that leave embers or residue after burning.

These fires are best extinguished by the quenching and

cooling effects of quantities of water or water fog. The entire mass must be cooled, since surface cooling would not effect combustibles below the surface. Smothering is not effective because the interior may smolder and rekindle later when air reaches those surfaces.

CLASS B. Fires in substances such as gasoline, oil, lubricating oil, diesel oil, tar, greases and other materials that leave no residue or embers. Here, the blanketing or smothering effect of the extinguishing agent is of primary importance. Rekindling of top surface is impossible if cover is thick enough to cut off supply of oxygen and act as a barrier against heat from hot bulkheads or other metal structures.

CLASS C. Fires in live electrical equipment, such as switchboard insulation, transformers or generators. Here, the extinguishing agents must be nonconducting so that electrical shock is not experienced by the fire fighter and damage to the electrical equipment is kept at a minimum.

SURVEY

Interruptions in production or service and destruction from fire can be minimized through good housekeeping, adequate physical barriers to prevent spread of fire, the elimination of fire causes, ample first aid, the manual automatic fire control and extinguishing equipment where products or process materials are combustible. In addition, the organization of a facility fire brigade, the establishment of a fire alarm service, and the assurance of thoroughly reliable and adequate water supply are essential.

Real fire protection starts with a survey of the facility and with realistic solutions to the so-called fire problem. The survey should give an accurate picture of the potential hazards, the means available for meeting them, and the areas of weakness. It would cover the following:

1. Physical factors such as the layout of the facility and the type of construction.
2. Hazards inherent in the type of operation, the location of dangerous substances or operations.
3. Alarm system.

4. Fire-fighting organization and equipment.
5. Fire prevention program.
6. Outside assistance.

A fire problem is a hypothetical fire that starts under adverse conditions in time and place within the facility. The question is, how can it be stopped before it gets out of hand. The solution of such a problem is not theory or guesswork, but the application of expert knowledge to a particular task. Management may obtain an impressive picture of the difficulties encountered in working out such problems by walking through a facility alone in the dead of night or on a Sunday. Management should stop and think of what would happen if and when an accidental or malicious fire got under way in the facility. How long would it be before someone found it? How would the finder report it? How would they get the alarm? Where would they connect their hoses? How long would all this take? During all this time, how fast would the fire be getting under way? The answering of problems such as these points the road to adequate facility protection.

CHECK LIST

The following check list is offered for use in reviewing fire protection measures at all facilities:

1. Housekeeping and elimination of potential fire hazards.
2. Proper storage of combustible materials.
3. Selection of proper fire-fighting equipment for various types of fire hazards and correct location of such equipment.
4. Regularly scheduled inspection of equipment by qualified personnel to determine if it is in good operating condition.
5. Employees trained in use of equipment and fully aware of their responsibilities in case of fire.
6. Arrangements with local fire departments to assist when needed—special training required for electrical fires.

HAZARDS AND PROTECTIVE MEASURES

The fire hazards created by normal production or service

methods are increased in time of alert or actual hostilities by risks arising out of the national defense emergency. The pressure to get out more work creates conditions that breed fires and make them spread. Taking chances with critical facilities is not only bad business judgment but bad defense planning.

Further, fire has always been the favorite tool of the saboteur. Time delay ignition devices give him a chance to escape. The destruction that follows hides his tracks. In addition to starting a fire in a vulnerable spot, the saboteur may effectively sabotage fire prevention devices. The saboteur counts on the weaknesses of the facility fire protection measures for his success. Management of a critical facility must assume that the facility may be a target for sabotage and must realize that advice is needed from experts in fire protection. Fire protection measures visualized by the average facility operator will not suffice. It is not that the saboteur is smarter than the average facility operator, but in such a case the saboteur is a specialist working against an amateur.

It is not possible to list every possible hazard, but some of the points to be watched are as follows: Maintain good order order and cleanliness. Watch shipping and receiving rooms to prevent accumulation of excess packing material and empty boxes. If paper, excelsior or other flammable material is used for packing, keep only one day's supply in the packing room. Keep this in a metal or metal-lined container with a self-closing lid held by a fusible link.

Provide standard waste cans for oil rags, scraps, wood and packing material, and see that these are burned daily or emptied into closed metal containers for safe disposal. This also applies to rags used for wiping paint and oil fillers. Do not permit an accumulation of waste combustible material near buildings, and keep dried grass, debris and brush well away from the walls. Make sure the roof is free of dead leaves and other trash.

Provide metal lockers for employees' clothing to minimize the danger from a hot pipe left in street clothes, or an oily rag in the overall pocket. See that "No Smoking" rules are rigidly enforced. Do not permit heating devices with open flames in areas where there are flammable materials, particularly volatile solvents or dusts of any kind. In such locations, use explosive-proof electric lamps and motors. Switches and starters should be guaranteed to be nonsparking, or should be installed in some place separate from flammables. Forbid key switches or pull chains.

Inspect all electrical wiring, and see that worn or frayed sections are replaced, and that motors are free of accumulations of dirt and oil. Do not permit open junction boxes or temporary wiring. If portable electric tools are used, see that they are properly grounded and that the cable is in good condition.

Make sure that power lines are not overloaded. This is particularly important in older plants in which new machines and more powerful motors have been installed progressively over a period of years.

Where steam pipes pass through or near woodwork, make sure there is sufficient clear space between them to prevent lint or light combustibles from accumulating there. If the back of the hand cannot be held in firm contact with the woodwork, the installation is unsafe. Do not attempt to remedy this by merely covering the wood with sheet metal. It will serve only to transmit heat. Use sheet asbestos and metal with an air gap between the wood and the insulation.

Insure adequate ventilation for coal storage areas, and do not permit coal to be stored around or against any wooden structure.

Watch out for lime. It must be stored where it cannot be reached by flood water or water from any other source, including that used to extinguish fires. When wet, lime generates great heat and may cause serious fires.

Do not permit railroad cars or other vehicles containing flammable or explosive materials to stand close to the plant. If they are consigned to the plant, have them emptied into safe storage immediately. If they are not for the plant, insist that the carrier remove them to a less hazardous location.

Screen openings from cupolas and other spark-producing equipment; keep sidewalk vault lights and window panes whole; keep plastering and sheathing in good repair so nothing can be thrown into the hollow space behind.

Pay particular attention to the storage of oils and flammable liquids. Bulk storage should be in a fireproof building detached from the main buildings, and no more than one day's requirements should be kept in the plant proper. In case of explosion in the oil house, burning oil may flow down the roadways, making it impossible for the firemen to reach the seat of the fire, and may pour through basement windows to set fire to the main plant, or into air-raid shelters. To prevent this, a substantial curb of earth or concrete should surround the oil house to provide a pool with sufficient capacity to contain the entire contents. The same hazard exists with acids and corrosive liquids, and the same precautions should be taken. Make sure that all fuel lines are grounded before pumping gasoline or other flammable liquids.

Make sure that all aisles are kept clean and clear to prevent spread of fire, as well as to provide ready exits. Keep stairways clear of all obstructions, and see that they are enclosed with fire-resistant partitions and self-enclosing doors, which must not be blocked open for any reason. This applies also to elevator and dumb waiter shafts.

Combustible or corrosive liquids commonly stored in glass vessels, particularly in hospitals and laboratories, should be kept on the floor to minimize the danger of their falling and breaking as the result of an explosion.

Keep doors, except the automatic type, bolted when closed, but bolts must be operable from both sides. Do not use locks, as the key may be missing in an emergency. The risk of fire passing from building to building through window openings is reduced if steel frames are used, the glazing is wired glass, the windows do not open, and each panel is limited to four square feet. Protect all openings from one building to another by fire doors.

Emergency exists must be clearly marked, and there should always be at least two exits from each room. In large buildings mark exit routes clearly. Small groups issuing from a number of rooms soon become large groups, and groups from different floors join and become a crowd on the stairs; be sure stairways lead to some place from which the crowd can disperse rapidly. The common practice of having emergency exits opening into narrow alleys is to be condemned. All doors should open outward, and panic bars should work freely. Keep external staircases, away from windows and doors if possible. If there are windows near or below any part of the staircase, equip them with wired glass.

Arrange combustible materials in storage so individual stacks are as small as possible, with clear spaces around each. They should not be piled so high that they are beyond easy reach of firemen. Keep adequate space between the tops of the piles and the ceiling. Alternate stacks of combustible and incombustible or less-hazardous materials will help check the spread of fire. Storage racks or shelves should be of incombustible construction.

Smoking still forms one of the greatest fire hazards, but, unless it is absolutely essential, total prohibition will only lead to the greater danger of illicit smoking. Some companies have had excellent results by permitting unrestricted smoking in specially marked off areas provided with fireproof ash receivers. Others forbid smoking entirely during the final half hour before quitting time, so any incipient fires caused by cigarette ends have had a chance to break out and be quenched. Do not use paper boxes filled with sawdust or shavings for "spit-kids"; when wet and oily they are regular little fire bombs. Instead, use sand in a metal container which can also serve as an ashtray.

All machines should be kept clean, and attention must be paid to lubrication to prevent overheating. Line shafting should be well hung and aligned, the hangers and bearings should be kept clear of woodwork, arranged to avoid soaking woodwork with oil, and be of a type that will permit easy cleaning. Provide static eliminators for belts in dusty or otherwise hazardous locations.

House acetylene generators in a separate, well-ventilated building, and keep gas cylinders, whether full or empty, out of the main building, except when in actual use. When welding operations are performed outside the regular welding shop, keep an extinguisher immediately available. Have a member of the plant fire department check that all combustibles are moved out of danger before welding begins.

Oil-cooled transformers represent a considerable risk of being punctured by bomb fragments and the oil catching fire. Protect them against blast by suitable walls and roofs, and a dry well for each unit. Filling should be rocks, graded from one-half inch to two inches, with sufficient capacity to absorb the entire oil content. This arrangement should quench effectively any burning oil draining into the well. Where this cannot be done, as in some indoor locations, build a curb at least six inches high and capable of holding all the oil.

When cables, pipes or other items pass through floors, close the opening through which they pass, preferably with asbestos packing, to prevent flames rising from one floor to another.

See that fire-resistant storage vaults are available for important plant records, and that no combustible materials other than the papers themselves are present. All shelving and partitions should be of steel. Keep vital papers in separate fire-resistant cabinets. Give definite instructions that, as soon as an alarm is given, all important papers are to be returned to the vault, and the door closed. Adequate fire protection in accordance with these standards will also serve to minimize fire damage resulting from direct enemy attack.

FIRE EXPOSURE PROTECTION

The plant should have adequate protection from fire hazards created by the proximity of adjacent buildings, other tenants in the same building and unusual exposure.

When required to obstruct the spread of a possible fire, protective installations and devices, such as blank masonry walls, fire doors, fire shutters, wired glass windows, fire breaks and automatic sprinkler systems should be provided.

Fire-resistant construction, especially of outside walls, roofs and separating floors, obstructs the spread of fire originating in adjacent buildings or from areas in the same building occupied by other tenants. Such fire-resistant construction should be considered and provided when required by the hazardous exposure of a building or structure.

The chemical treatment or use of specially prepared fire retardant paints on exposed construction offers a means of delaying the spreading of fire and provides a limited amount of protection. Fireproofing should not be considered as an adequate substitute for either fire-resistant construction or protective shielding.

FIRE-RESISTANT CONSTRUCTION

New structures being planned and those presently used which are capable of being remodeled should be made as fire resistant as is reasonably possible.

Fire Walls

All large production and storage areas should have the protection against the spread of fire afforded by fire walls. Such walls should be well maintained and all openings therein should be unobstructed and provided with fire doors.

Fire Doors

Fire doors should be in good operating condition and important fire walls should have two automatic fire doors of approved design at each opening. Fire doors should be provided at stair wells as a barrier against the passage of fire. Floor cutoffs should be provided at elevators, stairways and conveyances and should be as fire resistant as the floor itself.

Occupancy Fire Hazards

Fire hazards inherent to the type of structure and industrial process or operation should be eliminated as far as possible.

Inherent Plant Fire Hazards

Those common sources of fire, such as heat producing power unit installations should be eliminated by utilizing safeguards in accordance with recognized standards of the National Electric Code.

OPERATIONAL FIRE HAZARDS

Good housekeeping practices and procedures should be rigidly adhered to at all facilities and should include the following:

Waste Control

In plants where manufacturing processes produce large amounts of combustible waste, adequate measures should be in effect to speedily remove such accumulations. Waste containers should be emptied daily and immediately prior to the close of the day's operations. Even small amounts of combustible waste material should not be allowed to accumulate as they may impede fire-fighting operations or prevent the use of fire-fighting equipment in the event of fire.

Material Storage

In storage areas within the plant and in the plant yards, materials should be stored and arranged in such a manner that a fire hazard is not created and so that fire protection equipment is accessible at all times. Material should not be so stored as to alter compartmental arrangements specifically established as a fire protection means.

Yard Fire Hazards

Rubbish and other unnecessary combustible material should be regularly removed from yards and open storage areas. Grass and weeds in yards and open areas should be kept closely cut. Collected rubbish should be burned under proper supervision at designated sites sufficiently isolated to preclude exposure of flying fire brands and at prescribed times.

Welding Hazards

Welding and cutting fire hazards should be reduced by eliminating combustible material from the vicinity of permanent production-welding locations. A suitable routine should be established requiring the issuance of a permit for all maintenance welding and cutting operations not performed in designated welding or cutting rooms or areas. This procedure should provide for review by a competent maintenance supervisor of the hazards involved and of special safeguards required in such undertakings.

Fire Hazards Peculiar to Manufacturing Processes

Special precautions and protective measures should be taken to minimize special fire hazards created by unusual or peculiar processes of manufacturing and occupancy. Special fire hazards may be created by heat treatment, dip tanks, spray baths, drying ovens and engine testing. Particular attention should be given the storage and use of magnesium and magnesium alloys, rubber and other combustible materials and all operations and processes using flammable liquids, gases and explosive materials.

EMPLOYEE COOPERATION

No campaign of fire protection may succeed without the complete cooperation of all employees. Fire protection needs the help of every workman in the facility. The idea must be sold and interest maintained.

Remember that the attitude of management is very important in establishing the attitude of workers. If plant management shows a sincere regard for the fundamentals of fire protection and fire fighting, the men will be encouraged to follow suit. Therefore, do not smoke in no smoking zones, keep your office free of fire hazards, call attention at once to any hazard you see.

Every employee, male and female, should have some idea of the fundamentals of fire fighting. Bulletin boards and plant newspapers can be used, but best results are obtained by actual demonstration. Take small groups out to the parking lot or some other convenient location, and show them how to operate the various kinds of extinguishers, as well as the disastrous effects of using the wrong type, such as water on an oil fire. Frequently, demonstrators from equipment manufacturers are willing to stage an exhibition which makes a lasting impression.

One absolute requisite of employee participation is constant and free contact between all employees and the fire protection force. This is facilitated through training.

Fire Prevention Education

Continuous fire prevention instruction programs should be conducted for the education of all employees. All employees should be instructed in general fire prevention matters and trained to recognize likely fire hazards and to be constantly on the alert. Such instructions should include the following:

1. The safe practices for handling any hazards connected with his work.
2. Proper disposal of waste of a flammable nature such as rags, paper, or combustible liquids.
3. The importance of general cleanliness and orderly maintenance of the working area.
4. The importance of refraining from smoking in prohibited areas.
5. The need for keeping fire doors and fire extinguishers unobstructed.
6. Strict adherence to the rules and regulations concerning the plant's specific fire precautions.
7. Proper use of fire extinguishers. The principles of the automatic sprinkler system, so that no obstructions will be placed in the way of the lines.
8. Watchfulness for other employee errors that are contrary to fire safety.
9. How to sound the fire alarm, the operation of the various fire alarm devices, of the correct way to transmit such an alarm by telephone.
10. Proper conduct during the course of a fire.

Give special training to operators of equipment having unusually high fire hazards. Those handling flammable or dangerous liquids and chemicals must be made to appreciate the danger. Operators of boilers and furnaces should be thoroughly instructed in safe lighting procedures, proper operation of controls, and the importance of mechanical safety equipment. Men in charge of dryers or ovens evaporating flammable solvents must know how to keep them from becoming filled with explosive mixtures.

Evacuation Drill

The facility must be studied and plans established for evacuating each building as quickly as possible. Tests should be conducted so that the effectiveness of the plans can be evaluated. More than one escape route should be considered in the event that the most desirable is blocked.

A definite procedure should also be established and practiced to reduce the confusion that will arise in times of emergency. It will provide for the following:

1. The ceasing of work.
2. Securing of any machinery and the safeguarding of the area by eliminating potential hazards.
3. The opening and clearing of all escape routes.
4. Formation of an orderly group for departure and the designation of specific employees to command each.
5. The abandonment of all personal belongings as the attempt to obtain them will often result in a fatal delay.
6. Maintenance of silence.
7. The designation of those in charge of each evacuation group and those responsible for maintaining the exit route by keeping the employees moving and the exits open.
8. Searchers who will check the area for stragglers.

Plot Plan for Preparedness

A recurring problem in business and industrial fires is the ignorance of employees as to the location, as well as the use, of fire equipment provided against such emergencies. Even where periodic training programs are held, time passes and new employees are added with the result that the always unexpected fire is still met with confusion.

To meet this problem a simple layout plan should be prepared of each store and facility, showing the location of such things as extinguishers, shutoff valves, hoses and emergency exits. These sketches should be posted in the service and employees' areas at each location. Copies should be sent to the nearest fire department for their use in the event of a fire.

REGULATIONS

Fire Prevention Regulations

Fire prevention regulations specific to the individual plant should be conspicuously posted and actively enforced.

Smoking Regulations

Specific smoking regulations should be in effect at all facilities and such regulations prominently displayed. In areas where smoking constitutes a definite fire hazard due to occupancy or operations, it should be prohibited and adherence thereto strictly enforced. Smoking must be absolutely prohibited in the vicinity of flammable material of any kind, including hazardous areas subject to the accumulation of flammable vapors and combustible dust and vapors. Where smoking constitutes a hazard, it is especially important, in order to assure nonsmoking compliance therein, that nearby rooms or areas be provided for smoking.

Occupancy Regulations

Occupancy fire prevention regulations pertinent to the type of structure, its occupancy, and the inherent fire hazards therein should be posted and enforced in all buildings and allied structures. Operational regulations, adequate to prevent the occurrence and spreading of fire originating from a plant's manufacturing operations, should be established, posted and rigidly enforced. Such regulations should include suitable control of waste accumulation, material storage and yard hazards as well as the fire hazards of certain manufacturing processes—welding and cutting, heat treating, paint spraying and drying ovens, and engine treating.

EQUIPMENT AND SUPPLIES

Fire Protection Equipment

Facilities should be provided with normal fire protection devices and equipment adequate to meet a number of fires started simultaneously, and have available auxiliary equipment for use in event the regular equipment is destroyed.

Emergency Fire-Fighting Means

The destruction rendered by an explosion, or other plant disaster, may either curtail or cut off the flow from water mains, thereby reducing water pressure or entirely stopping hydrant and sprinkler water flow. Emergency water-supplying containers should be planned and established throughout the plant area. Sand piles and sand boxes should be frequently spotted throughout the facility and auxiliary self-sufficient fire pumps and extinguishers should be provided for use in case normal water supplies and other equipment are rendered inoperative.

Water Supply

The plant's water supply and the water supply system, including hydrants and water mains within the plant area, should provide adequate volumes of water at sufficient pressure for the simultaneous and effective operation of the sprinkler system and a number of hose streams capable of reaching the heart of any possible fire area. In addition, secondary sources of water supply should be available through such means as gravity storage tanks and fire pumps. Such tanks and pumps should be properly and adequately maintained. City water mains with hydrant outlets should be located immediately outside of the plant's yard area so that municipal fire department pumpers can be connected directly to them without adversely affecting operation of the plant's sprinkler system and diminishing the output of the plant's water supply system.

Automatic Sprinklers

Automatic sprinklers should be provided in all important buildings of combustible construction or in buildings of noncombustible construction over areas of combustible occupancy. All sprinkler systems should be adequate to control fire originating in the immediate area covered by the specific installation. In addition all sprinkler systems should have water-flow alarms to facilitate the discovery of fire and to supplement fire watchman service.

As a result of the experiences of recent riots, it is recommended that consideration be given to roof sprinklers. This was an area subject to extensive fire bomb attack, and in many cases the store did not know about the attack until the roof was involved in fire. Roof sprinklers might well have saved the store.

All sprinkler systems must be in perfect operating condition, that is, both the individual line valves and the main system sprinkler control valves should be sealed wide open, the water supply lines should be under adequate and constant pressure and the sprinkler heads should be in operating condition and frequently checked for tampering.

All members of the fire squads and brigade should be instructed in the operation of the automatic sprinkler system if the plant has one. They should have a thorough knowledge of water control valves and be properly instructed in their operation should one or more sprinkler heads have been opened by the fire.

In case of a fire in a building with automatic sprinkler systems, fire hose should always be ready for use even though the sprinkler systems may be adequate to control the fire without the use of a hose, for there is always the danger of a break in the sprinkler pipe.

Fire Extinguishers

Fire extinguishers and extinguisher racks should be located at appropriate places throughout the plant. Such locations should be easily available to all employees or the locations indicated by large directional signs. At each location there should be a variety of extinguishers. A color scheme should be used to indicate the type of extinguisher and the specific kind of fire which it will extinguish.

Other first aid fire equipment, such as water tanks, pumps, pails, sand boxes, ladders, axes and equipment conveyances should be provided in sufficient quantities and adequately distributed throughout the plant area.

Maintenance of Fire Equipment

Fire protection equipment should be maintained in condition for instant use. The fire chief or his representative should make regular inspections to confirm this readiness.

Frequency of Inspection

The plant's fire-fighting equipment and the valves and sprinkler system should be inspected at least weekly. Water pressures and usability of fire hydrants, hoses and all other fire-fighting equipment should be checked not less frequently than every three months.

Pertinent items which should be included in the checking of equipment are the following:

1. Hoses must be kept dry and drained, and the strength of the walls frequently checked to assure that they will withstand normal water pressures.
2. Hose nozzles must be cleaned, unobstructed, and in normal working condition.
3. Fire extinguishers should be full, in good working condition, and when tainted or corroded, checked for hydrostatic pressure, checked for correct ingredients content, located in designated spots and provided with lead wire seals.
4. Adequate heat to prevent freezing must be provided in buildings with wet pipe sprinklers and in dry pipe valve enclosures.
5. Fire pumps and gravity storage tanks must be kept full, leak proof, and in operating condition.
6. Auxiliary fire fighting equipment, such as pails, axes, ladders and equipment conveyances must not be used for other purposes and must be kept in readiness in a designated location.

To make an inspection routine effective, the fire chief should make a list of items which should be inspected. This will list, for example, every control valve on the fire protection system, and the inspector must report how the valve was found: open or shut and sealed or unsealed. The list will cover hydrants, hose equipment, the fire pump, the sprinkler tank and all special types of protection.

The list of items to be checked should be made, building by building, floor by floor by floor, department by department. For each department it would cover first aid fire appliances, small hose, fire doors, special hazards. Such listings will provide a valuable guide to inspectors who would be expected to make frequent checks. The inspectors should be capable and trustworthy men who will conscientiously check over equipment and hazards and record the conditions found on suitable blanks which are to be scrutinized by the fire chief and the plant manager.

FIRE ALARM AND COMMUNICATIONS

Speed is vital in fighting fire. Most fires start small; however, unless the proper fire control methods are brought to bear immediately, they may soon become large fires. Thus it is essential that they be brought under control as quickly as possible. For this reason it is imperative that every facility have an effective alarm system and adequate instructions for giving the alarm be provided to all personnel.

It must be stressed in employee instructions that the first person to discover a fire should give the local alarm. The method of sounding the alarm and the person to whom the fire is to be reported must be known to all employees. The report should give the location of the fire and an accurate description of the type of fire. The instructions should also designate who is responsible for notfying the public fire department.

When a facility has no fire detection system, fire watchman service should provide for patrolling the plant, especially during the non-operating periods. Watchmen assigned to this duty should be fully trained in looking for fires in most hazardous places, locating and using fire fighting equipment and in summoning assistance.

If plant guards are also used as fire watchmen, they should be especially trained in alertness to fire hazards and in measures to take in case of a fire emergency. Accurate records of watchmen patrols should be kept by means of an approved time clock system.

During hours when operating personnel are not present, buildings without sprinklers are a higher fire hazard and therefore should be protected by hourly foot patrols. Where sprinklers are used, they should be tied into a complete central station supervisory water-flow alarm system, with a direct communication alarm to the chief guard house.

Without an adequate communications system no emergency organization will be able to operate efficiently. The communications system must be such that it provides for receipt of all alarms on a twenty-four-hour, seven-day-a-week basis. The communications system must also include the facilities to relay immediately any alarms to the fire brigade, municipal fire department and any other agencies whose assistance may be required. The system must be reliable and must be tested frequently. In addition to the equipment, there must be complete procedures for notification of the fire brigades and other emergency organizations and responsibility for such notifications must be properly assigned.

FACILITY FIRE-FIGHTING ORGANIZATION

Regardless of the size of the facility, there should be an organization for fire prevention and for fire fighting. A plant fire brigade composed of a few selected, properly trained employees should be available at all times during plant operations to act immediately without panic in the event of a fire. The exact size and nature of the fire protection organization will vary. In very large plants or in plants having high fire hazard processes in operation, a full time highly specialized fire fighter department may be needed.

In smaller facilities regular employees are organized into emergency fire brigades, composed of fire companies or squads. Generally, however, certain employees should be selected within each department, section, floor area, or building for special training in the use of fire equipment and instruction as to what to do in case of fire. The assignment of these employees should be such that they will respond as a team within their area of the plant, and so they can act as a company of the plant fire brigade in other areas of the plant should the need arise.

Size

The number of men assigned to fire brigade and companies should be sufficient to control all normal-size fires originating within their areas. Thus it will depend on the following factors:

1. Size and character of the plant.
2. Nature of the fire hazard.
3. Availability of municipal fire-fighting equipment.

Irrespective of their size, the fire brigades should be composed of trained men from all departments who are fully familiar with the plant fire-fighting equipment and who are able to take over from operating departmental fire squads when any fire threatens to extend beyond the department in which it originated.

Selection

The members of the fire brigades or companies should be selected from among the regular employees of the various departments in which they work. In this way the men will be familiar with the surroundings and with the various processes and operations of the plant within which they might have to perform fire fighting activities.

The members of the fire organization should be selected with care, for they must not only be physically able to meet the demands of fire fighting but must have the mental ability and courage to do so.

Some of the personal factors to consider in selecting the fire personnel are the following:

1. Does he have the physical ability to perform fire-fighting tasks?
2. Does he have the mental ability to understand and apply the scientific methods and principles of fire fighting?
3. Does he have the emotional stability needed to face an emergency?
4. Does he have an interest in being a member of the fire brigade and a willingness to serve on it?
5. Is he available for fire duty at all times or does his work demand his constant attention or frequently take him out of the plant?

In this regard, it is usually more practical to have men who are not engaged in actual production. Such "nonproductive" employees have less demanding jobs and are more free to respond to a fire, without causing a stoppage of the production line.

It is desirable that the plant engineer as well as the maintenance crew be included in the fire-fighting organization. These men will have the technical knowledge in servicing and repairing fire-fighting equipment as well as operating various valves, pumps, fans and other machinery in the plant in the event of fire. The engineer will also have information concerning any special hazards which are present due to unusual manufacturing processes and will therefore be better able to assist the fire fighting activities in this connection.

Organization

The fire-fighting organization of a facility begins with squads of selected employees within each department, section, floor area, or building. They act as a team within their area and as a fire company of the fire brigade in other areas when needed. Special duties should be assigned to the various workmen in the departments, with definite instructions concerning the closing of valves, pulling electric switches, or other specific activities in the event of fire.

The fire brigade has facility-wide responsibility. It acts as a unit to aid the fire squad in its area. There should be a minimum of ten men assigned to a fire brigade, so that at least five men will always be on duty or on call at one time. These men should be designated with rank or numbers so that those with the most experience and training can take charge at a fire in the order of their succession in the absence of the captain. Each man should be assigned specific duties.

Command

The fire brigade should be headed by a fire chief who should report to and be responsible to an important executive in the plant operation. The chief should be qualified as a fire fighter and administrator. Among the desirable qualities of a good fire chief are the following: knowledge of the job, ability to handle men, energy, diplomacy, sincerity and willingness to do extra work when demanded.

The fire chief should have clearly defined authority on all matters pertaining to fire protection and in the fighting of fires. This authority should be understood by all personnel of the plant. He should be consulted prior to new construction or equipment changes and should advise management on fire and safety requirements. He should maintain a close working relationship with the local public fire department so that mutually satisfactory arrangements for assistance in the event of a fire or other disaster can be worked out beforehand. He should maintain an up-to-date roster of the fire brigade personnel, train and drill the members in the handling of equipment and in the fundamentals of fire fighting, supervise the testing of fire appliances, and maintain a high standard of fire prevention throughout the establishment.

An assistant fire chief should also be appointed. He will help the chief in the performance of his duties and act in his absence. He should have qualifications similar to those required of the chief.

A fire captain should be appointed to be in charge of each fire squad. He should be a responsible person with considerable ability in handling and directing the men in the work of the fire squad. These captains should also have a definite interest in fire prevention activities and should be willing to take special training so that they will be thoroughly familiar with the operation and handling of fire apparatus and appliances.

Training

Members of the fire squads should be trained in those fire hazards peculiar to their individual departments, in methods of their extinguishing, and in use of fire fighting equipment. Members of the plant's fire brigade should have extensive training in the control and extinguishing of fires which may originate anywhere in the plant. They should be fully familiar with the location and use of the plant's fire fighting equipment. In particular, this training should impress the employees with the importance of absolute discipline and the necessity for instant response to a fire alarm even though it may only be a training drill.

Fire Drills

Fire drills are an essential part of any fire training. The frequency of fire drills will depend upon the size of the plant and the nature of its fire hazards, but it is advisable to have drills for members of the fire brigade at least once a month, or more often if considered necessary. These drills should be held during the regular hours of employment. The value of such drills is that it permits the members of the fire brigade to perform as though a fire were actually in progress at a certain location. The location of these fire drills should be in different parts of the plant each time. Thus the men will become familiar with all conditions and be able to combat any conceivable emergency that may arise.

Drills should include making hose connections with hydrants, unreeling and stretching hose without kinks, coupling and uncoupling, attaching play pipes, carrying hose up ladders, over roofs and through the interior of the building. As a general rule, water should be turned on for all outdoor practice work, except during freezing weather. (This assumes rubber-lined hose; unlined linen hose should not be wet.)

Drills should be disciplined, moderately paced and accurate. Stress should be placed on accuracy and precision. To achieve this all drills must be supervised and evaluated. After the drill there should be a critique of the operation.

In addition to plant drills, there should be joint city-plant fire department drills. This will give the members of the facility force an opportunity to work with the regular fire department and thus develop coordination between the two. In addition such exercises, if conducted at the municipal training center, will expose the fire brigade to actual conditions which they will be called upon to face in an emergency. They should be exposed to various types of fires and be shown the proper way to extinguish them. The company should provide the extinguishers that will actually be used by the employees. They should be shown how to use them on actual fires. They should then be given the opportunity to themselves utilize the equipment they will have to use in an actual emergency. This will build confidence in themselves and familiarity with the equipment. In addition, the heat problems that develop should be experienced under the watchful protection of professional firemen.

The fire brigade personnel can also render important services prior to the emergency through a comprehensive fire prevention program, providing standby service during hazardous operations, and making scheduled inspections of fire protection equipment and other emergency equipment. Fire department personnel can be used for these inspections and standbys with the result that these members not only perform an important job in preventing fires and in insuring availability of equipment but they also gain a complete familiarity with various plant areas at the same time. All emergency personnel must be thoroughly familiar with the areas of the plant in order to minimize the confusion if they have to work in smoke and darkness or under other adverse conditions in an emergency.

A wide variety of fire equipment may be in a plant. This equipment could include sprinkler systems, automatic fire alarm systems, manual fire alarm systems, carbon dioxide or dry chemical systems, portable hand equipment or various types of emergency vehicles. Assigning the fire department or fire brigade members the responsibility for checking and maintaining this equipment serves a dual purpose as it takes care of the necessary checking of equipment and trains the personnel so they become familiar with the operation of the equipment. No situation can be worse in an emergency than having costly protective equipment which does not function properly or equipment which nobody in the area at the time of the emergency knows how to operate.

Inspections should be made to accomplish the following:

1. Obtain the correction of conditions creating an undue fire hazard.
2. Make sure that existing fire protection equipment is being properly maintained.
3. Recommend the necessary fire control equipment to meet the requirements of each specific shelter.
4. Assure the proper maintenance of features providing protection against spread of fire and, where possible, the adoption of such additional measures of this kind as may be necessary for reasonable protection to life and property.
5. Assure proper maintenance of exit facilities and the provision of sufficient exits as may be necessary for use in case of fire.
6. Impress on management that the advice and guidance of the fire departments be sought in connection with problems of fire prevention or fire control.

7. Get conditions involving accumulation of rubbish and other easily corrected hazards taken care of promptly.

8. Check for compliance with state and local building codes and fire prevention laws, ordinances and regulations.

FIRE-FIGHTING PROCEDURE

The first step in fire fighting is to learn of the existence of the fire. Rapid notification and response is of the utmost importance for the sooner the fire is attacked, the less will be required to subdue it, and thus the less the damage sustained. It is therefore essential that a warning system be organized and that a definite program for response be established. Each person must know his role and react as soon as alerted.

The first member of the fire squad at the blaze will take charge and begin to extinguish the blaze. The first fire captain will assume command of the fire fighting unit and remain in command until replaced by the facility chief or municipal fire chief.

If additional men are available and can be used advantageously in fighting the fire, the one in charge should put them to work as necessary in laying out fire hose, obtaining additional fire extinguishers from other departments, covering stock with waterproof material and turning off valves and electricity. The captain or fire chief in charge of the fire fighting operation should be the only person authorized to direct the closing of any valve that controls water for fighting the fire. Certain men should be delegated to close fire doors, shut down the power, operate skylights, or other necessary precautions to prevent the spread of the fire. If the public fire department has been called, the captain should dispatch a man to meet the fire apparatus upon arrival and direct the firemen to the scene.

Immediately after the fire has been controlled, the building should be ventilated to remove all smoke and harmful fumes. All fused sprinkler heads should be replaced at once in the automatic sprinkler system. This will restore the system as quickly as possible. All water mains and valves should be carefully inspected to see that no valves have been closed by mistake.

Fire extinguishers which have been used in fighting the fire should be placed on the flood for collection and removal for re-charging. Used extinguishers should never be rehung on their wall brackets, but should be replaced as quickly as possible after recharging.

All fire hose used should be immediately laid out to dry and then replaced on their racks. The hose should be inspected for injury of any kind and should not be put back for reuse if it is not in first class condition. Hose couplings should also be examined for injury. A watchman should be posted for several hours following the fire to make sure that it does not ignite again.

As soon as the immediate danger from the fire is over, start right in and clear up the mess. The salvage squad should be the first group to enter the buildings to make a quick survey of the damage and to make them safe for others to get to work. They will remove heavy, fallen beams, or those hanging danger-ously, and will examine the walls and shore them up if they seem in danger of falling. Make sure that all electrical power is shut off from the damaged area before anyone enters. The clean-up job may be left entirely to the salvage squads, or other em-ployees may be called in to assist, depending on the extent of the damage. Sometimes the entire plant personnel can be put to work on this job.

If the roof of the building has been destroyed or badly holed, it is important to secure machines against damage by weather. This may be done by covering each machine with tarpaulins or other water repellant material, by covering the roof openings with temporary boards, or by moving the machines to undamaged areas. In multi-story buildings it may be advisable to waterproof the floor of the upper story, and move machines downstairs. If the damage is extensive, try to get them under cover in nearby sheds, garages or warehouses. Check with neighboring plants to see if they can accommodate undamaged machines, and their operators, and get them back into production.

Board up broken windows and breached walls, and erect temporary doors, if needed. If the boiler plant is wrecked, and process or heating steam is required, an old steam locomotive or a steamboat can well be used.

Examine each machine carefully for obvious and hidden

damage. Clean off ash and other debris, empty and clean out the coolant tank and lubrication and hydraulic system. Remove the motors and check for wet or burned insulation, and for water or dirt in the bearings. Water can be removed from the windings by a hot air blower or by a low temperature bake. Check the starter and all wiring and electric equipment. Never apply full current until it is certain there will be no danger.

Experience with war damaged plants indicates that machine tools can withstand severe blast and fire without serious damage to their structure, but a careful check must be made for spindle alignment, and for straightness of beds and ways. Inspect rubber and leather belts for burning or excessive dryness caused by heat. Watch for working parts that show heat coloration. Replace these immediately or at least get an expert opinion as to their safe use.

Oil all bare metal surfaces, but remember that this will not prevent the spread of rust. Get after rust as fast as possible. Rubbing with oily rags will remove the deposits formed in the first day or so; after that, it may be necessary to use abrasives. Steel wool, or some of the common household cleaners will do the least harm, but if emery cloth has to be used, take it easy on working surfaces, and watch out for the dust getting into bearings.

Dismount grinding wheels and replace any that appear to be damaged by heat or water. Check all wheels for balance and ring, just as if they were new wheels.

If machines appear to be seriously damaged, consider the possibility of cannibalizing. You may be able to get one good machine out of two or three wrecks. Polish up and repaint fire-scarred machines as soon as possible. A messy machine tends to encourage sloppy work and has a depressing effect on the worker.

Dispose of debris as soon as possible, but do not be in too much of a hurry. There may be thousands of dollars' worth of small parts buried under ash and rubble, which could be sorted out, cleaned and used. Have an extinguisher handy when turning over piles of burned material; there may be smouldering stuff that will bust into flame as soon as the air reaches it.

Restore essential services, electricity, water and sewage, at the earliest possible moment, but be certain that damaged wiring has been repaired or replaced. String temporary wiring, water and air lines until permanent repairs can be made. Rig canvas screens where washroom walls are destroyed. Have telephone service restored quickly.

Investigation into the cause of the fire should then be made and steps taken to prevent similar fires in the future. If it resulted from carelessness, give plenty of publicity to the fact by posters and other means inside the plant. Strengthen the guard force by working extra shifts, if necessary, and caution guards to be particularly alert. The natural letdown following a serious fire, and the unusual number of persons entering the area for repair work may give an easy opening for a saboteur to commit even greater damage. If the plant is engaged in defense work, prohibit all photography inside or outside, except for such pictures as may be required for official purposes, insurance claims and the like.

A written report should be made within twenty-four hours to facilitate insurance claims and to assist the fire chief in determining high risk areas. If fires are reported frequently from one specific place it is an indication that something is wrong with the preventive measures, and proper steps can be taken. All equipment used in fighting the fire should be listed on the report, so the fire department can recharge used extinguishers and dry wet hose.

SUMMARY

In summary, a few of the points which should be emphasized in connection with industrial fire operations include the following:

1. One individual or group of individuals must have primary responsibility for fire protection in the company so that the fire protection requirements are not downgraded as a collateral duty.
2. Fire crews and other emergency personnel must be trained on a regularly scheduled basis.
3. The fire organization must be geared to the size and exposure of the particular plant involved.

4. Fire equipment must be provided, properly located and inspected and tested at regular intervals.
5. The fire departments in both government and industry must be organized to cooperate in an emergency.
6. Communications systems must be provided for reliable receipt of alarms and relay of information to emergency crews.
7. Adaquate preparation for day-to-day emergencies is the best way of reducing the possibility of major emergencies or of coping with them if they should occur.

Chapter 24

ESPIONAGE AND SABOTAGE

ESPIONAGE

ESPIONAGE MEANS SPYING, the obtaining of important information by covert or clandestine means. The very nature of many industries make it difficult to conceal many phases of their operations. In spite of this, it is desirable that these industries use discretion in the release of information. Perhaps one of the most insidious types of espionage is that of obtaining from various sources a vast mass of detailed, accurate data relative to the vital facilities of this country which may result in great loss to our industrial complex. These data may be developed piecemeal, through contributions of many agents whose fragmentary reports fit together like pieces of a jigsaw puzzle to complete a precise picture of the industrial structure.

Espionage agents specifically seek information such as the following:

1. Capacity, rate of production, industrial mobilization schedules, and details of orders on hand.
2. Specifications of products.
3. Test records of newly developed items or equipment.
4. Sources of raw materials and components.
5. Destination of completed products and transportation routes.
6. Data on production methods.
7. Critical points and possible methods of effective sabotage.
8. Measures in force for security and to prevent sabotage, such as, frequency of inspections by guards and their dependability.
9. Names of dissatisfied employees, former employees, and nonemployees who might be susceptible and utilized for subversive plans.

Espionage agents may be expected to use great ingenuity in obtaining information by the following:

1. Infiltrating plants as employees, visitors, inspectors, or by other means.
2. Obtaining information from employees by stealing, purchasing, or encouraging them to "talk shop."
3. Stealing information from records or other sources.
4. Using various means of reproducing documents, products, processes, equipment, or working models.
5. Using "fronts," such as commercial concerns, travel agencies, import-export associations, scientific organizations, insurance agencies, businessmen's groups, and other organizations to obtain confidential information or pertinent statistical information which can be translated into strategic information.
6. Using threats of danger to friends or relatives of an employee, to obtain information.
7. Using blackmail techniques by threatening to expose intimate and personal details concerning an individual.
8. Skillful extraction of information from employees or members of the family or close friends of an employee.
9. Picking up information at social gatherings, bars or restaurants.
10. Personal observation of production operations, test runs, shipment of finished product, or confidential papers.
11. Securing information from waste and carbon paper and other discarded records.
12. Attempting subversion by offers of money or by playing on the emotions such as love, hatred, desire for power, etc.

The spy is constantly engaged in target assessment. He is looking for weak points in industrial firms. A company's physical security may be excellent: control of documents; guarding of plant facilities; the wearing of identification badges when required. Yet, if he can develop a source of information (a person who will furnish him data) inside the plant, he will have at one stroke nullified much of this protective system.

Responsibility for Classified Information

Management is responsible for the security of classified information which may come into its possession as a result of contracts with the Army, Navy and Air Force. Security regulations furnished by the contracting authority must be compiled with in full. The security of industrial information of value is also a responsibility of management.

Protective Procedure

In general, espionage may be rendered ineffective or made more difficult by the application of protective measures such as a careful loyalty check of personnel, particularly before employment, prevention of unauthorized entry to the premises, special guarding and careful handling as well as safekeeping of classified material, controlled burning or shredding of waste paper, restriction of movement within the facility, and security education and training of employees and others who have information on the facility's activities.

An important means of safeguarding information is to establish administrative arrangements and methods which will make the espionage more difficult. Persons working on classified matters should be separated from employees who are under less strict surveillance and should be in areas where absolute security can be maintained. Safe combinations and keys to such restricted areas should be issued to the minimum number of persons. Safe combinations should be changed frequently and when employees knowing the combination leave the firm. Control of the number of copies of correspondence containing information of value must be exercised to assure that extra copies are not made and delivered to outside agents. Carbons and scrap paper may provide information and must be destroyed, if of possible value to the enemy. Janitors and charwomen have an opportunity to obtain uninterrupted access to information of general and possibly specific value and must be kept under surveillance.

SABOTAGE

Of all the subversive activities available to an enemy or a

malcontent, sabotage offers the widest range of targets, the best possibilities of covert action, and the most effective results. *Sabotage* means intentional destruction or disruption of productive capacity, usually by destruction or impairment of machinery, although sabotage of the work force's willingness to work, or to work efficiently, is also possible. Sabotage is as old as warfare itself and its methods have kept pace with the technical advance of science and industry.

Industrial sabotage is a basic doctrine of the Communist Party and other revolutionary bodies. The undeniable existence of this doctrine, the highly effective results which may be accomplished by the skillful employment of sabotage, and the known existence of substantial groups within this country available and willing to undertake such work, place this hazard higher upon the list of risks confronted by our industry than at any time in the history of this country.

In terms of trained manpoweer, equipment and munitions risked, a sabotage operation involves only negligible expenditures by the enemy, but the profit may be enormous if the target has been strategically selected. The disastrous consequences of an act of sabotage against a critical facility may be grossly disproportionate to the manpower, time, or material devoted to the act. It is only realistic, therefore, to assume that an attack on this country may be accompanied by one or a series of well-planned major sabotage efforts. In addition, however, throughout the period prior to the initial attack and thereafter, tremendous loss may be occasioned by a multitude of small acts, under the guise of accidents, whose cumulative effect may be of greater significance than an initial or subsequent major sabotage plan. It is suggested, therefore, that protection against these small acts be given primary consideration.

Actual sabotage motivated by political or military considerations is normally effected after the target industrial complex has been infiltrated and a complete, detailed picture of vulnerable points obtained as a result of the industrial espionage. This preplanned study determines the physical layout of the industrial complex, its productive processes, the types of materials used, the availability on the spot of explosives and inflammables intended for legitimate daily use, and many other useful types of information, depending on the particular plant and type of operation involved.

Recognition of an act of sabotage as such is often difficult as the ultimate target may not be apparent and the act itself frequently destroys evidence of sabotage. To employ effective countermeasures against the threat of sabotage, it is necessary to understand some of the methods and targets of the saboteur.

SABOTEURS

A saboteur may be a specially trained agent assigned a specific mission, or a disaffected native. Saboteurs may work alone or in groups. They may infiltrate military or industrial organizations as legitimate members, or they may work from the outside.

In time of war or heightened international tension, sabotage of important defense production may be attempted by those serving or sympathizing with a foreign power. The saboteur is not necessarily a foreign national or of foreign parentage. He may be a highly trained professional or a rank amateur. He may be a laborer, a machinist, a foreman, a topflight engineer, or even a member of management. He may be anyone, but one thing is certain, he is likely to be one of the least suspected members of the organization.

There are numerous motives that prompt, or can tomorrow prompt, the industrial saboteur to commit the crime of sabotage. Indeed the motives are as varied as the personalities that engage in these acts of destruction.

It goes without saying that money can be the motive impelling the individual performer of acts of sabotage toward his crime, but the paymaster, in turn, may be a foreign interest, powerful and unscrupulous competitors in the same business or industry, or the same sort of labor union attempting to break into the industry or upset existing labor contracts or to foment unrest or a strike situation.

Other motives will include those of bitterness or hatred on the part of disgruntled employees nursing real or imagined grievances against the employer. Such hate-inspired sabotage can in turn be motivated by love of the foreign ideology involved or misplaced nationalism, provincialism, class consciousness or religious or racial hatreds. Blackmail or the exploitation of fear, as in the case of an employee with relatives in the enemy country, may also be involved in motivating acts of sabotage by an employee. At times the act is committed to gain personal satisfaction or to achieve an aim or to further a cause.

All forms of labor unrest inevitably present the threat of sabotage. During these disturbing periods nerves are fraught; emotions flare. Mob rule spreads among the strikers; objective reasoning is abandoned, and a mere remark is sufficient motivation for crippling acts of sabotage.

An employee with a low IQ who becomes disgruntled with the plant or the employer is very apt to turn into a saboteur. The employee's desire to express his ill-feeling towards his supervisor or company, for whatever the reason, is not uncommon.

Discharged employees often get revenge on their "unfair" employer by sabotaging the plant. Too, when an employee is disappointed in his hopes of a promotion, he may resort to such acts to get even with the one who took his job.

Desire for recognition is another contributing factor to malicious acts. Overzealous employees desiring to display their loyalty and ability will often commit sabotage.

People who are guilty of other crimes will often commit sabotage to remove any implicating evidence. Thieves have set fire to buildings they have plundered; embezzlers have fired their offices in order to destroy fraudulent records.

Sad but true, there is also a class of saboteurs who so act because they are mentally deranged, such as pyromaniacs. The majority of firebugs are sexual psychopaths. They derive the same sensual pleasure from starting a fire as do others in performing the sexual act. These sexual psychopaths are not unlike average people in appearance and manner. Many are congenial and highly intelligent. For the most part they keep these perversions locked within themselves and, therefore, are almost impossible to detect in normal activities. Other more obvious pyromaniacal types are the imbeciles, morons and idiots who are often apprehended in starting fires. Epileptics and alcoholics, too, are common offenders.

It should thus be made explicit that the individual actor in the industrial sabotage drama need not be the lowly porter or clean up man, but can indeed be anyone in the industry or business, up to and including individuals in top management.

OBJECTIVES

The objectives of the saboteur in a particular facility may include one or more of the following:

1. Damaging buildings and equipment.
2. Damaging power, communications, water and sanitation systems.
3. Tampering with testing devices.
4. Tampering with drawings and formulae.
5. Infecting or polluting water and foodstuffs.
6. Tampering with ventilating systems or polluting the air supply.
7. Tampering with personnel safety devices and equipment and otherwise creating conditions which would injure personnel.
8. Damaging, spoiling, or destroying the product of the plant.
9. Sabotaging manpower by use of psychological methods.

A definite distinction must be made, however, between manpower sabotage by psychological means, such as the fomentation of strikes, "slowdowns," and the like, and legitimate labor activities. Manpower sabotage of this nature is extremely difficult to detect. One disloyal employee engaged in psychological sabotage may influence others who will thereupon, believing in good faith that a labor grievance exists, engage in strikes and other activities resulting in loss of production.

Sabotage tactics take maximum advantage of existing weaknesses in the security system, of accessibility because of location, of vulnerability through lack of adequate protective measures, and of the importance of the consequences of destruction. Targets whose destruction will initiate a chain reaction, either in further destruction or in immobilizing dependent processes, take a high priority.

TARGETS

In choosing their targets, saboteurs are influenced by two basic considerations that are analogous to those in a tactical situation, namely: the objective, and how best to attain it. Is the destruction of the target to be sufficient in itself, or is it but a contribution to a larger plan? The ultimate in sabotage is the complete and permanent destruction of the target. When this cannot be attained there are many lesser targets, and enough of these strategically grouped may result in the ultimate.

Target Analysis

In analyzing a sabotage target the saboteur considers the importance of the facility from a technical or military standpoint. Will its complete or partial destruction hinder or breach the overall defense? When complete destruction is outside the realm of possibility, what specific items of technical or military importance will have the most crippling effect on the mission of the installation? Examples of such items are the following:

1. Locomotives.
2. Transformers at power stations.
3. Dies in machine shops.
4. Pumps at waterworks.
5. Condensers at steam power plants.
6. Equipment that is hard to replace.
7. Special design equipment.

The capability of a target for self-destruction is always attractive to a saboteur. Heavy rotating machinery, such as turbo-electric generators, can be ruined by a disturbance of the shaft alignment or abrasives in the lubrication system. Other examples of self-destroying targets include ammunition and gasoline dumps, dams and warehouses containing inflammable stocks.

METHODS

The tools and methods of sabotage are limited only by the skill and ingenuity of the saboteur. A major sabotage effort may be undertaken after thorough study of the physical layout of the facility and its production processes by technical personnel fully qualified to select the most effective vehicle to strike one or more of the vulnerable parts of the facility. Sabotage may, on the other hand, be improvised by the saboteur relying solely upon his own knowledge of the facility and the materials available to him. Industrial engineers are well aware of the inherent sabotage possibilities through the use of products available in normal facility operation. An example of this is the sometime availability on the site of explosives intended for industrial purposes; product and process contamination by the use of additives and spoilers; incorrect cycle time phasing; tampering with control devices, operating equipment, etc. The saboteur, in such a case, may or may not possess or need a high degree of technical knowledge. Hence, the device or agent selected for sabotage may range from the crude or elementary to the ingenious or scientific.

For simplicity, the methods of sabotage may be generally classified as follows:

1. Mechanical—breakage, insertion of abrasives and other foreign bodies, failure to lubricate, maintain and repair, or omission of parts.
2. Chemical—the insertion or addition of destructive damaging or polluting chemicals in supplies, raw materials, equipment, product, or utility systems. In defense industries involved in the production of munitions, for example, the process can here work in reverse, by the substitution of ineffective, inoperative, or neutralizing chemicals to render the explosive harmless or minimize its effectiveness.
3. Explosive—damage or destruction by explosive devices or the detonation of explosive raw materials or supplies.
4. Fire—ordinary means of arson, including the use of incendiary devices, ignited by mechanical, chemical, electric or electronic means.
5. Electric and electronic—interfering with or interrupting power, jamming communications, interfering with electric and electronic processes.
6. Psychological—the fomentation of strikes, emotional unrest, prejudice, hatred and insecurity.
7. Pilferage—although not normally associated with indus-

trial sabotage and quite sufficient unto itself as one of the calamities suffered by industry, pilferage can be in fact a most important component of the saboteur's armory. The consist systematic theft of vital, costly, perhaps imported key components can do very great damage to a manufacturing plant. The same can be said of the theft of vital records, drawings, blueprints, formulas, processes, etc.

8. Passive sabotage—although "passive" might better be changed to "subtle," reference is here made to those particularly hard-to-to-detect acts of an almost nuisance nature, but which have the cumulative effect of impeding production and thus harming industry and aiding industry's and the nation's enemies. Although the good saboteur using fire or explosives attempts to make his act look like an accident, the saboteur resorting to "passive" sabotage creates little accidents here and there, giving normal acts just a little twist, in such a way as to create the appearance of normal carelessness or the usual accident. Examples run to great varieties of acts, all the way from the typical slowdown to such as the following: the passing of defective parts at inspection and, contrariwise, the rejection of acceptable components; tampering with control devices, meters, testing equipment, safety equipment, etc.; leaving switches open when they are supposed to be closed or vice versa; the falsification of control records, stock records, or the like, by such simple expedients as making entries on the wrong line, by insertion or omission of an extra digit or cipher, by placing the decimal point in the wrong place; the deliberate misfiling of important documents. Again, as the examples selected suggest, the possible acts of "passive sabotage" are as limitless as man's imagination and the circumstances surrounding the particular business or industry involved.

Still another classification of the methods of sabotage which may prove of assistance is the following based on the target:

1. Ruining fire protection equipment
   a. Damaging fire hose.

    b. Damaging or rendering inoperative the water control valves.

    c. Damaging the automatic releasing devices for sprinkler systems.

    d. Damaging or substituting liquids in fire extinguishers.

2. Obstructing the assistance of outside fire fighters

    a. Interference with communication systems.

    b. Attacking the watchman on duty at gates and other vital points.

    c. Disruption and congestion of traffic en route to the plant.

    d. Destroying the water supply system.

    e. Tampering with fire hydrants.

    f. Damaging fire fighter apparatus.

3. Planning arson with intent to destroy entire plant or bottlenecks of production.

    a. Secreting time bombs at vital points.

    b. Deliberate setting of delayed fires.

    c. Use of gasoline or other highly flammable fluids in the fuel oils used in the plant.

    d. Hurling incendiary or explosive bombs into the premises.

    e. Damage to oil storage or other flammable liquids.

    f. Starting fires on nearby premises which may spread to the plant.

4. Disabling of plant machinery

    a. Adding abrasives or gasoline or chemicals to the lubricants.

    b. Using chemicals to cause corrosion.

    c. Causing breakages by introducing foreign materials or otherwise.

    d. Interference with working parts to put them out of adjustment or cause breakage or undue wear.

    e. Introduction of foreign substances into the fuel used in the power units.

5. Interfering with electrical power

    a. Destruction or damage to the generating plant, including the dam or flume.

b. Breaking transmission lines by explosives or otherwise.

c. Destruction of or damage to transformer stations.

d. Damage to electrical switchboards or other key points.

e. Interference with electric wiring to cause short circuits, overloading or loose connections, with a view to causing a fire.

6. Interfering with the raw supplies necessary for industries
   a. Diverting the supplies at the source, such as by buying up the surpluses or hampering deliveries.
   b. Damaging raw materials at their source to render them unsuitable for manufacture.
   c. Damaging or destroying raw materials in transit.

7. Damaging materials in process of manufacture or in storage with the purpose of destroying them or producing an inferior product
   a. Arson or explosion as outlined in paragraph 3 above.
   b. Improper handling and inadequate protection against deterioration from undue heat or cold, water leaks and contamination by dirt or other foreign matter.
   c. Careless or deliberately improper processing and assembling, particularly with respect to munitions and instruments with delicate mechanisms.
   d. Rendering foodstuffs subject to rapid deterioration or unfit for consumption.

8. Damage or delay to finished products—it is when supplies have been completely manufactured and ready for use that the most costly acts of sabotage can be committed. This category includes such as the following:
   a. Disruption of shipping facilities by truck, rail, ship, or air.
   b. Introduction of chemicals or bombs to cause fire or explosion or damage while in transit.

9. The saboteur can also disable industry by attacking the personnel of the manufacturing plant by such means as the following:
   a. Bacterial infection or other polution of foodstuffs and water supplies.
   b. Spreading disease germs to incapacitate not only the

immediate plant personnel but also the general public.

c. Placing of poisonous substances in the fuel oil, ventilating systems, etc., to create a poisonous atmosphere.

10. Starting labor troubles: The labor saboteur will be active in the following:

a. Fomenting strikes and unrest.

b. Creating of personal antagonism and disputes between employees and employers and between groups of employees.

c. Deliberate slowing down of production operations.

d. Excess and disproportionate spoilage of work and waste of materials.

11. An indirect method of sabotage, more closely allied to espionage, is the theft and sale of important and confidential information by the following:

a. Disclosing confidential information acquired in the ordinary process of duty.

b. Theft of secret and confidential information from the files.

c. Theft or transfer of confidential plans and blueprints dealing with the design and production of war materials.

d. Disclosing secret information regarding formulas, materials, designed and developments.

SABOTAGE BY FIRE

Introduction

Incendiary sabotage is one of the most common and effective means employed by saboteurs because most structures and materials will burn. Moreover, if the fire is started at an appropriate time when the building is closed for the day and no one is around, it will cause much damage—perhaps even completely destroy the structure as well as any evidence of the cause of the fire. The materials necessary are easily available and possession in moderate quantities does not arouse suspicion. By using a delaying device, the saboteur can gain sufficient time to establish an alibi. Also, at times it is possible to set the fire in such a manner that it will appear to have been caused by natural means. However, the fires started by the morons and psychopaths are not as complicated, and the attempts at concealment are quite often crude and ineffective. It is probable that the fire itself will leave minimum identifiable traces of its cause. This will, of course, dampen the suspicions of the authorities and give the saboteur additional time to make a clean getaway and to plan other malicious acts.

The saboteur who commits sabotage by fire will frequently tamper with fire protective devices in order to delay attempts to fight the fire. Such sabotage is frequently accomplished by damage to fire hose, water valves and fire extinguishers, by preventing their free flow or by filling them with flammable material. Sprinkler systems, alarm systems, telephone lines and other communications may be disconnected to prevent notification of fire-fighting companies.

Incendiary Materials

Practically any flammable substance may be used to initiate a fire. The purpose is to bring the surrounding materials to the ignition point so that the heat generated will then, in turn, continue to enlarge the fire until everything within burning range has been consumed. Many cases of arson and sabotage have been caused simply by lighting a wad of paper with a match and placing it next to other flammable materials inside or outside the structure to be burned.

There are many materials which in combination form incendiary mixtures. The most common of these are the following:

PHOSPHOROUS. Phosphorous is a waxy, yellowish, translucent solid that burns spontaneously when exposed to air. In the dark, it will emit a luminous glow. It is stored in water, in which it is insoluble. When used to impregnate paper or cloth, it is first dissolved in carbon disulphide. Upon exposure to air the carbon disulphide slowly evaporates, leaving particles of phosphorous which burst into flame. Phosphorous is also used in explosive incendiaries where detonation scatters the particles over a wide area.

SODIUM. Sodium is a metallic element that ignites on contact with water. It is naturally shiny, resembling silver, but oxidizes quickly and becomes covered with a brownish patina. When placed in water, it reacts chemically to release the hydrogen particles contained in the water. The sodium acts as a strong catalyst, and through its reaction generates considerable heat. It is so intense that the hydrogen which is liberated will ignite. Sodium is usually stored in kerosene or oil to prevent its reaction when not wanted. It is most effective when combined with other chemicals and as an incendiary device is particularly useful in waterfront sabotage.

POTASSIUM. Potassium reacts in the same manner as sodium; it also ignites when brought into contact with water. It is a metal, silvery-white in appearance and is soft and pliable.

THERMIT. Thermit is a mixture of iron oxide and aluminum powder. It can be moulded into various shapes and is used extensively in airborne incendiary bombs. It is best triggered by magnesium tape and burns with an intense heat.

POTASSIUM PERMANGANATE. Potassium permanganate is recognized by its dark purple crystals. The contact of this compound with sulphuric acid and combustible materials or flammable vapors may result in a fire. Potassium permanganate when combined with glycerin is spontaneously combustible. Both are obtainable in any drug store. By using a capillary tube with a stop clock, an effective delayed action device can readily be made.

The methods in which the chemicals are used in setting a fire range from the most simple to the most complex. In almost all cases, the chemicals are placed near combustible materials that may be prevalent in the plant or business. The presence of the combustibles near the ignition source insures the eventual fire.

Many devices have been developed which utilize the characteristics of these ignition chemicals. Thus, since sodium when exposed to water ignites, it is merely necessary to place it so that in a normal manner it will in time be exposed to water. This may be done by making a small opening in the roof and placing the sodium close to materials that will burn. When rain seeps through, the sodium will ignite.

Phorphorous serves the purpose much better as it will ignite when only exposed to air. The arsonist has only to enclose the phosphorus in a moist wrapping and place it near accessible combustibles. When the moisture dries out, the phosphorous ignites the flammable material. The same effect is achieved by placing the phosphorous in a small can filled with water. When a few small holes are bored into the bottom of the can, the water slowly seeps out until the phosphorous is exposed to the atmosphere.

Manufactured Incendiary Devices

Manufactured incendiary devices are of two types. One is designed to deliver intense heat in a limited area and the other to scatter incendiary material over a large area by means of an explosion. Both types have been combined with success in aerial warfare, but for the covert action of a saboteur the separate and small types are more practical.

Both incendiary and explosive devices are of the same general construction and differ only in content and purpose. Each is made up of three parts, the delay device, the initiator and the main charge.

1. The delay mechanism is required to insure ignition at the appointed time, and to provide for the escape of the saboteur.
2. The initiator is to insure the complete and efficient ignition or explosion of the main charge.
3. The main charge is the incendiary or explosive material within the device.

By building magnets into a device, it can be affixed to a metallic target, such as a gasoline storage tank.

Improvised Incendiary Devices

The means of igniting a fire are so commonplace and accessible that improvisation of a sabotage device is exceedingly simple. A book of matches, a cigarette, a candle, or a can of gasoline are all available to anyone, and either by themselves or in combination can be used surreptitiously to start a fire. A book of matches triggered by a lit cigarette or a phosphorous impregnated piece of paper are simple devices and provide time for the saboteur to depart.

Slightly more complicated devices can be made on chemical and mechanical principles. Sulphuric acid can be separated in a container from a mixture of potassium chlorate and sugar by a paper partition. When the container is inverted so as to bring the acid in contact with the paper, the latter dissolves and the acid ignites the mixture. A variation of this can be made by separating a lead pipe into two sections by means of a brass or copper disc. One section is filled with sulphuric acid and the other with a solution of picric acid. Both ends are plugged with glass stoppers covered with wax. The time delay in each of these devices can be controlled by the thickness of the dividing disc.

Other simple incendiary devices can be made of mixtures of gasoline and oil, or napalm, rigged to be fired by a time fuse or by electrical or mechanical igniters.

Disguised Incendiary Devices

Since the entire incendiary device is actually an initiator and depends on a combustible target for its effect, it can be made small in size. This small size permits an incendiary device to be disguised as a commonplace item. Manufactured incendiary devices have been concealed in cakes of soap, pieces of imitation coal, cigarette packs, pencils and fountain pens. Even undisguised items, because of their size and their natural appearance as a component of some larger mechanism, do not attract particular attention. Chemical and clockwork contrivances are among the latter types.

Mechanical Delay Devices

Mechanical delay devices are used normally in connection with dry cell electric batteries. The basic idea in these mechanisms can be well represented by the use of an ordinary pocket watch. By removing the minute hand, setting a small screw crystal to a depth that it will contact the hour hand but not the watch face, and using this screw and the main stem as contact points, the watch becomes a timing device with a twelve-hour span. This same principle is employed in the majority of mechanical delay devices.

INVESTIGATION OF INCENDIARY SABOTAGE

Before a person can qualify to investigate the cause of fires, he must have special training and adequate knowledge and experience regarding fires. The job is not one for a novice. The investigator must show the criminal intent of the saboteur by establishing the point of origin of the fire, the time the fire was started, what method was used to start the fire and whether or not any flammable liquid or other substance was used to accelerate the rate of burning. Such information as the color of the smoke, the intensity of the fire, odor of burning materials, whether or not more than one fire was started, and the rate of spread of the flames are important. After the fire is out, the area is carefully scrutinized for further evidence such as the charred surface of burned wood and the coloration of steel beams, in order to establish further the point of origin of the fire and the intensity of the flame.

Plant personnel can be of great assistance in this investigation if they make it a point to observe and make a note of any incident pertaining to the fire. Among those factors are the following:

1. Where and how the fire began, how it spread, any unique odors and the color of the smoke.
2. What were the circumstances under which the fire was discovered.
3. Who was in the area or loitering near it at the time of, and immediately preceding, the fire.
4. Who witnessed the fire and what was their reaction, appearance, comment.
5. The condition of the area, the building, all entrances and exits, all fire fighting equipment.

SABOTAGE BY EXPLOSIVES

Introduction

The use of explosives by a saboteur may accomplish several

objectives. First, explosives can totally and completely destroy a target which cannot be destroyed by other means. Second, explosives can be used to instill fear and confusion in the civilian population, causing panic, absence from work and disruption of the normal routine of living. Third, it is possible to use explosives in such a manner that the cause cannot easily be determined due to the complete destruction of the target, followed perhaps by a consuming fire and the erasure of all telltale evidence from the scene. The most probable targets are the heavy construction of power and transportation facilities. Small quantities of explosives may trigger a chain reaction or destroy an extremely vital adjunct of an installation.

One problem to the saboteur in explosive sabotage is the difficulty of surreptitiously bringing the explosive to the target. For example, only approximately three pounds of an explosive can be concealed on a person.

Certain explosives can be moulded into forms of items of everyday usage and carried to the target. A saboteur may use any ingenious method to accomplish this mission. Then, too, explosives are used extensively in mining, agricultural and some industrial operations. Also they are not difficult to produce, and the ingredients are readily procurable. These factors work to the advantage of the saboteur.

Common Explosives

Explosives are classified according to their sensitivity and as to the velocity of the expanding gases released, i.e., according to the time it takes common explosives to burn or detonate. They are classified as low explosives or high explosives.

Explosives of high velocity (high explosives) are nitroglycerin, trinitrotoluene, nitrostarch, picric acid, dynamite and plastic explosives. Black powder is an example of a low velocity or low explosive.

High explosives, such as nitroglycerin, detonate with a ham - merlike blow of shattering or cutting effect. A low explosive like black powder, on the other hand, while powerful in effect, produces a more pushing or heaving motion against any retaining object. Consequently, the saboteur selects his weapon much as a surgeon chooses his instrument—the one most ideal for the task at hand.

Low Explosives

1. Black powder is the oldest known explosive, but its use has declined due to the development of more efficient types of explosives. It has many different forms, ranging in size from very small granular forms to compressed cylinders more than two inches in diameter. In appearance it is shiny black.

    Black powder must be handled with care as it is sensitive to heat and is easily ignited by friction or sparks. On ignition it gives off a dense white smoke. One of the products of the combustion is hydrogen sulphide, which has a characteristic "rotten egg" odor. Black powder is one of the very few explosives that can be destroyed by immersion in water as the nitrate content is completely soluble. It burns freely in the open air and must be confined for an explosive effect. It is used in pipe bombs and other improvised devices.

2. Smokeless powder is not a powder and is only smokeless in comparison to black powder. It is made by treating plant fibers (cotton or wood) with nitric and sulphuric acids to form nitrocellulose. It may be used in pipe bombs and similar arrangements in the same manner as black powder, and generally has a more powerful effect. It can be formed in flakes, strips, pellets, sheets, or perforated cylindrical grains. It is cut by machine and therefore has uniform size, which helps to distinguish it. In its natural state it is amber in color but is usually coated with graphite which gives it a metallic gray appearance. Upon ignition it gives off a light colored smoke, not nearly as dense as that of black powder. When not confined, it burns more slowly than black powder and is, therefore, not quite as sensitive to handle in the open. It can be ignited by sparks or flame and although it cannot be permanently destroyed by immersion in water, it will not burn when wet. It is best destroyed by spread-

ing it out in thin trails on the ground and igniting it with a piece af safety fuse.

High Explosives

1. Nitroglycerin is a viscous, syrupy, oily, liquid that is heavier than water. It is colorless when absolutely pure, but ordinarily has a pale yellowish color similar to olive oil. It has a sharp, sweet taste and in large doses is toxic and may cause death. It is a most dangerous explosive to handle because of its senstivity and is physiologically unpleasant as even a single drop on the skin will cause violent headache and nausea. A liquid suspected of being nitroglycerin should be treated with extreme caution. It should not be moved until its identity is established, and then only when it has been absorbed in sawdust or otherwise neutralized as explained below. If reddish-brown fumes are observed over the surface of a material suspected to be nitroglycerin, *extreme caution* should be exercised. This condtion indicates decomposition and in this state sensitivity to shock is increased many times. There are two simple tests for nitroglycerin.

   Hammer test. A strip of ordinary newspaper, twelve inches long and one-fourth inch wide, is dipped into the suspected liquid to a depth of approximately one-fourth inch. The wet end is then placed on a metal block, far removed from the suspicious liquid, and struck sharply with a hammer. If the liquid is nitroglycerin there will be a loud, sharp report. In making this test care should be taken not to exceed the quantities mentioned.

   Flame test. A strip of newspaper, twelve inches long and one-half inch wide, is dipped into the supected liquid to a depth of approximately one-half inch. It should then be removed from the suspected liquid and ignited. Nitroglycerin burns with a peculiar green flame.

   If a suspected liquid is found to be nitroglycerin it must not be moved in its liquid state. Sufficient sawdust to absorb the nitroglycerin should be brought to the spot and the liquid poured into it. A thick mat of newspapers should be prepared on the ground at a distance from any structure. The nitroglycerin soaked sawdust can then be removed and spread lightly on the newspapers. A six foot fuse of twisted or rolled newspaper can then be employed to light the mat. The person lighting the fuse should take cover for although this method of disposal is considered safe, nitroglycerin is very unpredictable. Nitroglycerin can also be destroyed by decomposing it through chemical action. The formula for this decomposing solution is as follows:

one half gallon of water,
One-half gallon of methyl alcohol.
Mix with two pounds of sodium sulphide.

This mixture should be slowly poured into the container of explosive and allowed to stand.

In emergency cases, alcohol may be used, but only in an emergency.

2. Dynamite is the most widely used commercial explosive. Commercial dynamite is prepared by absorbing nitroglycerine in a porous or absorbent material, such as woodpulp, wood flour, sawdust, or other materials. Dynamite is made in many different strengths, generally ranging from 25 per cent (meaning 25 per cent explosive) up to a form called blasting gelatin which is considered 100 per cent explosive. Gelatin-type dynamite has a puttylike form and its explosion is extremely violent. The most ordinary types of dynamite will be 25 or 40 per cent. The strength is marked on the outside of the wrapper.

Military dynamite is an explosive composition fixed in a binder but contains no nitroglycerin. Both commercial and military dynamite are used in the form of cylindrical sticks or cartridges wrapped in heavy waxed paper. The sticks range from six to twelve inches in length and from one and one-fourth to one and three-fourths inches in diameter depending on the type and manufacturer. A blasting cap detonator is necessary to cause an explosion.

Dynamite can be handled with safety provided it is in good condition. However, if it is wet, or the nitroglycerin has seeped through the wrapper, extreme care should be exercised. If dynamite is stored in one position for a long period, however, the nitroglycerin tends to seep to the lower side and becomes sensitive. Military dynamite is relatively insensitive as compared to the commercial types. Dynamite can be disposed of either by detonation or burning. It is burned in much the same manner as nitroglycerin-soaked sawdust, by spreading the sticks at least a foot apart on a mat of newspapers. The dynamite, the mat and the fuse should be soaked with kerosene, and the person lighting the fuse should observe the same precautions given above.

Its convenience, availability and effectiveness make dynamite a favorite explosive for the saboteur. The high velocity of its explosion makes it unnecessary to confine it to make an effective bomb.

3. Trinitrotoluene (TNT) is a crystalline powder or flakes pressed into half pound and one pound blocks for ease in handling. It is a light straw color. The blocks are incased in a fiber container to prevent crumbling and to render them waterproof. Each block is provided with a cap well to receive an electric or nonelectric blasting cap. TNT is quite insensitive to blows and friction, and in fact will seldom detonate on impact with a rifle bullet. Because of the safety of handling and storage there are few problems concerning its disposal. It can be destroyed by burning in a manner similar to that used with dynamite. The blocks should not be broken but should be well spaced on the burning mat. As with other explosives, all precautions should be taken by destroying it at a distance from personnel or structures.

4. Nitrostarch is slightly less powerful than TNT and is more sensitive to flame, friction and impact. Accordingly, more care must be used in handling this explosive. It is gray in color, and is packed in blocks similar to TNT, except that the blocks are of one quarter and one half pound

size. It is also provided with a cap well, with the position marked on the outside of the wrapper. Nitrostarch blocks may be disposed of by detonation or burning.

5. Composition C is a plastic explosive. The different types are designated by a numeral following the letter C, such as C2 and C3, and C4. The higher numerals represent further development of the original Composition C. The explosive, a gray or white puttylike substance, is pressed into blocks, two inches square by eleven inches long, and enclosed in a glazed paper or plastic wrapper. It has about the same sensitivity as TNT but is more powerful. Plastic explosives should be destroyed by detonation.

Initiating Devices

There are two general types of initiating devices: those which produce flame for use with low explosives, and those which detonate for use with high explosives.

FLAME PRODUCING DEVICES

1. A safety fuse consists of a powder core that is wrapped in paper or cloth fiber and usually is waterproofed. It is the medium through which flame is conveyed for the direct firing of a low explosive, such as black powder, or for indirect firing, as in the ignition of a blasting cap to detonate dynamite.

2. A miner's squib is a thin paper tube of powder sealed at one end with a wax plug, and the other end contains a fuse or wick. The wax plug is pinched off at the time the squib is used. The fuse burns slowly and when it reaches the powder a flame shoots out the open end.

3. An electric squib consists of an aluminum tube one and one-half inches long with leg wires protruding from one end. When an electric current is applied to the leg wires, the firing element flashes, ruptures the tube, and sends an intense flame into the explosive.

DETONATING DEVICES

1. Blasting caps and squibs must, in themselves, be regard-

ed as explosives and should be handled with caution. A blasting cap (nonelectric) is a small tube closed at one end and loaded with a charge of an explosive, which is capable of being detonated by the spit or sparks from a safety fuse. In appearance, blasting caps are metallic cylinders, about two inches long and about the diameter of an ordinary lead pencil. In use, a length of safety fuse is inserted into the open end and the edge crimped lightly. The explosive material used in these caps is sensitive to heat and shock and must be handled with caution. They are generally packed in shock-absorbent material.

2. The electric blasting cap consists of a metallic cylinder similar to the fuse type with wires inserted into the metallic case and connected in series to a short piece of resistive wire within the cap and adjacent to the explosive material. As the electric current is fed through the wires, the resistive wire heats and ignites the explosive material. They have the same sensitivity as nonelectric caps and great care must be used in storing and handling them. They come in various numerically designated sizes.

3. Safety fuse, detonating fuse, or detonating cord or simply fuse as it is sometimes called, has the appearance of a smooth surfaced clothesline rope about one-fourth inch in diameter. It has a wrapped and waterproofed powder core and cannot be extinguished by immersing it in water. A burning fuse emits a cloud of white smoke. The burned portion of a safety fuse will show a discoloration with the exception of the heavily coated types. It requires a blasting cap to detonate and the extreme violence of the action is sufficient to detonate a high explosive in contact with it. Fuse itself is nonexplosive and may be handled without fear of injury except from burning. It is usually wrapped around or taped to the charge to be detonated. It is used for the simultaneous firing of a number of shots at some distance apart.

Timing Devices

After the explosive charge is placed in position by the saboteur, it may be detonated by any of a number of methods. Usually, a time delay device is employed to give sufficient time to get out of the area before the explosion occurs. These delay devices may include a time clock, a slow burning fuse, or a chemical which slowly eats away a thin retaining wall of metal inside the reaction chamber, and which will cause the explosion to take place when the chemical comes in contact with reacting agents on the other side of the metal retaining wall. This latter method is more often employed in incendiary sabotage, however.

Some favorite timing devices are used repeatedly because of the ease of their construction, procurement and use. Some of these more common timing devices are the following:

TIME PENCILS. Explosive bombs may be activated by time pencils operated by mechanical or chemical actions.

CLOCKWORK DELAYS. Like the time pencils, these clockwork delays are instruments which can delay explosions up to forty days. After the time of delay has expired, these timework mechanisms release a striker which fires attached detonator caps, or close an electric circuit which detonates the cap.

ALTIMETER SWITCH. The altimeter switch is a weapon used to sabotage aircraft after they reach a predetermined altitude. It is a combination bomb containing detonator, battery and explosive.

RIGID LIMPET. This sabotage device is expressly made for the sabotage of boats with steel hulls. It contains high explosive, is held to the boat by strong magnets attached to the device, and is detonated by a chemical delay attached to one end.

MAGNETIC CLAM. This is a commonly used device, easily applied to iron or steel surfaces through the use of a bar magnet attached to the bottom of the device. It is usually activated by a time pencil.

Sabotage Bombs

The popular conception of a bomb as being a black sphere about the size of a bowling ball equipped with a sputtering fuse is not likely to be encountered in sabotage by explosion. Because an explosive bomb itself is the unit of destruction and is not dependent upon outside aid as is an incendiary bomb, it is normally larger than an incendiary bomb. However, the same ingenuity of disguise is applicable as in the case of incendiary bomb.

Amateur saboteurs, such as the disgruntled employee, the "wildcat" labor agitator, or the homicidal maniac, would tend to use explosive materials that are most readily obtainable. For example, black powder, unfortunately, is in widespread use commercially and may be obtained without much difficulty. This type of explosive burns rather than detonates.

A typical bomb of the black-powder type might consist of a length of iron pipe capped at both ends with a fuse inserted through one of the caps. Upon ignition, the powder burns, expands, the pipe bursts, and pieces of the metal will be thrown to a considerable distance with damage to people and property.

The professional saboteur is more likely to make use of high explosives, such as dynamite. The high explosives achieve their destructive effect as a result of shock rather than sudden burning. Dynamite is generally ignited by the use of blasting caps. These may be set off either by a fuse or electricity. By the use of delayed action devices an inoffensive-looking package may be made to explode hours or days after being placed by the saboteur.

It must be recognized that the professional saboteur, and in many cases the skilled amateur, can make the appearance of his explosive material normal and harmless. Five sticks of dynamite taped together and equipped with a blasting cap would make a capable bomb, but upon sight would excite suspicion and concern. The same five sticks of dynamite stuffed in a suitcase with a dry cell battery and a clockwork delay device would be just as destructive, but would not attract attention. A lump of plastic explosive coated with a mixture of shellac and coal dust would be unnoticed in a load of coal. The possible combinations of explosive, activator, delay device, and outside container are many.

A bomb can be concealed within a bundle of clothing, a pair of shoes, a package of lunch, a thermos bottle, a suitcase, a camera. Thus it may be seen that the means of concealing explosive devices are almost limitless.

Bombs may be triggered or detonated in various ways. For example, a weight-type trigger bomb is actuated by some normal act of the person finding it. A chemical mechanism bomb is constructed in such a way that, after placement and removal of a simple arming pin, tilting or upsetting of the package starts a chemical action which serves to detonate the bomb. Electrically actuated bombs may be constructed in a variety of ways. For example, a package containing a bomb can be placed in a dark room with an electrical cord plugged into a socket. When the light is turned on the electrical device explodes the bomb.

An open or unconcealed bomb is generally directed at materials rather than persons. This bomb usually consists of the pipe-type bomb or several sticks of dynamite bound together and actuated by a lighted fuse. Due to the characteristics of the burning fuse, it is dangerous and usually inadvisable to attempt to put the fuse out. Upon discovery, it is advised that the area be cleared and that steps be taken to minimize damage. As a general rule, time bombs are concealed at the point of intended explosion or sent by mail.

MECHANICAL SABOTAGE

This type of sabotage interferes with proper operation of machinery and equipment, or results in the manufacture of inferior products. Mechanical sabotage can be carried out without special knowledge or training, and the necessary means are frequently indigenous to the target.

The methods of mechanical sabotage are usually within one of the following classifications; however, the field is wide, and they may be committed either singly or in combination.

Breakage

If breakage is employed as sabotage, it is usually directed against critical parts of machinery which are difficult to replace and which will stop the production processes. Expensive and delicate apparatus, instruments, tools and other equipment may also be the object of sabotage by breakage. Other objects of breakage may be gear mechanisms, fuel pumps, cam shafts, or any vital part which will impair the working of machinery and cause harmful delay. Breakage is usually directed against the moving parts of a machine or motor, or against delicate control or measuring devices. The simple act of dropping a wrench into the moving parts of a machine or breaking a pressure gauge will cause a halt to service or production.

Abrasives

This method of sabotage is most often accomplished by adding abrasives such as emery, sand, powdered glass, or carborundum to fuels and lubricants, or directly into the moving parts of a machine. The powdered abrasive materials circulate through motors or machinery, causing undue wear, costly damage, eventual breakdown and delay because of necessary repairs or replacement of the vital parts involved.

Contamination

Not only may lubricants, fuels and various chemicals and solutions which are used in production processes be contaminated by the addition of foreign substances or chemicals, but commercial and public water supplies may also be contaminated so as to make them unfit for drinking or other purposes. Contamination of foodstuffs can make them inedible, even without bacteriological aspects.

The addition of contaminants to motors and other machinery may cause breakdowns due to bearing failure or may freeze the pistons to the cylinder walls of engines. When chemicals are diluted by liquids of similar appearance and viscosity, the act may go unnoticed until the end of the process at which time the product will be faulty. Contamination of the fuel of an internal combustion engine is one of the most common methods of mechanical sabotage. Ordinary granulated sugar poured into a gasoline tank will immobilize a motor and necessitate overhaul.

Substitution

Items, such as raw materials, processing solutions, measuring gauges, patterns, or blueprints, may be altered by substituting faulty materials or erroneous information. The substitution of false information, wrong materials, faulty equipment, or the wrong measurements or amounts of materials may be quite effective in sabotaging engineering jobs or plant production. This substitution may be done at any point along the production line or where materials are stored and may consequently be difficult to trace to the one who performed the act. Even when discovered the worker can claim that he made a normal "mistake."

Acts of Omission

An act of omission is the failure to do something which one has a duty to do. In failing to perform his duty a worker can sabotage machinery and processes just as effectively as though he performed a deliberate act of sabotage. Modern industrial methods are quite complicated. By omitting a vital step in a production process or by neglecting to turn a valve at the proper time, a worker can effectively sabotage machinery and products. Even though the failure can be traced to a certain man, criminal intent is difficult to prove in such cases. The harm has been done.

The infinite variety and the simplicity of methods for committing an act of mechanical sabotage make it extremely difficult to detect or prevent. Constant vigilance, alert supervision and frequent inspection by all concerned are the most effective defenses.

PSYCHOLOGICAL SABOTAGE

Normally when psychological sabotage is mentioned one thinks of strikes. But in addition to the deliberate fomentation of strikes as such, the industrial saboteur has also in his repertoire the confusion and strife that can be stirred up over jurisdictional, regional, factional, communal, racial, religious, political and personal differences, to name only some potentially emotional issues that can be exploited. The encouragement of boycotts is another example. The dissemination of false, misleading, or inflammatory propaganda or rumors can have the effect of creating work stoppages or slowdowns or bad morale.

Another means of agitating crowds of employees to the explosion point can be the display of "atrocity-type" photographs of workers killed in industrial accidents, injured in strike violence, or even discharged from employment. Similarly, the staging of public funerals and the turning of such affairs into demonstrations can be exploited to inflame emotions and bring about psychological sabotage. The actual manufacturing of martyrs can sometimes be accomplished by thus exaggerating the significance of otherwise minor or routine incidents. The use of "hired" crowds from outside the industry itself—unemployed drifters, drunks, or drug addicts, for example—may help exploit such cases.

The creation of panic among industrial workers can be an exceedingly potent weapon by which the industrial saboteur can attack the defense effort. Panic can be created by an act of violence and by exploiting the ignorance, fear and lack of confidence of the workers in their leadership. Psychological sabotage is most difficult to control or combat because it deals in intangibles and takes full advantage of normal human frailties. In its simplest form it is the implanting of a doubt or fear or discontent in the mind of an individual. It depends on natural garrulousness for exaggeration and multiplication.

Psychological sabotage may be employed effectively on a local scale to corrupt a unit or facility. It attempts to stop or slow down production by causing unrest and dissatisfaction among the employees of a facility.

BACTERIOLOGICAL AND CHEMICAL SABOTAGE

It is unlikely that security personnel will be faced with a purely local problem of bacteriological sabotage. It is more probable that such a weapon would be used on a wide area basis, and that technical and professional advice on defense would be available to combat the specific situation. However, the possibility of an attack by waterborne or airborne poison or bacteria exists. Such sabotage efforts might take the form of introducing chemical, biological, or radiological elements into air, food, or water supply. Likely targets for chemical sabotage are installations employing highly skilled technicians.

Not only are chemicals highly effective and capable of affecting employees to the extent that their productive efforts may be totally impaired, but in addition such an attack has a tremendous psychological impact. The fear and panic induced by a successful attack on a few facilities might be out of all proportion to the casualties involved.

It is important that both industrial and governmental facilities be prepared, therefore, to institute protective measures should an emergency develop or a potential attack appear imminent. The security force must be flexible enough to enforce the instructions of experts in the field, in addition to carrying on their normal functions.

Food Handlers

Personnel handling food at a plant's cafeteria or dining hall should be hired with at least as much care as other plant employees. This is especially important in view of the opportunity of food poisoning sabotage.

Water

The drinking water in key facilities should be checked frequently by personnel whose loyalty has been thoroughly investigated. Such checks are necessary as a precaution to minimize the effect of bacteriological sabotage.

Water samples should be taken from within the plant even if regular tests are made by the municipal authorities or at the plant's own filtration or pumping plants.

Water supplied from the plant's own deep wells is peculiarly vulnerable to contamination since normally it may not be chlorinated. Frequent potability tests of such supplies should be performed.

METHODS OF ATTACK ON SPECIFIC TARGETS

The following specific targets are vulnerable to one or more methods of sabotage:

Natural Resources

Mines may be sabotaged by causing cave-ins or flooding of the shafts or tunnels. Forests may be destroyed by incendiaries; fruit trees may be killed by an induced blight. Farm produce is vulnerable to parasites and various blights, and on a smaller scale by the diversion of water used for irrigation.

Army, Navy and Air Force Installations

Any action against an armed forces installation that disrupts or prevents full accomplishments of its mission constitutes a potential threat. Sabotage actions directed towards the disaffection of personnel, destruction of ammunition or fuel supplies, and the disruption of communications are common to any of the armed services. Other targets are peculiar to the particular service, such as drydocks and repair facilities to the Navy, and complex flight and navigation equipment to the Air Force.

Industry

Industry presents innumerable possibilities for explosive, incendiary and mechanical sabotage and is especially vulnerable to acts that will initiate a chain reaction. The following are examples of vulnerable points of industrial machinery and the means by which they can be damaged:

1. The drainage of oil or the blocking of lubrication pipelines.
2. Introduction of abrasives.
3. Missetting or damaging process control instruments.
4. Introduction of small tools or other pieces of metal into moving gears.
5. Explosive charges placed to have a shattering effect when detonated.

Warehouses and Supply Depots

Materiel in storage is subject to ordinary explosive or incendiary sabotage. There is also an opportunity for delayed sabotage by the introduction of abrasives or adulterants into the items stockpiled. This latter type of sabotage will not normally be discovered until the materiel is put to use, and is difficult to detect or trace.

Transportation

The propelling machinery and cargoes of land, sea, and air transportation are subject to acts of sabotage similar to those mentioned in *Industry* above. In addition, rail transportation can be sabotaged indirectly by damaging switches, rails, roadbeds and various structural adjuncts, such as bridges, tunnels and shop facilities.

PREVENTIVE MEASURES

Physical sabotage is essentially an inside job, or requires the assistance, knowingly or unknowingly, of someone inside. Hence the principal protective measures must be designed to limit the entry of continued presence of saboteurs or their assistants.

Nothing is more basic in this direction than the careful recruiting and selection of employees in general, including pre-employment investigations of all key personnel. A thorough background investigation of all porters and janitors, messengers and maintenance men is equally important, as they are in a good position to gather information to which they are not entitled.

Closely related to this and of very great importance is the building and maintaining of a high level of employee morale. Regular checks of employee morale and prompt action to correct any deficiencies or situations adversely affecting the mental health of the employee is essential.

Avoid labor turnover as far as possible. Every employee who leaves the plant carries with him some knowledge of its layout and operations. This is invaluable to saboteurs who commonly obtain employment just to gain such information. Preserve the records of every employee who quits or is discharged, and answer inquiries from other plants concerning former employees promptly and fully, after reference to the records.

To reduce turnover, the exit interview has proved useful. A company official interviews the employee and tries to find out his reason for leaving, and whether it is possible to keep him. This also helps to show up the activities of some saboteurs, particularly those who are engaged in psychological sabotage. Such persons will frequently stop short of committing an act which would expose them to prosecution and, when their disruptive activities are detected, will quit before any action is taken against them. They then seek employment elsewhere and start all over again. In the exit interview, make tactful inquiry about the man's future plans; if his answerse are vague, make a note on his record.

The personnel office and the security office should take a

joint interest in employee absenteeism. Any absences not sub-
stantiated by good reasons should be promptly investigated. An
employee may fake a reason for being absent from work in order
to make contact with subversives or enemy agents.

There should also be routine checks of "influential" employees. Be alert to which men are becoming prominent in plant
organizations ranging from hobby and sport to religious and
political groups. Know who is running for office and who has been
elected in the various organizations. A quick look through personnel and security files usually tells all that needs to be known.
Of course go farther into backgrounds, if it appears to be warranted.

The employees should be kept as fully informed as possible,
by means of bulletins, leaflets, lantern slides, etc., of dangers to
ge guarded against, protective measures available, and steps to
be taken to meet situations indicating sabotage or resulting from
acts of sabotage. Such training measures should be related to
emergency plans in effect for other disasters, such as those
brought about by floods, bombings, or earth quakes.

Encourage employees to report suspicious activities of fellow
workers to the plant security department. Ask them to report
this information directly. When the lead is received, immediately open the case, Check the suspect's personnel file
and his security file. Check and learn whether the charges
are founded in fact. Keep complete records of the case in
locked security files for later reference.

There is a difference of opinion as to the advisability of publication of information concerning possible acts of sabotage.
Some believe that such publicity may jeopardize their chances
of apprehending the culprit and will in addition cause unwanted
outside publicity. Others believe that their employees should be
kept advised of such conditions, for in this way they will become
more alert to the danger and more observant of the acts of others.
Both approaches have merit. The one to be followed in a particular situation will depend upon the facts of that case.

Rules and regulations governing the actions of employees
while on the job or on plant premises must be established and
enforced impartially. Strict perimeter security should be main-

tained by the guard force. The identity of all persons entering and leaving the plant should be checked, and the movement of packages, materials, merchandise, or other deliveries, as well as private vehicles and trucks, should be controlled and inspected.

Tight security should be maintained on critical and vulnerable areas in the plant and areas where explosives, flammable liquids, or gases are stored or handled, as well as restricted areas where only certain persons are allowed.

All classified material should be handled according to government regulations, with particular attention being paid to storage, logging documents in and out, handling only by authorized personnel, restriction on reproduction of classified material, destruction of classified waste, etc.

Strict security of safe combinations and keys is mandatory. A record of the persons knowing safe combinations and holding keys should be maintained. Keys must be turned in upon transfer or discharge of such personnel, and the locks and combinations changed in such cases.

Careful investigation must be made of all unusual incidents, serious accidents and acts of known sabotage and even of those minor suspicious acts or incidents that may be part of a chain of acts of "passive sabotage" as described above. Such inquiry may lead to identification of the actors involved in such a chain of preplanned incidents and of their motives and may also forestall repetition or continuation of the acts.

Specially valuable precision instruments and tools must be kept in a locked tool crib and issued only on a charge-out ticket. All damage to such tools is to be reported immediately and investigated carefully, as a possible act of sabotage.

Firm, quick administrative and legal action should be taken against those found guilty of acts of industrial sabotage or, on a lesser scale, against those clearly established as reasonable suspect. Such action should be taken irrespective of personal connections, trade-union affiliations, or other repercussions.

Advanced detailed surveys must be made to identify all points vulnerable to sabotage, to set forth protective measures appropriate in the circumstances, to determine whether controls previously instituted are in fact being carried out, and to determine the need for corrective measures or additional safe-guards.

Creation of a counter-espionage organization is indicated wherever the risk of industrial sabotage appears real and substantial, as a means of keeping informed as to hostile elements, subversive labor union activities, underground cells, etc. This may include the development of well-posted informants or sources of information, attendance at outside meetings or rallies, regular review of literature emanating from hostile organizations, and perhaps the payment of incentive awards for prior useful warning of forthcoming acts of sabotage.

Effective employment of all the preceding and other counter measures presupposes the existence of a well-educated, experienced, trained, effective plant protection or security department or unit. Such an organization must be adequately mannered, well paid, properly motivated and of excellent morale. It must, especially in such matters as sabotage, communicate effectively with top management and have confidence that its sound recommendations will receive deserved consideration.

Another method is to establish top security or sensitive areas where the saboteur will find his most important targets. Some of these areas may be given extreme protection by examining all packages passing into them. Further, no package would be delivered to the sensitive area until its origin is certified. This can be controlled by not permitting the delivery of any package not covered by a requisition. Any box coming in without a requisition should be examined, opened and the contents identified before delivery.

Store and guard dangerous chemicals carefully as well as any particularly scarce or valuable materials and explosives, and inform guards of their locations. Purchase food supplies from reliable dealers only, and investigate all cases of food poisoning within the plant. Protect water supplies against pollution, or interruption in case of fire. Prohibit the use of liquor anywhere on plant premises, and see that smoking rules are rigorously enforced. Where smoking is permitted, it is suggested that it be forbidden during the last half-hour of work, so a butt cannot act as a delayed action fuse.

PLAN TO MINIMIZE SABOTAGE DAMAGE

The plans should be made which cover the action to be taken to prevent or minimize destruction of property and life where a sabotage attempt has occurred or an attempt at sabotage is suspected. These measures include the following:

1. Clearing the building or danger area of occupants.
2. Establishing a guard force outside the danger area.
3. Obtaining the services of a competent explosive expert.
4. Obtaining portable x-ray equipment when deemed necessary by the technical expert for detection of suspected explosives in packages, containers, etc.
5. Obtaining mattresses to be used as protection from flying fragments.
6. Insuring that adequate fire-extinguishing equipment is available.
7. Removing all inflammable materials from the surrounding area.
8. Shutting off all power, gas and fuel lines leading into the area.
9. Avoiding any article that may be connected with the bomb or which may act as a trigger mechanism.
10. Arranging for medical aid for the injured.

HANDLING BOMBS AND BOMB THREATS

In any discussion of the handling, disarming or disposal of sabotage bombs it must be realized that the exterior appearance of a known or suspected bomb gives little or no indication of the explosive used or the manner of construction. Both of these key factors are largely dependent upon the availability of materials and the technical skill of the saboteur. In view of the infinite varieties possible, it is obvious that no set procedure can be established for their handling. However, the primary consideration is the safety of life and property and there are certain basic rules which must be followed. Wherever the possibility of a sabotage bomb exists, the staff of the facility should be checked to determine who, if anyone, is qualified to handle the bomb. There must be a prearranged plan for coping with such an emergency so that the following steps may be carried out quickly and in many cases concurrently:

1. Immediately notify the security force headquarters which is responsible for establishing an operations headquarters.
2. Clear the area of all personnel and establish a guard around the danger zone. When it is necessary to evacuate a building, notify all persons therein that they shall leave and proceed quietly and in good order to the street. It is of utmost importance that nothing contained in the announcement generate the slightest element of panic. This notice should avoid the word "bomb," and stress the fact that employees are not to go to any building shelter as in an air raid, and that it is not a civil defense drill.

   If facilities are not available for a public address system type of announcement, selected employees such as telephone operators, security personnel or building service personnel should be predesignated to make the announcement personally.

   The greatest hazard to life and property in these bomb threats is panic. A poorly worded announcement or a frightened voice over a public address system or the telephone may be the necessary impetus to generate a panic resulting in injury or death.

   Once evacuation is completed, all doors serving as major entrances to the building should be locked and guards posted at such doors to prevent entrance or re-entrance of all persons except individuals concerned with the search of the building.
3. Shut off power, gas and fuel lines leading into the danger area.
4. Notify the police and fire department.
5. Secure mattresses or sand bags for use as protective shields and barricades. Sandbags may also be used in confining and directing the force of an explosion.
6. Send for technical help, military or police bomb disposal unit, or employees of the facility experienced in this

work. Except in an emergency presenting immediate danger to life, inexperienced personnel should not handle or attempt to dispose of any explosive material. Inexperienced personnel should take the above precautionary steps and await the arrival of qualified bomb disposal personnel.

7. Remove flammable materials and small objects from the surrounding area. However, anything that might be connected with the bomb or might act as a trigger mechanism must not be touched.

8. Arrange for the use of portable x-ray or fluoroscopic equipment which will be used by technical personnel only.

9. Search. The basic purpose of a search in case of bomb treats is to either discover the bomb or make certain that it is not within the building. Since persons who plan such destructive devices are often mentally disturbed, there are no "likely" places in which a bomb may be found. Therefore, the entire premises should be searched as thoroughly as possible, as soon as evacuation of the building is completed. Search parties should be organized by the security director. They will be assigned specific portions of the building, directed to report completion of the search in person at headquarters, informed that they will be held responsible for the thoroughness of the bomb search in their assigned area, and notified of the signal to be used for reassembly at headquarters in the event the bomb is found or for other reasons.

All search party personnel should be instructed to conduct their search based upon the assumption the bomb is present in the building and must be found. They will be cautioned to work alone in a room or hallway, being careful not to set off any triggering device. Light switches will not be touched, flashlights will be switched on before the start of search and not turned off until search is completed. They must use extreme care in examining movable objects for strings, wires and springs before disturbing them. They will also be warned not to move, touch or disturb the bomb in the event it is located, but to notify the headquarters for dispatch of the bomb disposal technician, guarding the area at a safe distance until such crew arrives.

In evaluating the areas to be searched, remember that a bomb will most likely be placed where it will do the most damage, or where a secondary effect of its explosion will create havoc. Such vulnerable spots might be near flammable materials, boiler rooms, delicate machinery, power generating equipment, critical structural points in the building, cafeterias or lunchrooms, or any place where large numbers of people gather.

10. In actual bomb discovery cases the security director should be guided by the report and recommendation of the bomb disposal technician. He shall take immediate steps to guard against any vibration in the area of discovery, or the use of magnetic tools or radio transmission.

Unless there is imminent danger to personnel in the immediate area, pulling the fuse should not be attempted. The risk of losing an individual life may be unjustified. If it is decided that the risk is justified, cut the fuse. Stamping it with the foot will not hinder its ignition. If knife or scissors are not available for cutting, pull it with speed and firmness. Do not hold the fuse in the hand after pulling it free. Throw it in a safe direction to avoid injuries from the cap detonator that may be attached to it.

If there is not time to pull the fuse, drop flat on the floor and shield the head and eyes. Advantages should be taken of any available cover, however scant, due to the possibility of the blast and heat passing over the body.

It is a fallacy to assume that a bomb should be immersed in water. If it should be actuated by an electrical timing device, there is a possibility that the water may complete the circuit by shorting it. Many types of fuses are water proof. Bombs that contain metallic sodium will burst into flame in the presence of water. One of the most satisfactory liquids to be used for this purpose is light machine oil which will clog the clock or timing works without causing a short circuit.

If a decision is made to leave the bomb undisturbed in its original position, the logical precaution to take is to attempt to deflect the blast, thereby decreasing its damaging effect. All objects immediately surrounding the bomb or touching it should be left undisturbed.

Since an explosion creates large amounts of gas and flame, all doors and windows should be opened wide to permit the escape of as much of this as possible, thus reducing the effect of the shock waves that will inevitably occur.

Materials such as sandbags and mattresses (minus springs) should be built around the bomb in such a way as to direct the blast toward the least vulnerable areas. These deflective materials should be built up so as to retain the greatest flexibility rather than a rigid structure. Flexibility permits absorption of some of the blast effect. This material should not ordinarily be placed over the bomb, especially in the case of the bomb being located on a second or third floor.

Simultaneously with the above operation, another crew may be safeguarding the physical facilities. All flammable materials that would otherwise add fuel to an explosion followed by fire should be removed. Pillars or supports of the building subject to blast should be protected with sandbags or mattresses. All power, gas, compressed air and steam or fuel lines should be cut off. Valuable machinery may be safeguarded by removal or covering in the same manner as the support structure of the building.

Fire fighting equipment should be positioned so that not a moment will be lost in quenching any resulting fire. Care should be taken, however, that the equipment is not placed in a dangerous area where it might be damaged by blast. The plant security officer who is in charge of the situation should see that the public authorities are notified. The bomb should be removed and destroyed.

11. When the search is completed, the security director should make necessary arrangements for the prompt return of employees and others to the building. The police personnel assigned will return to their regular duties. The security director should notify the chief of police by telephone of all the circumstances of the bomb threat. This telephone notification should be followed up as promptly as possible by a full written report.

12. In the event any damage results from a bomb explosion the security director shall take whatever action is necessary to put out any fires, to reduce the extent of damage, and protect employees and other persons from any hazards in the affected area of the building. He will also assign sufficient security personnel to interview witnesses, prevent thrill or curosity seekers from overrunning the bomb scene, secure evidence, conduct a general investigation into the circumstances of the damage and report all facts to him.

Chapter 25

THEFT

INTRODUCTION

T HE PROBLEM of internal theft is not only pressing and immediate but staggering in its scope and importance. Indeed it is impossible to obtain a truly accurate picture of the magnitude of the problem. This is because of the reluctance of many companies to report or prosecute because of the fear of adverse public relations or costly civil suits. In other instances the records of the company are inadequate and thus do not reflect the loss due to theft.

However, more than the loss of merchandise or trade secrets is here involved. Ignoring theft encourages laxity and more serious defalcations. Moreover, uncontrolled thievery will adversely affect employee morale and production. Indeed, employers must note the direct relationship between employee morale and employee dishonesty. Recognizing the connection is not enough. Management must do all that it can to dispel employee restlessness and discontent. This means that management should interest itself in the family and financial problems of its employees and do all that it can to develop high morale, for it is axiomatic that bad morale is the first indication of theft.

THEFT DETECTION

Theft can be detected by utilizing the following methods:
1. Check inventory shrinkage figures carefully from period to period. Substantial, systematic thefts may be the reason for a big rise in the shrinkage figure. Shrinkage figures for the industry, available from some trade associations, also serve as a useful yardstick to indicate when losses are excessive.
2. Look for evidence of stealing outside the plant too. Maybe

the products are being sold through unauthorized channels, or being offered at prices substantially lower than cost. It is probably the work of an employee who stole for purposes of resale. Because management knows so little about its losses, this is very often the first notice of theft.

Some plants go so far as to cultivate the acquaintance of pawnbrokers and scrapdealeers. It is one good way to insure quick reports of suspicious material that has been brought in.

3. Investigate abnormally high tool replacements. Check repeat orders for tools and materials indicating systematic theft or loose control.

4. Make spot inventory checks of tempting or easy-to-steal materials such as radio tubes, electric wire, small hand tools, antibiotic capsules. It is possible to spot skullduggery by concentrating on products that are most vulnerable to theft.

5. Keep close watch over danger areas, areas where thefts are most likely to occur. For example, sheet aluminum coming into the plant is hard to steal once it gets inside the building, so the place to watch it is at unloading docks and related handling areas at the point of arrival.

ANALYZING THEFTS

Not only must the existence of thefts be recognized, but an analysis of the character of thefts and the employees committing the offenses should be analyzed. Many plants have discovered that thefts have ups and downs during layoffs, vacations, on a seasonable basis. A large aircraft manufacturer, for example, had large losses of electric wire just prior to the Christmas season, and shortages of paint, rope and compasses for use in employees' boats during early spring. Also, it is good to analyze thefts to learn which work group is responsible. One company found that new employees tended to steal tools and long-seniority employees took materials. The type of product stolen and those engaged in the crime will suggest the most effective preventive methods.

THEFT PREVENTION PROCEDURES

Prevention first, detection and apprehension second, is the approach to the problem of thievery. By preventing theft, management not only eliminates many distasteful scenes involving employees who have been "caught in the act," but this also puts management on record as being interested in the efficiency and effectiveness of its personnel and minimizes the costly procedure of firing an experienced employee for stealing and the necessity of training another person for the job.

In order to prevent loss from theft, the following actions should be considered:

1. Educate the employees
2. Alert the supervisory personnel.
3. Screen job applicants carefully.
4. Improve the guard force.
5. Encourage prompt reporting of thefts.
6. Establish a control of all purchases, the receipt and storage of material. Improve inventory and auditing procedures.
7. Control and protect tools and supplies. Lock up valuable metals, materials, or equipment at shift end. Cribs should be provided for storing metals such as nickel anodes, babbitt, and solder. Stow away valuable tools.
8. Lend tools to employees for home use. Oftentimes employees steal because they feel they are entitled to a break in using plant tools and equipment and in buying company products. Many plants give employees a break by lending tools, selling or giving away materials, and selling slightly irregular products at greatly reduced prices. Some even go so far as to open up after hours for employees who want to use company tools and facilities.
9. Maintain careful checks on trash, scrap and salvage disposal. Supervise scrap trucks while being loaded. Check truck weights, both gross and tare. Lock up broken or scrapped tools so employees cannot use them for good replacements.

10. Institute a good central lock and key control system.
11. Check the package pass system.
12. Maintain a close inspection of all vehicles entering and leaving the plant area.
13. Improve the physical protection of the plant.
14. Control movement of employees in the plant after hours.
15. Obtain union cooperation by enlisting their support and advising them of your policy against offenders.
16. Increase insurance coverage. You may have the wrong kind of insurance, too little insurance, or no insurance at all. You may find you are not covered with your standard burglary policy if a loss occurs by key entry to the premises—no matter who the key carrier is. If you have a "deductible" policy, you cannot collect on small thefts, which can add up to considerable total.
17. Investigate suppliers' financial stability, credit rating, net worth, etc. This is important because an unstable company, or one of poor credit standing, could possibly fail in the middle of a production run and materials would not be available until new sources were obtained. From the standpoint of fraud, this investigation would determine that the particular supplier is an existing company.
18. Establish separate receiving and shipping areas. No foot traffic by employees should be allowed in these areas. Guards should be assigned to these areas to double-check all receipts and shipments and to observe all employees and all truckers. A record should be kept by the guards which shows the name of the trucking company making the pickup, the driver's name and chauffeur's license number, the make of the truck, color and the vehicle license number, both cab and trailer.

SUPERVISORS

Key men must not set a bad example by themselves engaging in petty pilferage or by winking at the practice on the part of others. Such men, as so often happens, tend to forget that every move they make is observed with interested attention by their subordinates. Their leadership obligation must be impressed upon them. In addition, they must recognize that it is their responsibility to prevent thefts by their employees and to report those who are guilty of a violation. To accomplish this it is important to overcome their feeling that theft in their department reflects poorly on them, their dislike in turning in an employee, and their belief that it is not an important problem.

One word of caution must be given. Encourage foremen to report losses or strange doings, but discourage them from becoming private eyes on company time. Catching thieves is no job for the amateur.

INVENTORIES

Accurate inventories are essential. Unless plant people are trained in taking a physical inventory, their lack of experience and knowledge will result in duplications, omissions, or improper counts. Use an outside firm of auditors to supervise the inventory taking. This will insure technical supervision by people who are concerned only with doing the job right. It will eliminate collusion among employees who want to cover up thefts.

Whenever possible, inventories should be taken by employees from other departments. In this way you are better able to forestall cover-ups of theft.

Insist that departments with an unsatisfactory loss experience take frequent recounts during the inventory taking period. Spot inventory checks should also be made at unannounced times.

Check inventory of precision tools monthly. Advance notice will give employees a chance to replace tools they "borrowed."

CONTROL OF TOOLS

Control of tools owned by the company is comparatively simple if fixed responsibility is placed on the employee who uses the tools. The tool crib should be locked and placed out of traffic. A window should be used for issuing tools and replacing those that may be broken. The tool room attendant should be held responsible not only for the maintenance of tools, but must also keep a card index of all tools issued to employees.

If the employee is transferred or leaves, the personnel department should not issue his last pay check or complete the transfer until a clearance is issued by the tool crib attendant.

Marking the tools with the company name or symbol will prevent the employee from claiming that the tools were his personal property. In machine shops, where large tools are supplied but not necessarily issued through the tool crib, the employee who uses such tools should be given responsibility for them. With this added responsibility, he will maintain a careful watch over them.

Quite frequently the complaint is heard that the employee had his tools stolen. To prevent such an alibi, the plant must provide suitable locked lockers in which he may store his assigned tools overnight or during lunch hours and breaks.

CONTROL OF STOCK

Those firms that must maintain a large supply of usable components and parts that are intended for incorporation into the finished product are presented with a constant threat of shortages. The first step toward the control of this situation is to cut down as much as possible the number of parts issued to the production line. Stock room partitions should be well constructed and should reach from floor to ceiling. They should never be left unattended and employees should not be allowed access to fill their needs at will.

The methods department can determine the schedule for the run on any particular section of the line. If the number of parts required is determined, it can easily be established how many parts will be required to run the line, say, until lunch break and excess parts will not have been exposed. Each position can be provided with a bin that can be locked by the operator during a break. This bin can then be refilled at the lunch break with enough parts to carry the operation for the balance of the work period.

All rejects should be returned and allowances made for the exact number of replacements required. If no provision is made for rejects, the control will die through the abuse of the reject alibi.

PROMPT REPORTING OF THEFTS

Continual emphasis should be placed on the prompt reporting of thefts. The security office should maintain a written record of all thefts occurring on the plant premises. The following information is of value in such a report:

1. Name and department of the person reporting the theft.
2. Names of all witnesses, supervisors, or other persons concerned or having knowledge of the theft.
3. Whether the property is owned by the company or by an individual, and if individually owned, by whom?
4. Approximate time and date of theft.
5. Place from which stolen.
6. Description of property stolen, including serial number, catalogue number, color, type of material, size, shape, age, etc.
7. Approximate value of property.
8. The method or means by which the merchandise was stolen, if known, including any conditions or circumstances which may have aided the theft.
9. Description and disposition of any evidence found at the place of theft.
10. Any other information that will aid in the investigation.

PROSECUTION

You can do something about stealing. You can make it tough to steal. You can let your employees know you mean business. Ignoring a theft is absurd, in fact it is equivalent to licensing theft. More than that, it is destructive to employee respect for the company and morale.

Some advocate criminal prosecution for all thefts, but most plants prosecute only the big ones. Nevertheless, all agree that occasional prosecution is a good deterrent to future crime. When the employees know you mean business, they think twice about stealing.

Prosecution for theft has certain definite beneficial effects, among which are the following:

1. Enhances respect for management.
2. Improves employee morale.

3. Encourages honesty and discourages others from criminal activity.
4. Attracts a better grade of workers.
5. Identifies thieves and thus protects others from their theft.
6. Protects the liquidity of the company and thus the investment of the stockholders and the jobs and salaries of the employees.
7. Helps the thief himself.

Most employee thieves are basically honest but have gradually slipped into a pattern of crime. The prosecution of a culprit will snap him out of the pattern, whereas merely firing will encourage him to continue the activities with a new and unsuspecting employer, for he has in effect "gotten away with it."

If the guilty employee is not prosecuted, he should be fired no matter how small the "take." Indeed the company policy in this regard should be made known to employees as soon as they are hired. Do not bow to pressure from veteran, political, or religious organizations; you are bound to get it in some cases.

UNDERCOVER OPERATIONS

The nature of sabotage, espionage and theft is such that consideration must be given to the development of an intelligence gathering organization. Thus it is that modern plant protection requires both uniformed and plainclothed agents, each charged with different protection tasks. In most cases the two complement each other. In many, they rely upon each other for maximum security.

The uniformed guard acts as an ever-present deterrent to petty theft, protects the plant from outside vandalism and intrusion and symbolizes the awareness of management that protection is necessary. For large scale larceny protection, of course, only undercover agents normally fill the bill. Unknown and unsuspected, they can normally discover more about criminal operations in a fraction of the time required for uniformed guards to unearth the criminals.

The undercover agent is ostensibly a production line worker, mail clerk, sales girl, or what have you. Like other workers, he collects a salary from the company for the job, but he is also a trained observer and actor. He picks up a fact here, a rumor there, carefully building an information background one piece at a time. Ideally, the agent becomes an "active" member of the conspiracy to obtain complete details on its membership and operations.

Surveillance also plays an important part in such operations. In this instance the agent, as the name suggests, keeps locations or subjects under constant surveillance in order to observe criminal activity and to ascertain those who are implicated. Part of such an operation can consist of an investigation of employee expenditure inconsistent with their known financial resources, as conspicuous overexpenditure, is often the first clue to the pilferer and embezzler. Credit checks of employees, particularly those in sensitive positions (cashiers, bookkeepers, money-handlers, shipping and receiving clerks) should be made through recognized credit exchanges from time to time. In all such operations it is of paramount importance to avoid giving the impression that a spy system is being set up. Instead, it should be stressed that security will improve a company's income, profit and employee benefits.

The most valuable informers are plant employees. Employees are familiar with all phases of operation, and they are cognizant of weak points where products and materials may be stolen. One of their best qualifications is that they become familiar with the personality, family problems, money difficulties and the character of their fellow employees.

It is necessary to understand the motivation for informing. Human beings who inform on the activities of their fellows will have one of the three following motives, or a combination of them: (1) loyalty to country or employer; (2) personal gain, money, recognition or advancement; or (3) revenge, or the desire to see other people in a "stew."

Of the three types, the person motivated by loyalty is the one to be most desired and utilized. The employee who informs for personal gain must be picked with the utmost caution. This is also true of the revenge type, who will have no scruples against supplying entirely misleading or false information just to see someone else suffer. Of these last two categories, the most reliable is the type who desires a monetary gain in addition to his regular wages, but who will, at the same time, give straight information.

In selecting an informer, careful consideration must be given to just how valuable he will be. This requires an evaluation of his work, location and mobility. Further, he must be alert, observant and intelligent enough to obtain pertinent information. Before selection, a careful investigation must be made into the background and associations of the individual.

THEFT INVESTIGATION

The general attitude of investigators should be that of a firm determination to get to the bottom of each case with which they are confronted and then to present their findings in an honest, fair and factual report, without bias or prejudices. Such reports will serve to establish guilt, determine future action to be taken, recover stolen property and act as a deterrent to others from similar activities.

The first step is to determine if in fact a theft has been committed. This is done by examining all controls that affect the loss to learn whether the procedure has broken down, if the loss is merely a bookkeeping error. Check these points:

Was the material or item involved actually brought into the plant? When? What happened to it? Who received it? What papers were signed proving that it entered the plant?

If material, was it placed in the store room or stockpile? If tools, did they actually get into the tool crib?

Where was item when last seen? Was it a stock item?

What is the number of items used daily in production? If parts, how many are normally used in a day, a week, a month?

Who has access to the material or item? How is it handled? Who is responsible for it?

If a theft has in fact taken place, no uniform pattern of procedure can be outlined, but the following steps are suggestive of those that can be taken to apprehend the thief:

1. Determine what is missing. Get a complete description of the stolen articles.
2. Determine from where, when and how the property was

stolen.

3. Determine whether the theft was probably for personal gain (resale) or for personal use, and if for resale, where it would most probably be disposed of.

4. Question supervisors and employees working in the area and check the files for records of similar losses in order to obtain all the information possible. Also check the local police, if it is considered advisable.

5. Get complete up to date data on each employee, including key personnel in affected departments.

6. Check with supervisors of the affected departments and obtain their views and suggestions as to who is the thief.

7. It may be advantageous to institute an undercover operation and surveillance. At other times, because of the nature of the thefts, ultraviolet powders, crayons, pastes, or inks might be used to identify the culprit.

8. Question employees and try to obtain any leads possible.

9. Learn as much as possible about the character, habits and living conditions of the suspected person. For this purpose the personnel files, police sources, banks and credit unions, credit and loan companies, or other sources may be used to advantage. Particular attention should be paid to evidence that the individual gambles or has overextended his credit and is living beyond his means.

10. Watch for the reactions of all suspects to the progress of your investigation. Those involved will become wary and cautious. Their manner and conduct will hint that they are worried. The supervisor or foreman is in a position to observe this and can be of great assistance in applying pressure indirectly. He can, for instance, casually state that something is about to break.

11. Then "put the heat on." Enforce management controls to the limit. Insist that all procedures be rigidly followed. All controls and procedures should be checked and double-checked. Be sure to close the loophole that permitted the theft. Spot inventory checks should be made of various departments—even night checks should be made of watchmen and guards. All entrances and

exists should be inspected to make sure they are properly locked

12. Now question all employees who work with the suspects. You should determine what they have observed, what they know, and what they believe about the case. You will be surprised how much the employees know about each other, but they will only divulge it when they consider it necessary to protect their own neck.

13. Call the prime suspects into your office if you have sufficient data to lead you to believe they may be the guilty ones. Go over their entire occupational background. Show in your conversation that you are well aware and fully informed regarding their habits, mode of living, and how involved they are financially. By careful handling at this stage the suspect will reach the conclusion that he has been under observation and that management knows much more than it is saying. He will feel the game is up, that there is nothing to be gained by lying.

Remember two points. Never accuse any employee directly. Let him do the talking. Let him answer the questions. Just watch his reaction. It is not difficult to see the emotional strain a person is under if he is afraid of something.

Force will get you nowhere. Your man will clam up if you start to get tough. He will think you do not really have anything on him, that you are trying to scare him into talking. If the employee confesses, his confession should be reduced to writing. Have the suspect sign each separate statement made. Also include the time, date and place the statement was given and all witnesses' signatures. If a statement takes more than one sheet of paper, each sheet should be signed.

14. Consult the legal representative of your company to protect your plant against any possible action by the employee that may be against your interests. A wrong step at this point has often resulted in a suit by the employee against the company. He can claim false ar-

rest, duress. You will not only lose your man, but may find yourself a defendant in a court action.

Reports should be made to the local police, especially in theft cases wherein the articles stolen are readily identifiable by serial numbers or other peculiar characteristics. These articles may turn up later in local pawn shops or be recovered through police investigations in other cases.

If the evidence against the suspect is adequate and the company wishes to prosecute the case, the complaint should be signed in the name of the company and the following items delivered to the prosecuting attorney:

1. Copy of the suspect's signed statement.
2. The merchandise involved (if recovered).
3. Copies of statements of witnesses to the theft and investigator's findings.

In the case of the professional thief or employee with a criminal record, it is another matter entirely. After a preliminary inquiry the local authorities should be called in without delay. Then the insurance or bonding company should be notified of the situation.

If management intends to prosecute, it should also call in the local law enforcement agencies. The sooner the law is advised of the situation the better.

Chapter 26

STRIKES AND DEMONSTRATIONS

INTRODUCTION

THE BASIC FREEDOMS guaranteed by the Constitution of the United States give workmen the right to refuse to work under conditions believed to be unfair. These same freedoms give the employer the right to defend his employment policies and work standards. Such disputes at time will become most bitter. This is because a labor quarrel is concerned with fundamental human rights. It is a contest between the workers, who seeks a living wage, and management, which is endeavoring to obtain a fair return on its invested capital. To both sides, then, the issue involves standards of living, prospects for security and, to a certain extent, the very right to live.

Preservation of public order during a strike is the joint responsibility of the disputing parties and the police organizations having jurisdiction of the places involved. This requires a great deal more than simple respect for legal process. It implies a positive duty on the part of the disputants first, to act from a sincere resolution to avoid the kind of situation which converts a strike into a battle, and then to cooperate with public enforcement authorities who must actually police the public aspects of the dispute.

Not all strikes and labor disturbances result in violence, but most, at some stage, manifest a potential for it. The violence erupts when (1) bitterness exists between a large number of strikers and the struck employer; (2) violence is deliberately employed (or encouraged), either as a direct strike weapon or as a basis to justify counter-force or restraint, and (3) an incident or misunderstanding momentarily inflames the persons then and there involved. The last of these is the most troublesome as it is the hardest to anticipate. A garbled announcement,

416

an insulting remark, or the appearance of some person who is symbolic in the dispute can easily provoke a sudden outburst. Such flareups can be devastating if not immediately suppressed because they trigger the pent-up aggressions of all persons on the scene and can, in a terrifying brief moment, swell to riot proportions until the charged emotions are spent.

OBJECTIVES

The first step is to recognize the objectives of the parties involved. Labor will resort to a cessation of operation of a plant or industry in order to pressure management into granting demands. To accomplish this, the union will resort to the following:

1. Picketing premises.
2. Preventing and discouraging the movement of raw materials or finished products into or out of the plant.
4. Subjecting employees who enter to embarrassment and harassment.
5. Slow down the operation of the industry.

Management on the other hand has as its objective the maintenance of operation of plant or industry. To do so, it will try the following:

1. Encouraging employees to resume work.
2. Maintaining free ingress and egress of raw materials and finished products.
3. Replacing striking employees.
4. Attempting to limit picketing and other union activity by injunctions.
5. Maintain at least partial production.

The police have as their objective the following:

1. Maintenance of law and order.
2. Protect life and property of all parties.
3. Protect civil rights of all involved in the public interest.

The police can realize these objectives only if they are instilled with the attitude of neutrality and responsibilities of their duties. This does not mean that they will blind themselves to crime or wait until injunctions or restraining orders are issued by a court.

Independent police action is not only permissible but necessary when violence erupts or force is used. It is not necessary that police await a judicial decree to quell a disturbance. If assaults on the person or trespass against property are committed, or if thoroughfares are blocked, or the public denied any right of access, then prompt action is essential. The parties may still have recourse to the courts for further clarification of their rights in an injunctive order or denial, and may argue in the calm of a courtroom the need for judicial intervention. Police will also be guided by the order, if any, but they need not wait for it in the face of violence. The public right to peace is paramount and must be enforced against parties to a private dispute.

PLANNING

Mass picketing of an industrial facility, accompanied by a work stoppage, is one of the most serious types of actions that confronts the professional security officer today. To meet this challenge there must be a complete prearranged flexible program approved by company officials and under the control and direction of the security department which can be immediately activated. It should be coordinated with that of the police authorities. The police should also be asked to review and approve the plan and operational tactics.

The detailed approved plan should be known to only selected top officials of the company. It should be in written form and in sufficient quantity for immediate distribution to all affected individuals when the need arises. This plan will cover the following:

Security Force

CHAIN OF COMMAND. There must be established a specific chain of command with particular individuals assigned to perform designated duties and with well defined authority. Arrangements must be made for a command post.

MANPOWER DETERMINATION. The plan must provide for necessary manpower for all emergencies. There is no set rule of thumb as to the number of uniformed guards to be assigned to any plant or area during a strike. The factors to be considered are the value of the property within those confines, the threat to its security, the potential for violence, the need to protect workers who will continue to be employed. Based upon the evaluation of these factors, a decision can be reached on the number of persons needed. There is no such thing as being overmanned in the case of industrial property protection; this is one of the cases where it is far better to be safe than sorry.

AREA OF AUTHORITY. All guards should fully understand their jurisdictional rights and restrictions both on and off company property. Security forces should confine their attention to police activity within the confines of the struck facility, such as directing interior traffic, protecting sensitive locations and preventing trespass on the employer's property. The perimeters, entrances and access routes should be handled exclusively by the police.

Company or contract guards should be cautioned to have no contact, verbal or otherwise, with the strikers. Everything possible should be done to avoid and prevent clashes between strikers and guards. A buffer zone should be set up between the legal property boundary and some arbitrary line further inside. No vehicles should park or persons gather in that zone. This has the double advantage of further separating hostile parties and of clearly exposing for prompt apprehension anyone committing an illegal act in the zone.

WEAPONS. Side arms should not be worn or displayed during a strike. First, it is not likely that use of a deadly weapon will be needed. While crimes may grow out of labor disturbances, strikers are not lawbreakers. Should a serious crime be committed, the police present will be responsible for apprehension. Second, the chances that a guard may become involved in a melee and be disarmed are great enough to warrant caution. Allowing a deadly weapon thus to fall into unauthorized hands would not be in the public interest. To afford a means of personal protection, the guard might be supplied with hand gas dispensers or a standard police type baton is suitable, but the caution to be added here is that it not be needlessly displayed.

TRAINING. Special attention should be given in security force training to the unique problems of labor relations. This should include a knowledge of the rights and interests of both parties to such disputes, the duty of law enforcement, the problems and the methods to be employed. Much can be learned from films of other labor disturbances in the same community. The sight of others in action will convey more of the chaotic, fast paced strike action and the pressures it produces than mere discussion. It will also portray the manner in which typical strike connected violence is perpetrated.

COMMUNICATIONS are vital. Guards should be linked with each other and with the security chief by walkie-talkie radios, and the network should also link key management people in an effective communications system.

Routes

In selecting routes to premises within several police jurisdictions (a plant cut by county or city lines, for example), preference should be given to the route through the jurisdiction with the largest force as it is better able to meet the great manpower demands of such disturbance. If the enterprise has a system of interior roads, all vehicular movement from place to place should be by these roads as much as possible. This may require that cars admitted at one point be allowed to move to a distant interior location for passenger discharge and parking. In this way contact between workers and the strikers is reduced and thus the likelihood of violence is also diminished.

Entrances

As the gates and entrances are the points of contact between strikers and nonstrikers, these are sites of potential violence. Therefore, the employer should examine his premises to determine the fewest open access ways necessary and should close all those not needed to continue operation during the strike. In this way, maximum use may be made of the limited security and police power available as they can be concentrated at these points rather than spread over all the entrances.

Car Pools

If the struck enterprise has many employees who drive to work, they should be encouraged to form car pools to reduce the traffic flow.

Security Measures

Adequate provisions must be made to secure and protect all company property including motor vehicles and related equipment and other facilities that are outside the main plant. In the industry this would include unattended central dial offices which contain switching equipment, power entrances, manholes, repeater stations, microwave towers, equipment yards and a multitude of other facilities including telephone poles, wire and cable.

Building Security

A great deal of vulnerable glass in plants with labor trouble often results in many broken windows. Unprotected windows are vulnerable to rocks and bombs. Security oriented architects should design into building exteriors some modern form of the old-fashioned shutter. If a shutter is impractical, one-half-inch to three-quarter-inch mesh burglar screen welded to a stout steel frame will cover and protect the glass while letting air and sunlight pass. The shutters or screens could be detachable and used only when trouble arises.

Lighting

Permanent and portable floodlights should give twenty-four hour exterior visibility. During the nighttime hours, portable floodlights should augment the normal exterior lighting, to insure that the perimeter is clearly visible at all times and at all points.

Patrol

Patrol officers (either contract or company guards) should maintain a roving patrol within the fenced area. Other guards should be stationed at the company's gates.

Delineation of Boundaries

Strikers, for all technical purposes, are no longer employees once a strike has been undertaken. If found on the property after that, they could and should be considered as trespassers. Determine the property boundaries. Post signs which would convey the massage that the area lying beyond is private property of the company and trespassers will be prosecuted. Make certain in the disaster planning that a plot plan of the boundaries is displayed on bulletin boards and discussed in meetings involving foremen or leadmen.

Injunctions

Special consideration must be given to the legal problems that will arise during a strike and the legal action that can be taken to aid in maintaining security. Since the potential for violence increases in proportion to the size of the group of pickets, it is advisable either to reach a voluntary agreement limiting the number of pickets or to take legal action to obtain a restraining order. It is also possible that other methods of harassment may be employed that will call for similar restrictive legal action.

Care of Employees

Every effort must be expended to protect the autos of employees parked on company premises. The welfare of non-striking employees must be provided for. This should include, where necessary, temporary housing, food, clothing, medical services, transportation and financial assistance.

Briefing

Through management all non-striking employees including regular employees and temporary employees must be fully instructed about their duties and responsibilities. They should be reminded that employees have the constitutional right to engage in an economic work stoppage and to engage in peaceful picketing for a lawful objective. In addition, non-striking employees should be instructed that they should not engage in arguments, name calling or exchange of insults with pickets, should not issue threats of reprisal to picketing employees, should not expect them to return to work by promises of benefits and should not engage in any activity that might be construed as a violation of fair labor practices.

Log

Maintenance of a full and complete log or record of pertinent activities is the principal function of a security organization during a work stoppage.

Cameras

Using zoom lenses, maintain constant motion picture coverage of the pickets. At the first blockade or stoppage of vehicles, or intervention with personnel entering or leaving the plant, these films should be turned over to the firm's attorney for his use in obtaining an injunction against illegal activity of the strikers.

If violence flares, people may be arrested. Then the question of who did or did not do what will be confusing. Several officers should carry cameras which should be as simple as possible but capable of taking good flash pictures of moving people. Several men with movie cameras should be stationed every seventy-five to one-hundred feet on the building roof to record any disturbance. The presence of cameras will often deter wrongdoing. This is true even if the cameras are empty or the photographers inexperienced, for, naturally, the strikers are ignorant of that fact.

Illegal Acts

Unfortunately, during work stoppages emotions frequently run high, sometimes leading to unruly and even illegal activities. In view of this, the company security department should be particularly alert for incidents of threat of violence, actual violence, blocking entrances to company properties, blocking public thoroughfares, damage to company property and other incidents of illegal activity. If incidents of this nature are reported, the company legal counsel should be promptly informed in order that necessary relief through legal procedures can be undertaken.

Arrests

If illegal actions occur within the employer's premises and he has a security force, initial response, including apprehension, will generally be by that force. The legal liabilities and consequences of action taken by the employer are his responsibility. If security guards apprehend, there must be one who will swear a complaint affidavit to permit police to assume custody. Such action should be avoided if at all possible except where the person apprehended is patently attempting or committing a felony or a misdemeanor with serious possible consequences.

COORDINATION AND LIAISON

Coordination of effort is important in planning as well as in meeting the actual problems posed by a strike. Therefore, liaison with officials within the company and outside of the company is proper and should be encouraged.

Advance plans should provide for the prompt notification of and coordinated emergency action by all officials and key personnel. Local law enforcement officials, legal counsel, suppliers of material and services and guard forces should be promptly informed of the work stoppage and the picketing. The manager of industrial security should have freedom of access to top management people and the legal section, not only to ask their assistance but to give them help in turn. If conditions require the legal department to take court recourse during the height of the problem, they should notify the security department and obtain its advice. Liaison with the police department is a "must" and the exchange of field intelligence reports would be invaluable in planning manpower distribution from both sides.

Because the employer will have regular names and designators for gates, lots, roads and other interior locations known to the striking workers as well, police should be given a detailed map or plan of the premises in which all such data are set forth. Police commands and orders can then use the same language and the risk of error or misunderstanding will be lessened. If the employer maintains a security organization, the strike deployment can also be indicated so that the police commander knows the location of all enforcement resources.

Not to be excluded is liaison with top union officials because of the need to call to the proper authorities' attention violations or possible impending difficulties that these officials might be able to prevent.

During the strike all employees should also be kept currently informed on at least a daily basis of all pertinent developments and of the general progress of union and company negotiations. Coordinating meetings should be held between the security department and company officials, at least on a daily basis and if needed, more frequently. in order that complete and accurate information can be exchanged concerning the progress of negotiations, serious incidents, extent of work being performed, public reactions to the work stoppage and other pertinent data.

INTELLIGENCE

A distinct advantage to plant security in handling the problems of a strike is foreknowledge of the strike. Such advance warning allows the gathering of intelligence well before the actual disturbance. This permits the formulation of plans of action if and when a disturbance occurs and also alternative plans to cover exigencies. The following general areas of inquiry and suggested check list should prove of assistance in procuring this information.

Initial Strike Report

The first information needed is knowledge that there is or will be a labor dispute. Every facility should have a definite plan for reporting the possibility of a strike to its security department long before it commences. Further provision should be made for notifying the local police of the strike. This plan should assign the responsibility for making such a report and designate the individual who is to receive such reports.

The report to the police should contain the following:

1. Name, busines address and telephone number of employer.
2. Name and address of union, union local number, affiliation, and telephone number.
3. Kind of business.
4. Number and occupation of employees involved in the dispute.
6. Date strike declared.
7. Number and occupation of employees who will continue to work.
8. Number of pickets.
9. Trouble anticipated.
10. Kind of strike (sympathy, wildcat, lockout, secondary).

11. Any additional factors which would aid in determining the number and kind of police details required.

Detailed Dispute Report

To evaluate the potential problem properly, it is fundamental that as much as possible be known by security personnel about the parties to the dispute, their disagreement and the relationship between them. The following check list will aid in obtaining the pertinent data:

I. Disagreement
   A. When does the current contract expire?
     Date ————————————— Hour —————————
   B. When will the strike begin?
     Date ————————————— Hour —————————
   C. Reason for dispute.
     1. Economic (salaries, hours, conditions)
     2. Improper conduct of one of the parties.
   D. Status of negotiations if possible.
   E. Expected duration.
II. Employer
   A. Identification
     1. Name.
     2. Address.
     3. Telephone.
     4. Liaison representatives.
     5. Home address.
     6. Telephone.
     7. Products manufactured.
   B. Nature of employer
     1. Single location.
     2. Multiple locations.
     3. Subsidiaries.
     4. Parent.
   C. Plans
     1. Will the struck facilities be operated?
     2. What will be the extent of operations?
     3. Total number employees and an estimate of workers who will report despite strike.
     4. Time of arrival and departure of employees who will not strike.
     5. Transit facilities and routes used.
     6. Entrances and exits used by employees.

7. Time when merchandise is to be received or shipped.
8. Meal periods for employees and whether they eat on the premises.
9. Exits, entrances and routes used by employees during meal periods.
10. Special hazards or other conditions effecting police duty (e.g., loose debris that might be used as missiles).

D. Physical Plan, location and nature of struck premises.
E. Security
    1. Security or protection officer.
    2. Address.
    3. Telephone
    4. General security plan.
F. General Attitude

III. Union
  A. Idenification
    1. Name and local.
    2. District or region.
    3. National.
    4. International.
  B. Union profile
    1. History of union conduct in past strikes.
    2. Character of members.
      a. Skill.
      b. Seniority.
      c. Wage level.
      d. Education.
  C. Officials
    1. Name of president, secretary, strike manager.
    2. Address and telephone (business and home).
    3. Picket captains (address and telephone).
    4. Command post (address and telephone).
  D. Equipment
    1. Sound trucks.
    2. Public address systems.
    3. Permits for use required.
  E. Picketing
    1. Number of pickets.
    2. Locations.
    3. Times.
    4. Captains.
  F. Tactics
    1. Marches
      a. Caravans.

          b. Motorcades.

          c. Parades.

      2. Harassment.

      3. Boycotts.

  G. Financial arrangements

      1. Feeding.

      2. Housing.

      3. Benefits.

  H. General attitude

IV. Relations Between Parties

  A. History of previous relations.

  B. Character of recent relations—strained or amicable.

  C. Existence of alleged grave wrongs that popular sentiment will demand be corrected.

  D. Other influences, groups or individuals that would benefit by a prolonged and violent strike—their structure, leadership, aims, tactics, power.

Physical Survey

If it appears that there may be violence, a survey should be made showing the following:

1. Location of the plant, number of exits and entrances, loading patforms, etc.
2. Other buildings or location which might be affected by the dispute.
3. Time of arrival and departure of employees who will not strike.
4. Transit facilities and routes used.
5. Entrances and exists used by employees.
6. Meal periods for employees, and whether they eat on the premises.
7. Exits, entrances and routes used by employees during meal periods.
8. Time when merchandise is to be received or shipped.
9. Special hazards.

Evaluation of Reports

Once the basic information has been gathered, it must be assembled and evaluated and a preliminary assessment made of the possible severity and duration of the strike. In addition to basic identifying information, it should evaluate the following:

1. Nature of dispute.
2. Nature and history of the relationship between the parties.
3. Nature and temperament of parties
   a. Employer.
   b. Union.
4. Capacity of parties for disorder
   a. Employer.
   b. Union.
5. Degree of threat to public place
   a. On part of employer.
   b. On part of union.
6. Need for police action to protect the facility and the public.

PRE-STRIKE CONFERENCE

The security director should contact and confer with both the police and the union before there is an open conflict. This call made prior to the time of the dispute, is at a time when the pressures that accompany a strike have not yet developed. Such meetings are concerned only with public order. That tenor should immediately be set. The responsibility of all parties should be understood. All should pledge their cooperation in maintaining the peace. If possible, agreement should be reached on the methods to be used.

Ground Rules

Experience shows that if all parties concerned know exactly what is going to happen on the street at the scene of a strike, there is little likelihood of misunderstanding which might lead to trouble. It should be established that the following ground rules will apply:

1. That force or violence will not be tolerated.
2. That the law will be enforced with strict impartiality.
3. That the rights of the public using the streets and sidewalk will be protected.
4. That unlawful conditions or acts which lead to disorder will not be tolerated.
5. That the employment of professional bullies and thugs will not be sanctioned.

6. That activities of professional agitators will not be allowed.
7. That no parties to the dispute may use language or manner of address which is offensive to public decency or may provoke violence.
8. That the rights of striking employees to conduct orderly picketing will be fully protected in accordance with the circumstances and conditions existing at the location.
9. That the number of pickets to be permitted should be established and adhered to.
10. That striking employees may picket in the vicinity or in front of the place of employment to do the following:
    a. Persuade those still employed to strike.
    b. Persuade those considering employment not to do so.
    c. Inform customers about the labor dispute.

It should also be agreed between the parties that each will be responsible for supervising his own partisans and enforcing the agreement among them at the scene of the dispute.

Mutual Trust

Even more important than establishing ground rules is the establishment of mutual trust. Experience shows that the one thing that determines which will prevail, peace and tranquility or strife and turmoil, during a labor dispute is distrust by all parties concerned. Management does not trust labor, labor does not not trust management and neither trust the police.

Coordination

Arrangements should be made for maintenance of close relations during the disturbance. The union, management and the police should designate definite individuals who will be responsible for keeping the others advised and for receiving communications. In a large organization with a formal security force, the director or manager of security will usually represent the employer on this matter. The strike committee chairman or chief picket captain can act for the union.

CRITICAL PERIODS

Certain critical periods exist in any strike situation. The first few days are critical periods. At this time the tone of the strike is not determined, and the labor and management strategy and the number of pickets to be utilized are not known. At the very outset both labor and management will endeavor to use the police for their best interests in carrying out their objectives.

It is also during this period that drinking may occur at the scene. Union officials and management will quickly assess the type of enforcement that is exhibited at the scene. If they feel that the police are unbiased and intend to maintain law and order, there will be little or no trouble. By the same token, if they recognize that the police department is biased, and not too sure of itself and its duties, it follows that an attempt may be made to take advantage of this apparent weakness.

Each succeeding Monday is a critical period, as usually a show of strength is made at this time. The second Monday is more critical, due to failure of strikers to receive their usual pay checks. One of the most critical periods arises when some strikers start back to work, as this often develops into physical clashes between strike-breakers and persons on the picket line.

PICKETING

Whenever a union calls a strike, it is the normal procedure to establish pickets or a picket line outside of the struck facility. Picketing is patrolling by union members or sympathizers near an employer's place of business to publicize the existence of a labor dispute, to persuade workers to join the work stoppage, and to discourage customers from buying or using the employer's goods or services. The right to picket peacefully and truthfully is one of organized labor's lawful means of advertising its grievances to the public, and, as such, is guaranteed by the Constitution as an example of freedom of speech. Thus, peaceful picketing is permitted during either the daytime or nighttime. It is the duty of law enforcement to protect the rights of pickets to establish picket lines and to uphold their right to picket legally.

Illegal Picketing

The constitutional guarantee of freedom of speech extends no further than to confer upon workmen the right to publicize the facts of an industrial controversy by peaceful and truthful means. In case of a resort to acts of violence, physical intimidation, use of threats, force, or false statement, picketing loses its character as an appeal to reason and becomes a weapon of illegal coercion. Picketing becomes illegal if it blocks streets and roads or interferes with the free and immediate use of the sidewalk or with ingress and egress to any place of business.

Placards

Placards of any size may be carried while picketing. The use of placards in labor disputes in congested areas where pedestrian traffic is heavy, such as downtown areas, poses a problem. It is suggested that the union official be made aware of the inconvenience to the public. The usual practice is to call on the uinon concerned and request that sandwich-type signs (the width of the body) be used.

If it is alleged that placards or banners are of a defamatory nature, redress is in the civil courts.

Mass Picketing Problems

When a strike develops, the union encourages full turnout of striking personnel to be present at the strike location. This then becomes a crowd control problem for law enforcement.

In some cases, the union might encourage mass picketing in front of personnel gates in order to keep these gates closed to those persons desiring to use them. It is the responsibility of the police commander to have his personnel deployed prior to this buildup. The officers detailed must keep the gates from becoming jammed and must keep an open passageway for those persons desiring to move in or out of the plant.

Officers must also observe closely the strategy in the picket line. Orders are passed through the line in whispers, and the picket line may move very slowly and surreptitiously closer to the gate. Officers must keep pickets moving. Under no condition allow the gate to become closed. Officers must move the pickets back before the passageway becomes jammed. It is difficult to disperse the crowd once a gate is closed, and an attempt to reopen the entrance may lead to a clash between the pickets and officers.

Another problem of mass picketing is the traffic hazard caused by a crowd overflowing into the street. The officers must keep these people as close to the curbing as possible so a traffic hazard will not be created and pickets will not be injured by passing vehicles.

Handling Movement Through Picket Line

One of the principal causes of trouble at a strike scene is the movement of pedestrians and motor vehicles. For this reason, at the very outset of the dispute, the parties must be informed of their rights and the legal limitations on their action. The pickets must realize that all persons and vehicles have the right to enter and leave at will. It is the duty of police officers to see that persons have the opportunity to enter and leave the plant, and to keep the peace when this is done.

The suggested method is to allow a person—the union official or his delegate—to talk to the vehicle driver. This talk should occur in the presence of an officer. Pickets should have the right to present their case to the truck driver—that they are on strike and have established a picket line. They should be allowed a reasonable time to present their case so the driver may make his decision. If advice is requested from the officer at the location, he must be careful to notify the driver that he may stay out, or enter the plant. Preferably, this should be done in the presence of the picket captain. Under no circumstances should the officer advise the driver to enter the plant. The driver must make his own decision whether or not to enter the plant. If the driver has decided to enter the plant, it should be made clear that upon his leaving the premises, a convoy will not be provided for him. In order to avoid creating a traffic hazard, care should be taken that vehicles are not stopped in the street. Arrange for the vehicle to park, so the normal flow of traffic will not be interrupted.

If the driver decides to enter, the picket captain should be asked to clear his pickets from the path so that the vehicle may enter. Allow the pickets adequate time to disburse. If the persons still refuse to move, labor leaders and possibly the central labor council should be contacted and told of the problem and asked to alleviate it so that police action is not necessary. If the picket captain fails to comply with the request, or the pickets do not respond to his orders, the officers should open the lines without hesitation and allow the movement of persons or vehicles. The line should be broken only temporarily. When broken, sufficient clearance should be made to allow for safe passage.

When the picket lines are opened by officers, the officers should face the pickets rather than the persons or vehicles entering the gate. This action affords the officers the opportunity to observe the actions of the pickets and prevents the possibility of assault being committed, or damage being done to vehicles entering the plants. This affords an opportunity also for the officers to identify any pickets who might commit a crime.

If the picket line is small, the officer's caution to the group to "watch the cars" will, in most cases, suffice. The use of hand signals by the officer, normal procedure in working traffic, is always interpreted as directing the vehicle to enter struck premises. This should be avoided in most instances.

Trains Entering a Strike Location

One of the most serious problems which faces law enforcement at a strike location is that of trains entering the location. It should be made clear here that railroads are involved in interstate commerce. Therefore, deliveries of materials will be made to a plant and finished products will be removed. In many cases this creates misunderstandings with representatives of labor, as they feel law enforcement is being partial to both management and railroad. In the past a procedure was developed by the railroad to accomplish this movement of trains. The railroad representative always notifies the station or commander in advance, and establishes the time the shift will be made. Special agents of the railroad are detailed to the location to assist with the railroad movement. Inasmuch as railway personnel are unionized, they disembark at the location and a supervisory train crew takes over in order to move the cars into the plant.

The commander at this time notifies the picket captain that the train will be moved in, and requests his cooperation in removing the pickets from the railroad tracks. If the pickets refuse to move, the railroad special agent will require the strike commander to remove the pickets. Arrangements should be made beforehand with the special agent to allow the picket captain and pickets time to discuss the situation before the special agent requests the officers to remove the strikers. This precludes charges of the union that labor did not have an opportunity to present its case, and that officers are siding with management and the railroads.

In the event it is necessary to remove pickets, officers should be certain that they are removed from the track only so the train can be moved in. Pickets should not be removed from adjoining railroad property, as this is a problem for the railway special agents.

SPECIAL PROBLEMS

Security personnel and police officers assigned to a labor-management dispute should be alert and observe unusual acts requiring police action such as the following:

1. Putting sugar in gas tanks.
2. Cutting ignition wires, etc.
3. Breaking into railroad cars.
4. Cutting tires.
5. Threats.
6. One side following members of the other side to their homes and threatening them or members of their family.
7. "Palming" or concealing in the sleeve sharp objects such as can openers which are scraped along the sides of vehicles passing a bunched picket group. Sometimes, flat rocks or metal plates are used in the same way to fracture glass windows or headlights.
8. Scattering tacks or sharpened devices on the road or in entrances to produce flat tires. Children's play jacks have been used in this way after the points were filed. Nails have been similarly used after "pretzel folding" a trouser pocket hole deliberately made.
9. Lock-step picketing which prevents a continuous bar-in a vice. Often the items will be allowed to slip through

rier and does not permit passage through the line.

10. Falling or lying in front of vehicles or in entrance ways. This is often combined with feigned injuries from an alleged accident.

11. Inciting to riot or encouraging a breach of the peace through specific, inflammatory words, or actions (including placards).

12. Abandoning motor vehicle or deliberately blocking a thoroughfare. Frequently a car is driven to or through a gate and then deliberately stalled. Other vehicles follow and the area is effectively sealed off. A well-prepared enterprise will have equipment on hand at entry gates to move such vehicles. Police should also consider including a wrecker or similar vehicles in their on-site supplies.

13. Hurling missiles or explosives at the struck premises or non-striking personnel, or even police.

14. Making, carrying or using unlawful implements such as slingshots, billy clubs, switchblade knives, etc.

15. Assault on nonstriker or police officer from the crowd and when the assaulted person's attention is directed elsewhere.

16. Employment of bullies and thugs.

17. Activities of professional agitators.

Drinking at a Strike Scene

Drinking and intoxication are common problems at the scene of a strike, particularly the first day or two. Without exception, drinking always leads to trouble when carried on during labor-management disputes. Drinking is usually not condoned by either side at the scene of a labor-management dispute, as drunkenness tends to hamper negotiations and gives the general public an adverse impression of those involved in the dispute. Security personnel should be alert for any drinking and take immediate steps to restrict it.

DUTIES OF THE POLICE

At a strike location, the duties of police are the same as

in any other situation requiring police attention. It is their duty to see that persons have the right to enter and leave at will, if they so desire. It is also their duty to see that strikers have the right to picket the plant in a legal manner. The only purpose for which officers are detailed to a strike scene is to maintain the peace. The issues of the strike are of no concern to the officers.

Although the police do not have legal authority to establish a strike zone, they should enlist the aid of the union in restricting the operations of the picket so as not to block sidewalks unduly. Picketing strikers should be kept on the move. They should not be permitted to collect in standing crowds or remain loitering on the sidewalks, streets, or any other public place and should not be permitted to picket in a manner that will block any entrance being used and hinder or impede passage of pedestrians or vehicles.

Many cases of strike violence have been initiated by nonstrikers. Therefore, the fewer persons present at the strike scene, the less chance of group action. Keeping the general public away from the strike scene prevents trouble and allows working room for the police when and if violence should break out.

SUMMARY OF RULES FOR POLICE

Things to Do

1. Be absolutely impartial and neutral.
2. Limit your conversation primarily to the picket line captain and ranking company official.
   a. Get to know the other disputants and converse with them when you believe it is advantageous.
   b. However, keep conversation to a minimum.
3. Keep the general public away from the dispute, thereby lessening tension.
4. Place responsibility by issuing instructions to either the picket line captain or the ranking company official.
5. Be aware of professional agitators. They will often put the police in a position where they appear to be taking sides.
6. Whenever possible, use the Intelligence Section to ob-

tain information pertaining to the dispute.

7. Forward all information to the commands primarily concerned with labor disputes.
8. Give verbal instructions when asked directions by a disputant.
9. In handling trucks and automobiles, break the line only temporarily. Make sure of sufficient clearance to allow the auto to pass. Beware of the use of regular hand signals as they give the impression you are directing cars to enter. The best policy is to have a union representative direct his men to clear the entrance. Let a union official in police presence talk to the drivers of all trucks. A driver then makes the decision to enter or leave.
10. Union officials should be notified of and given an opportunity to take proper action in the case of the drinking of alcoholic beverages by the pickets.
11. All bars in the area should be given special attention.
12. Arrange for periodic relief of police on the line.

Things Not to Do

1. Do not give an impression, by an overt act such as waving, smirks or any gesture, that would make others believe you are biased.
2. Do not become provoked by name calling or derogatory remarks directed at you.
3. Do not at any time go to the scene of a dispute in numbers to obtain information, thereby creating restlessness among the disputants and the onlookers.
4. Do not drive on the property of management.
5. Do not under any circumstances talk over the merits of the dispute with anybody.
6. Do not give any advice pertaining to injunctions.
7. Do not discuss an injunction with anyone involved in the dispute. This procedure is civil in nature and should be treated as such. Advise them to contact an attorney for advice.
8. Do not physically assist or escort a disputant where it

would give the impression you are taking sides in the dispute.

9. Do not lead a person or vehicle with police vehicle when asked for directions.

10. Do not drink coffee or eat in establishments that are frequented by the disputants.

11. Do not perform a police task immediately after conversing with a disputant. This gives the impression you are taking orders from him.

12. Do not go on company property unless such action is necessary to enforce the law.

13. Do not use toilet facilities on company property.

14. Do not accept gifts—donuts and coffee. This gives the impression of partiality.

15. Do not let plant guards assist the police on public property.

16. Do not by outward indications of sympathy, such as smirks, gestures or wearing a campaign button on your uniform, cause resentment.

Chapter 27

MANAGEMENT AND AD HOC
DEMONSTRATORS

INTRODUCTION

Of LATE MANY DEMONSTRATIONS and acts of civil disobedience by various and sundry *ad hoc* groups have been directed against private facilities. Often the demonstrators are justified in the action they take, but quite often the alleged cause for the protest was merely verbal subterfuge without substance in fact. Indeed, the whole campaign is undertaken to embarrass and harass the agency, school or company and to gain personal recognition for the "cause" and leaders of the demonstration.

It is therefore important that management be familiar with the tactics of such groups, and that they know what is legal and illegal and what are the duties and responsibilities of the police.

The first fact which must be recognized, is that with increasing frequency demonstrations and civil disobedience are resorted to as weapons of social conflict. Ironically, the most effective and professional the local police, the more understanding and cooperative the business, the more vulnerable the industry and the jurisdiction is to the exercise of the tactical weapon of civil disobedience. This is not to say that such a department or business itself is vulnerable in the sense that it will suffer any loss of prestige or will fail to cope properly with the problem. This is not true. On the contrary, the area is selected because of the very fact that law and order is likely to be maintained. In such a climate the demonstrator feels safe, secure and free to use these destructive tactics. The business is selected because it is more likely to be sympathetic to the demands made by the protestors.

NONVIOLENT, DIRECT-ACTION DEMONSTRATIONS

A demonstration is a public assemblage of persons exhibiting sympathy with or opposition to some political, economic or social condition or movement. The intent of the demonstrators is to persuade by focusing attention on a problem or problems, and the persons or establishments against which action is directed, and to publicize the procedure or beliefs of the organizations and persons participating.

Demonstrations must be distinguished from crowds and mobs in that they are as follows:

1. Organized.
2. Have leadership.
3. Participants are well disciplined.
4. Participants are orderly and nonviolent.
5. Action is legal.

The means used by the demonstrators to express their views are the picket lines, parades, meetings and rallies. Demonstrations may be staged at a single location, or may be held simultaneously at various points. Indeed, numerous simultaneous demonstrations may be held throughout the nation, in fact, throughout the world.

Rather than looking on a peaceful or lawful demonstration with fear and horror, it should be considered as a safety valve serving to prevent a riot. *He who makes peaceful demonstration impossible makes violet revolution inevitable.*

Indeed heavyhanded efforts to suppress a public demonstration or failure to protect the demonstrators from opposition elements within the community will only create a climate of frustration and violence which will eventually culminate in riots with subsequent loss of life and property.

CIVIL DISOBEDIENCE

Civil disobedience has been defined as "refusing to obey the relations between the citizen and the state as regulated by law."

Another definition declares that civil disobedience is the deliberate public violation of the law, with every expectation of arrest, as a protest and in order to dramatize one's sympathy with or opposition to some political, economic or social condition or movement. It is thus apparent the term *civil* is highly inappropriate. "Criminal" disobedience would be far more accurate.

Civil disobedience may be based upon belief in the unconstitutionally or immorality of the violated law, or upon the desire to influence official action or major changes in our society. In addition, other groups, with less noble purposes, use the trappings of nonviolent civil disobedience to further their own ends. The understanding and identification of the problem are essential for the alert police executive who would guide department policy in this most sensitive area.

TACTICS OF THE DEMONSTRATORS

Basically a demonstration or act of civil disobedience is an attempt to apply social pressure against a selected target. The ultimate goal is to force a confrontation over an emotional issue, whether it be civil rights, poverty, the draft, or the war. The next step is to create a situation where the community or company has "backed down" or can be made to appear to have capitulated. Once this has been achieved the inevitable victory celebration will be held with resulting national or even international news coverage of the event.

Whenever a protest demonstration or acts of civil disobedience are being planned, there is a sequential development of events that normally occur. During the initial or preparatory stage the organization or group develops background information about a particular segment of community life, such as the education system, employment, public accommodations, job opportunities, housing, community social structure or other areas of alleged discriminatory practice. Frequently an organization will infiltrate its members, adherents or sympathizers into an industry, area or function in order to gather information or create a situation. If this is unsuccessful, they will try to develop informants within the target organization.

The second or testing stage consists of developing the issue or cause, i.e., discrimination, free speech, etc., to serve as a bargaining point or basis for protest. The issue is phrased in terms that possess a high moral appeal that will justify whatever means the demonstrators may see fit to use. It must be something that catches and challenges attention without great effort. It is normally in terms of "civil," "moral," or "constitutional" rights so as to appeal to the broadcast base and to give the aura of respectability and righteousness to the group's activities.

The organization will at this stage of development frequently request a meeting with the group with which it is in disagreement and submit the information they have obtained through the previously mentioned methods, and at this time, particularly in a militant organization, the *quota* or *demand* is made. Should the problem still remain unresolved, attempts are frequently made by the organization to enter into negotiations. If they are recognized for purposes of bargaining, their stature and standing is tremendously increased, particularly among their own group. Such recognition will also serve as a lever for broadening their actions and activities in related fields. Lawsuits may be filed if the organization is devoted to the principles of court test cases or if legal remedial action appears to have a favorable opportunity of resolving the issue.

At any given point in the chronology mentioned, or even before its inception, direct action may occur either as a result of a spontaneous occurrence which tends to arouse quick resentment or anger or whenever the development pattern reaches a point where the protest organization feels that it is being ignored or it is losing face. Normally, a number of minor incidents will be provoked by the demonstrators to insure a steady flow of free publicity in advance for the demonstration date. Direct action may take the form of picketing, boycotts, entry and exit blockades, or premises capture or occupation. Picketing techniques are designed to harrass, block, or encumber an operation. Perimeter picketing or blockades are used to establish control around an area and isolate the area through a blockade as a support action for the main picketing. Entry and exit blockages are accomplished by walking pickets or by standing, sitting or lying within the entry or exits. Premises capture involves stand-ins, sit-ins, lie-ins, kneel-ins, walk-ins, chain-ins, and similar "ins." It may be accomplished by conducting prayer meetings on court house steps, city halls, or other public locations.

Rally meetings are designed to broaden the base of popular support, instill enthusiasm, obtain additional publicity and as a forum for prominent speakers in the organization who are brought to the community. Organizational meetings, during which varous decisions are ostensibly submitted and voted upon, are usually held at established intervals which increase in frequency in a project that is in an active stage. There are usually strategy meetings held by the leadership prior to or immediately after organizational meetings at which the actual decisions are made, and it is imperative in an intelligence operation that information be forthcoming from these meetings.

There is a rapid development in the training of members in the principles of civil disobedience, its aims and purposes, together with the techniques which have been developed in previous demonstrations or actions. The group will try to enlist the aid of certain types of persons who can be utilized effectively in any direct action campaign because of their energy, enthusiasm, or the appearance of respectibility and righteousness they give to the movement. Among these are the following:

1. Children, particularly in the front of a group, particularly infants in their mother's arms.
2. Women.
3. Youths, particularly adolescents.
4. Elderly persons, particularly cripples.
5. Military personnel in uniform.
6. Clergy in clerical garb.

Attempts will be made to influence employees and gain their sympathy and support. The methods may be nothing more than conversation with the employees at the facility, or as they enter or leave, or by telephone calls to their homes. At times it will take the form of constant shadowing, threats of violence or actual violence. The employee may be asked to merely give his moral or verbal support or he may be recruited to commit acts of sabotage or espionage.

Civil disobedience envisions a protest based upon noncompliance, without active resistence. Disobedience to directives against congregating or grouping and also the disobedience of ordinarily accepted regulations relative to parking, parading and littering are the most common methods used. The ultimate purpose of passive resistance is to create a logistical problem which will focus direct and continuing attention on a given area and require substantial police manpower and equipment to restore order. Frequently the removal of these individuals is also used as the basis for charges of police brutality, alleging twisted arms, legs, improper handling and similar charges.

The basic drive behind all of this activity is to generate the maximum publicity. Indeed this is an area where considerable friction has been generated between law enforcement agencies, the victims and the press. The press has given the advance publicity to the demonstrations with the teaser of possible civil disobedience. This usually results in large crowds turning out for the event, but more important, it has encouraged the demonstrators to produce an event worthy of extensive coverage by the television and news media. Responsible community leaders or company officials find it difficult to counter such irresponsible tacics and are sometimes triggered into making a foolish remark that only serves to inflate the publicity over the affair further.

MANAGEMENT ACTION

How is such an assault to be met? If the cause for complaint is justified, then the solution is to correct the deficiency. This act alone will hit at the heart of the movement, even though it will not satisfy some of the group who are not really interested in solutions but instead desire constant conflict.

The effectiveness of this approach is based upon sound principles of crowd control, for one of the most striking traits of the inner life of a crowd is the feeling of being persecuted. Thus the crowd nurtures a peculiar angry sensitiveness and irritability directed against those it has once and forever nominated as enemies. These can behave in any manner, harsh or conciliatory, cold or sympathetic, severe or mild, whatever they do will be interpreted as springing from an intention to destroy the crowd, openly or by stealth.

In order to understand this feeling of hostility and persection it is necessary to start from the basic fact that the crowd, once formed, wants to grow rapidly. It is difficult to exaggerate the power and determination with which it spreads. As long as it feels that it is growing, it regards anything which opposes its growth as constricting. It can be dispersed and scattered by police, but this has only a temporary effect, like a hand moving through a swarm of mosquitoes. However, it can also be attacked from within by meeting the demands which led to its formation. Its weaker adherents then drop away and others on the point of joining turn back. Thus an attack from outside strengthens the crowd, as those who have been physically scattered are more strongly drawn together again. An attack from within, on the other hand, is really dangerous, for it causes the group's power to crumble visibly.

Thus, although the correction of the shortcomings will not placate the leaders because it takes away their cause and thus weakens them, it does satisfy those who were seriously striving for the improvement.

What of those instances where absurd demands are made that cannot be met? How is a slanderous attack to be met? There is only one weapon to use against falsehoods. It is the one weapon the demogogue cannot tolerate—truth!

To successfully launch such a counter-offensive the company should pursue the basic tactics outlined in the preceding chapter on strikes, for basically a strike is a demonstration. Thus many of the security principles are identified, but there are certain elements that do deserve special consideration.

Intelligence

It is essential that a company that is to be the target of an *ad hoc* attack learn of the assault and the methods to be employed before any move is made by the outsiders. Only such knowledge will enable the company to organize and plan to meet the threat. The gathering of such information is the task of the security department. It should devise a plan for the gathering of such information and for coordinating its intelligence operations with those of the local police authorities who will be involved in the action. For it must be recognized that the police are charged with the responsibilty of maintaining law and order. They can best do this if they have the confidence of all parties, so that the conflict, when it comes, will be conducted in a lawful manner. They also must have advance warning of such conflict and thus have developed contacts unavailable to local industry which is seldom in contact with the various protest groups. Thus, they will be in a position to learn long before industry of any planned demonstrations.

Intelligence can be obtained by the employment of numerous methods (see *Riots, Revolts and Insurrections* for a detailed treatment of the subject). In the situation here under consideration a large scale cloak and dagger operation is not needed. Rather, as in the case of strikes covered in the preceding chapter, such intelligence is often available through direct contact with the demonstrator command. At times the problem will be to learn who is the actual leader of the demonstrators. Often those who are publicly paraded as the leaders are merely the puppets of the true leaders who, for various reasons, wish to remain unknown and unseen.

In still other instances the demonstration is directed by public group planning. Such open participation and discussion makes the gathering of intelligence data a relatively simple task. An observer attends all of their public planning and coordinating meetings. Personnel selected for such an assignment must be personable, intelligent and, above all, possess a sense of humor. Whenever possible they should be of such an age and be so dressed that they are not conspicuous while among the demonstrators. All handbills, magazines and publications supporting the demonstration goals must be carefully analyzed for clues concerning shifts in tactics, changes in leadership and indications of what group or groups appear to be providing the funds and overall direction. Very often these handbills and other publications contain the detailed demonstration plans since this is one of the principal methods used by the organizing cadre to reach all members of the participating groups. What they too often overlook is that the organizing cadres of the demonstrators make a number of test probes in the community while evolving their final strategy. Therefore, almost any contact with the demonstration groups must be considered as a scouting or probing operation to test the types of countermeasures that may be employed. It is during this period that the organizing cadre itself will be most exposed to public view because they will be virtually anonymous during the actual demonstration itself. It is therefore of the utmost importance that reports be forwarded to the central intelligence gathering point where these seemingly casual contacts can be properly evaluated. All information from whatever source must be gathered at one central point so that it can be studied, evaluated and disseminated to those who must plan for the action.

Obviously the intelligence operation must be continuous. Often during the course of a demonstration, tactics and plans of the *ad hoc* committee will change. Indeed in instances leadership will pass from one individual of the group to another, with a resulting change in the action of the group. Management must know these facts as soon as possible to make any alteration in its plan that may be required.

The intelligence operation should not only be directed to learning the plans of the *ad hoc* committee, it should also cover the nature of the leadership and the participants, their temperament, experience, potential for violence, their aims and their true motivation.

Part of the company's intelligence operation at the scene is the photographing of the participants and their conduct. These photos are of value as evidence of illegal activity but also to counter any adverse publicity that may flow from *ad hoc* produced pictures of the events.

Indeed cameras are the best defensive weapon to be employed against the *ad hoc* groups. The knowledge that their actions are being photographed discourages individuals from performing acts that could incite a crowd to violence. The camera also robs any group of the feeling of anonymity. It is this feeling of anonymity that is essential to encourage massive civil disobedience or other direct action tactics. Photographs are by far the best evidence in any subsequent court action and serve to demonstrate graphically to the jury that appearances can be quite deceiving (e.g., when they compare the evidence photographs with the clean-shaven, respectably dressed defendants in the court room). Speeches are a staple diet of any demonstration and often the delivery is somewhat extemporaneous in nature. Therefore all speeches should be recorded both as a preventive measure and a basis for future legal action should the speaker incite the crowd to perform illegal acts.

Part of the intelligence operations at the scene of the demonstration is the listing of all license numbers. This will prove invaluable in many ways. It helps to identify the participants and their sympathizers. At times it is the key to the identity of the secret leadership and the interests who are financing the operation. A collateral benefit that can be obtained is in insuring that vehicles from various tax-supported agencies are not used to provide free transportation for the demonstrators.

Contact of Leaders

Contact should be made with the leaders and agreement reached on the basic ground rules for the demonstration. Demonstration leaders should be encouraged to maintain their own discipline and reduce the possibility of a confrontation between uniformed officers and the crowd. Individuals for each group should be designated to receive communications from the other party. The swift reliable flow of communication is of the utmost importance in preventing unfounded rumors from spreading, in lessening confusion and in reducing tension should a misunderstanding arise or an incident occur.

It is particularly important that these leaders be present whenever a crowd is assembled, as at that critical time a misunderstanding may cause the crowd to degenerate into a mob.

Legal Position

At the earliest possible time it must be determined what legal action will be taken if there is a violation of the law. What laws will be enforced, which will not? This is particularly important in the case of trespass, as the owner of the property has the power to render the act legal by consenting to the entry.

It must here be recognized that if arrests are made there will be trials, and employees may be essential witnesses. The company must therefore be prepared to pursue the matter through to its completion and to give all the support required to make that prosecution a success.

Once all the substantive issues regarding the legal position of both the community and the company have been established and responsibilities assigned, it is essential to make the possible legal consequences publicly known. Unless this is done many of the demonstrators may be mislead into believing that the "moral right" of their cause will protect them from normal civil or criminal prosecution. Most of the demonstrators do not consider themselves as lawbreakers and are often profoundly shocked when confronted with the reality of criminal prosecution for their acts.

Public Relations

Once it is learned that an attack will be made on a company, the nature of the tactics and the propaganda campaign must be learned. If legitimate complaints are made, they should be corrected. If a false slanderous campaign is planned, it should be met by a campaign of truth which, if possible, is launched prior to the inauguration of the attack. Thus a claim of bias in hiring could be met by the release of an open letter to the Fair Employment Practice Commission affirming one's status as an equal opportunity employer and stating certain actions which are now practiced and others which will be undertaken, including disclosure to the FEPC of a breakdown of staff composition, showing particularly the number of minorities in the company's employ by racial groups and by various areas. In addition, the willingness of the company to meet with any of the various minority groups to discuss progress and entertain their suggestions leading to an increase in the number of minority employees should be proclaimed. Further, it should be made clear that no agreement will be made with any nongovernmental agency nor will the company capitulate to any illegal pressures by any groups.

This open letter to the FEPC should be widely distributed, and, indeed, published in a full page advertisement in the major newspapers. The company's public relations campaign must constantly keep ahead of the attacks. It must not be limited to a mere counterpunching technique, but must be aggressive to keep the opposition constantly off balance.

All company officials and employees should be specifically instructed not to engage in conversation with the group. Such conversations will be directed to obtain embarrassing statements. It will undoubtedly be secretly recorded by the group.

Always remember that the *ad hoc* group wants publicity. To frustrate them in this regard a campaign should be launched to create a climate within the company that will discourage the development of anti-demonstration sentiment and actions, such as name-calling, egg throwing or physical encounter. The development of "martyrs" is often either a direct or indirect goal of any demonstration and there are countless cases where the original demonstration received only lukewarm public support, but a stupid act of violence against one of the demonstrators gave it a weight and public support it could not have otherwise achieved. It is here that the greatest danger to the community exists since the dynamics of crowd psychology operate in even the seemingly most peaceful of demonstrations.

To achieve these ends the company should designate a specific individual to handle all press releases. He and his staff must be in close and constant communication with management, intelligence and all public agencies involved in the action.

Basic Strategy

Any defensive measures taken must be designed to deny a crisis issue to the organizers of the demonstration since the lack of a crisis issue will cause their support to melt away. Likewise, a demonstration that fails in its principal objective, which is to force a confrontation, immediately loses both its news value and financial support.

To achieve these ends is not a simple task, for even the slightest miscalculation or misunderstanding can cause an entire community to erupt. It is even more difficult when dealing with fanatic individuals whose personal ambitions are dependent upon the success of the demonstration since it must be expected that they will resort to any means to gain their ends. Thus at times they have deliberately destroyed their own headquarters, bombed their own homes, indeed murdered their own followers.

Bargaining Conferences with Demonstrators

The decison whether or not to hold a conference will depend on many factors. At times the demands of the demonstrators may be patently beyond all reason. In such an instance a meeting will merely give them a degree of recognition and an opportunity to gain instant status and headlines as a result of the confrontation.

In other instances although the demands of the demonstrators may seem reasonable, it will be found that as soon as they are recognized and a meeting arranged, they will claim that such a conference is an admission on the part of the adversary of the merits of their charges. Still others will increase their demands once a conference is begun. Those who are motivated by personal ambitions will not be content with anything less than a dramatic confrontation in the streets, for then they will maximize their publicity. Thus any conference with them is worse than a waste of time for nothing constructive will come of it.

There are instances where the *ad hoc* group is genuinely interested in achieving definite reasonable goals. In such instances a conference is desirable, for it affords an opportunity to discuss the problems and the possible solutions rationally and intelligently.

What action to take will thus depend upon the facts of each case, which in turn will depend upon accurate intelligence. Naturally once the protestors have lost public support, there should be no meetings except on the company's terms. Whenever it is decided to conduct a conference, the company should insist that the demonstrators provide an agenda and should inform them that the subjects on it must be matters that really can be productive. Any meetings with the demonstrators should be recorded so that no distortion of the truth can be made. Two tape recordings should be used so that both parties will have a complete taped transcription of the meeting. Such a procedure not only assures an accurate record of the proceedings, but it tends to restrain the demonstrators.

Chapter 28

RIOTS

INTRODUCTION

"**W**HAT CAN YOU DO to protect your store or facility during a riot?" This is a question that is becoming more and more the concern of countless merchants and industrialists throughout the country. Now we know that the problem can present itself anywhere at anytime. Any city can have a riot of the magnitude and proportion of those which occurred in Detroit, Rochester, Harlem, Los Angeles, Cleveland and Chicago. As we all know, such disasters are generally accompanied by looting and by wanton destruction of property. The retailer, because of the exposure and appeal of his merchandise and the location of his facilities, will often suffer heavy losses in a situation of this type.

Having recognized this, we must consider what steps can be taken to protect commercial and industrial facilities. Any thought given to this subject and any positive preventive action taken is bound to be beneficial, and any preparation made, no matter how slight, will help protect against loss of life and property in the event of a riot.

IMPORTANCE OF STAYING CALM

The conduct of human beings in a state of riot is unpredictable. The situation is fluid and constantly changing. Top management will fall under these same influences. They will have definite feelings which will vacillate on whether to close down operatians entirely or partially. Naturally, the protection of human lives, employees and customers, comes first, but on the other hand, a business cannot be abandoned merely because of panic.

Thus, even though there may be a dispute as to methods, it is definite that the wrong answer is to throw one's hands up in despair, complaining that nothing can be done unless the police or national guard arrives on the scene in the nick of time. In such a crisis the important thing is to avoid panic and to exercise good judgment. If one runs scared, he has lost.

OWNER'S RESPONSIBILITY

Under normal operating conditions the merchant and industrialist have available to them countless services supplied by the municipality, police utilities and others. This would include police and fire protection, ambulance and medical services for the sick or injured, maintenance and repair of telephone lines, gas, electric and water lines, transportation of goods and people, messenger service and the ability to make purchases of necessities, such as foods and drugs.

When riots occur, municipal and private facilities—including police, fire and public utilities—are stretched to the breaking point, are spread so thin that it becomes unlikely that any one individual company or business will be able to obtain substantial help from these sources. Therefore, the merchant and manufacturer may have to perform many of these functions themselves if they are to be done.

Similarly, in recent riots private security agencies have not had the manpower to meet the substantial increase in demand. Guards were at a premium. The situation has been much the same with central-station alarms and patrol and guard services. These firms were hampered by insufficient help, and when they did have help it was often found that it was impossible to enter the riot area.

As a result of these recent experiences, it is generally recognized that during a riot the prime responsibility for protection of any establishment falls on the owner. It is up to the owner and his own staff to make the hard decisions on how best to protect the property and to what extent to risk their lives in the protection of that business.

PLANNING

How well a business survives a riot may depend on how careful and sound was the preparation for it. The type and extent of the programs that are needed will differ radically. While a small establishment would need a minimum of preparation to secure the place of business, a large complex establishment with numerous plants, stores, warehouses and service buildings would require extensive and detailed planning for effective protection.

Another factor is the nature of the enterprise. One inescapable lesson of the riots is the fact that some businesses represent high risks in riot situations and will need all the security measures applicable. The businesses attacked first were liquor stores and pawn shops, and these were also the most thoroughly looted. After these came markets (initially attacked, apparently, for liquor and cigarettes), furniture stores, appliance stores and clothing stores. On the other hand, industrial plants, schools, libraries and banks suffer relatively little damage. The reason for the distinction is to be found in the motivation of the rioters. In these recent riots the rioters have displayed some acquisitive motivation. Thus the rioters give top priority to merchandise that can be used or is readily resalable. Liquor, furniture and clothing can be used. Clothing can be sold on the street.

This does not mean that in the future the same drives will control. Further, it is possible that subversive groups will use the confusion and chaos of a riot to attack and destroy industrial plants or obtain classified information. Thus although much can and must be learned from past events, it cannot bind security to changes and distinctions that may arise.

While planning, it must be kept in mind that all civil disturbances are not of the magnitude of the riots in Rochester, Harlem, Los Angeles, Cleveland or Chicago. There have been instances of limited violence—the hurling of molotov cocktails or bricks through store windows. That such gradations exist must be first recognized and then taken into consideration in drafting appropriate countermeasures to meet each class of problem.

In preparing the plan it is advisable to coordinate and cooperate with similar organizations. Many of the problems will be the same in other firms, and the broad and varied experience of the larger group will prove helpful and will furnish a basis for unified action.

Also, it is of the utmost importance in planning to determine whether a private guard service is willing to supply guards and protect a business in the event of widespread civil disorder.

There should be full coordination between the plant emergency head and the surrounding city and county emergency agencies, so that there will be a minimum or overlapping or duplication of effort. This will also insure that the equipment is standardized and that the training of personnel is carried out along the same lines and using the same methods of operation.

Some of the organizations with which the emergency plans should be coordinated are the fire department, police department, hospitals (including doctors and medical services), the Red Cross, the local civil defense agency, transportation facilities, water and power department and the telephone company and other communication services.

The plan in order to be effective should be as simple as possible and should provide for one or more alternate plans of action in the event the first plan proves unworkable due to the peculiarity of the emergency. Elsewhere detailed discussion of disaster plans will be found. As they are designed to meet all disasters, manmade as well as natural, they apply to a riot situation, but for assistance, an actual plan for a large retail chain follows.

Riot Plan for the Company

I. Introduction

In view of the number of riots which have errupted of late throughout the country with the staggering loss of life and property, and the ever present danger of future riots which might directly threaten The Company and its personnel, the following plan has been developed. Such a plan is essential as:

Police and fire departments may be incapable of providing aid during a major riot;

Private security organizations may be short-handed and ineffective; and

Riot areas may be sealed off, so that entry is impossible for even emergency crews.

The purpose of the plan is first and foremost to protect the lives of all personnel and customers. Secondly, it protects the employees' livelihoods by preserving the facilities of the Company.

All planning and policy decisions both before and during a civil disorder should be evolved in close cooperation with civil authorities. THEY ARE THE EXPERTS.

II. Basic Policy

The policy of The Company is to continue operations wherever possible, but never to compromise the safety of customers, personnel or property. Further, The Company recognizes that riots and other civil disorders are usually caused by a relatively small percentage of the community or neighborhood in which they occur. Many innocent persons, including peaceable Negroes or members of other minority groups, frequently suffer the worst. It is tempting to blame even the innocent victims for the sins of a few. Yet they should be the objects of aid rather than of blame. They may be in dire need of food, shelter, clothing and medical help following a riot. It is therefore the policy of The Company to cooperate with officials in helping the victims of riots, in restoring normalcy and in planning constructive programs to remove the causes of civil disorders.

III. Advanced Preparation

A. INTRODUCTION

In order effectively to meet any challenge it is essential that all procedures be well established that every employee should know his responsibility and have received training so that he can perform his duties effectively.

B. RESPONSIBILITY

The primary responsibility for preparation at each facility rests with the manager. He must do the following:

1. Contact the local police and fire departments to inform them of the Company's policy and plans. Further supply them with the names and telephone numbers of Company personnel who can make a decision during a crisis.

2. Ascertain the willingness of the employees to

participate in the protection of the facility in the event of a riot. Assign definite responsibilities for merchandise, cash and record removal; fire fighting; for evacuation duties, etc., to members of emergency squads. Impress upon all personnel that personal safety is more important than property protection.

3. Compile an up to date list of both business and residence telephone numbers including the following:

    a. Police.

    b. Fire Department.

    c. Manager of the facility.

    d. Assistant Manager of the facility.

    e. District Manager.

    f. Security Director.

    g. General Manager.

    h. Glass repairman.

    i. Electrician.

    j. Fire extinguisher maintenance.

Alternates for all personnel should be listed and their telephone numbers given. All key personnel should be given a copy of this emergency call list. Additional copies should be posted at critical locations on the facility for immediate access Establish "pyramid" telephone system (each person phoning several others) to advise employees of store closing. Establish system for notifying emergency crews.

4. Establish procedure for evacuation of store if emergency occurs during working hours. Assign specific responsibility to given employees and provide personnel drills and training to avoid panics and confusion.

5. Establish procedure for protecting customers and personnel if evacuation is dangerous or impossible.

6. Plan for emergency transportation for evacuation. Assignment of emergency car pools. Assign specific exits to which cars are to be brought.

7. Establish first aid facilities and first aid training for members of emergency squads.

8. Provide for identification by special badges, arm

bands, etc., of persons assigned to emergency duty. Arrange with police department to have authorized personnel conducted through police lines if necessary and practical for them to get to the store. Provide emergency transportation to store for emergency squads.

9. Arrangements for food and other provisions for emergency personnel on duty.

10. Plan for emergency evacuation of personnel remaining on duty.

11. Consult and coordinate plans with other businesses and merchants associations.

12. Establish procedure for emergency diversion of merchandise deliveries.

13. Review insurance coverage and provisions covering riots.

C. FIRE PREVENTION

1. *Preparation*

"Molotov Cocktails" and other incendiaries will provide the greatest property threat in the event of civil disturbance. Because of expected roving bands of rioters, this danger may not be confined to a "riot area." Each location manager should check the equipment and facilities available to him and his personnel to prevent and/or fight fires, and to assure emergency evacuation of the building for customers and employees. These include:

a. Normal prescribed complement of fire extinguishers, all filled and situated at proper locations with instructions for use. Locations of fire extinguishers clearly marked by signs and warning notices regarding types of fires for which extinguishers may be used.

b. A hatch to provide accessibility to the roof, and a ladder or fiber rope (at least ¾" diameter) with knots at 18" intervals, sufficiently long to provide emergency evacuation from roof to ground.

c. A garden hose (at least ⅝" inside diameter), stored for ready accessibility, sufficiently long to run from faucet connection, out roof hatch and provide a stream of water to any point on the roof. Be sure adjustable nozzles and screw-type faucet connections are available for connecting hoses.

d. All trash stored in proper receptacles.

e. All access aisles to outside doors clear of merchandise, equipment, etc.

2. *Fire Fighting Training*

Each location manager should contact his local fire department and arrange for instructions and demonstations in fire fiighting techniques to be given to his key personnel. Each location manager should select people to be trained and have them on hand for such instruction. Key men, assistant managers, department heads and foremen should be included in such training. Following completion of such instruction, all male regular full-time and part-time employees should be given basic instructions by those who attended fire department demonstrations regarding use of extinguishers and their location, and general emergency assignments.

D. EMERGENCY LIGHTING

Since civil disturbances have resulted in power failure for some areas, provisions should be made for emergency lighting of interior areas, and of exterior areas where necessary and practical. Each location manager should survey his minimum lighting requirements for the safety of his personnel. If present emergency lighting is not adequate, the manager should request additional lighting equipment. Long burning flood lights using "B" batteries are recommended for area lighting. In addition, sufficient flashlights in good working order for use by key personnel should be kept on hand.

E. EMPLOYEE TRANSPORTATION

In some instances, it may not be desirable for employees to drive their own cars to or from work, nor to park their cars in the neighborhood or on the parking lot, even though the location continues to operate. Store Managers, or men in charge, should work out an employee transportation system with their District Operations Managers whenever a civil disturbance is imminent. Managers of other facilities should work out such a program with the Director of Warehousing and Transportation. The Security Director will assist in such arrangements upon request. The employee transportation program should include arrangements for:

1. Picking up employees at a predetermined location outside the disturbed area and bringing them to the work location.

2. Taking them from the work location to the predetermined area outside the disturbed area.

3. Assuring the employees have been given adequate instructions.

F. PUBLIC RELATIONS

1. The policy of the Company is to hire and promote all qualified applicants regardless of race, creed, or color. This policy and its practical application should be extensively publicized.

2. The Company's prime goal is the service of its customers and to supply them with the best merchandise at the most economical price. This policy must be made known and any and all complaints promptly investigated and appropriate action taken to correct any problems.

3. The Company is interested in the community in which it does business and participates in community betterment projects and youth programs of education and recreation. It encourages similar activity on the part of all employees. This should be known in the community.

4. All employees should at all times refrain from making any remarks which will insult or embarrass or antagonize any individual or group. This is not only good business, it is good manners.

IV. *Riot Operations*

A. OWNER'S RESPONSIBILITY

Under normal operating conditions, The Company has available to it countless services supplied by the municipality.

This includes police and fire protection, ambulance and medical services for the sick or injured.

When riots occur, municipal and private facilities— including police, fire and public utilities—are stretched to the breaking point, are spread so thin that it is unlikely that The Company will be able to obtain substantial help from these sources. Therefore, The Company must be prepared to perform many of these functions, if they are to be done.

We must recognize that during a riot, the prime responsibility for protection of our establishment falls upon our own organization.

B. STAGES

As the location and severity of the riot will have varying effects on the various facilities of The Company, the response must also be graduated. Thus, the plan provides for degrees of response, each successively more extensive, designed to meet the stage of intensity of the riot.

Circumstances will determine which of the plans will be used at a location in the event of any emergency. In some instances, it may be necessary to start by putting one plan into effect and, if circumstances alter, shift to another plan. The stages are:

1. An outbreak *threatens* within the general marketing area (normally a 2-mile radius) and news broadcasts indicate that trouble is imminent. - OR - An outbreak actually occurs, but it is in excess of 2 miles distance.

2. An outbreak actually occurs within the marketing area (within 2 miles) and mobs are approaching the location. Public authorities indicate they cannot control the situation.

3. An outbreak *occurs* in the immediate vicinity of the location or comes within a half mile of the location and there is a threat to persons and property. - OR - When public authorities advise that persons and property are in immediate danger.

C. ACTIVATION OF THE PLAN

The person in charge of a location will place the appropriate plan into effect when:

1. Notified by any of the following officials to activate the plan:

NAME   POSITION

a.

b.

c.

d.

2. Without orders when the circumstances as outlined above have actually occurred, in such an instance, the following shall be immediately notified:

NAME   POSITION

a.

b.

c.

d.

D. CONTROL POINT

A Control Point will be established by the General Manager in the event civil disturbances become critical in any of the areas where The Company has stores or facilities.

The location and phone contact of the Control Point and the composition of its members, will be announced when the Control Point is established.

The following will immediately report to the Control Point and perform their preassigned duties:

NAME

a.

b.

c.

d.

Once the Control Point is established and announced to all locations, all communications from the locations should be directed to the Control Point, and all instructions will be issued to locations through the Control Point. Until a Control Point is established, however, any communications are directed to the District Operations Manager or Director of Store Operations.

E. Procedure

STAGE

	(1)	(2)	(3)
CONDITIONS	An outbreak threatens in the general marketing area of the store. News broadcasts indicate trouble is imminent - OR - An outbreak occurs but is in excess of 2 miles distance.	An outbreak actually occurs within the marketing area and mobs are approaching the location. Public authorities indicate they cannot control situation.	An outbreak occurs in the immediate vicinity of the location and there is a threat to persons and property - OR - When public authorities advise persons and property are in immediate danger.
ALERT	Man-in-charge calls following individuals until *one* is reached and explains situation as it exists. 1) Store Manager, or facility manager as appropriate, if not on duty. 2) District Operations Manager, if a store. 3) Director of Store Operations or Director of Whsg & Transportation, or Director of Manufacturing, as appropriate. 4) Director of Security. 5) Personnel Director. 6) General Manager. Man-in-charge advises plan he is following and any special additional action he is taking, or	(Same as "1")	(Same as "1" except that man-in-charge *also* notifies local police)

E. PROCEDURE *(continued)*

STAGE

	(1)	(2)	(3)
ALERT *(continued)*	changes he is making to plan, based on local situation. Asks individual contacted for any additional instructions and advises phone number to call if necessary to reach the location. Notifies location personnel to refrian from using that phone in order to keep open for emergencies. Individual first contacted notifies all other individuals listed.		
MANNING	Man-in-charge calls meeting(s) of location personnel just as soon as possible after giving alert as above. Advises of situation. Check all individuals on where they live and how they travel. Permit those who wish to leave to punch out. Alert all key men by phone, if not on duty, to stand by and await further instructions, or call them in if substitute staffing is required. Determine full extent of substitute staffing that will be required. All guards should be stationed *inside* the facility.	Man-in-charge provides for safe departure of those employees leaving for home. Make arrangements to move cars, belonging to employees remaining on duty, out of parking lot to a safe area. Manager reports to location, if not already there and location is operating. Obtain names of employees who volunteer to continue to work. Issue Manager's calling cards as ID cards to all volunteers. Write on back of card employee's name, working hours and that he is employed by Com-	Man-in-charge sends home all employees except for a minimum number of volunteers who have agreed to remain. Move all employees' cars out of parking lot to safe place, except those necessary to transport employees remaining in store. When safety of members remaining is threatened, attempt to obtain protection of public authorities. If protection not available and safety is threatened, evacuate and lock location. Leave on lights in location and parking lot. In any

E. Procedure *(continued)*

STAGE

	(1)	(2)	(3)
MANNING *(continued)*		pany. Sign name. Employee will use such card if stopped by public authorities. Notify Office or Control Point of personnel situation.	event, location is to be completely evacuated when ordered by public authorities or by direction of Company.
FIRE DEFENSE	Open hatch to roof and run out hose for use in event of emergency. Designated stores and locations should ready other hoses. Spot fire extinguishers at critical locations and be sure employees know where they can be found. Conduct a quick review of fire defensive measures with location personnel and agree on warning signals. Provide heavy knotted rope or ladder on roof for descent of spotters in emergency. Designate volunteer employees who will keep roof watch if situation worsens.	(Same as "1")	(Same as "1") *In addition*, man-in-charge assigns at least 2 volunteer employees to roof watch with instructions to keep concealed as much as possible from street and neighboring buildings behind parapet. Test hose and wet down roof thoroughly. Move at least 2 fire extinguishers to roof. These men, or replacements, are to remain as long as the location is manned and the situation exists.
CLOSING	Location is to remain open for business during regular hours unless ordered to close by public authorities or by direction of Company. When closing loca-	(Same as "1") Instead of leaving a key man or foreman and one other volunteer, Manager and two other male volunteers remain in loca-	(Same as "2")

E. PROCEDURE (*continued*)

STAGE

	(1)	(2)	(3)
CLOSING (*continued*)	tion for any reason, including routine closing at night, leave interior lights and parking lot lights on. Send employee to roof with hose to thoroughly wet down roof approximately half hour before closing. If a reliable neighbor is nearby, leave Manager's phone number and request neighbor to call if destructive action is taken by rioters against location. Notify Office or Control Point when closing location early because of order by public authorities. Leave a key man or foreman and one other male volunteer in location after normal closing as security and alert.	tion after normal closing for the night as security and alert.	
GENERAL SECURITY	Secure and lock all store doors except customer entrances. At other facilities secure and lock doors other than those actually in use for deliveries, etc. Check back room, rest rooms and other locations for loiterers, or other unauthorized persons in locations of store not de-	(Same as "1")	(Same as "1") *In addition*, when customers leave store, lock customer entrances. At facilities, complete any loading or unloading operation and lock all doors.

E. PROCEDURE *(continued)*

STAGE

	(1)	(2)	(3)
GENERAL SECURITY *(continued)*	voted to business. Keep careful watch for suspicious individuals entering store. Request Office or Control Point for Security Agent.		
CURRENCY valuable & "attractive" merchandise	Place all currency, checks and pertinent Company records in safe. Currency and checks in compartment with variable time lock. Be sure all checkstands operating do not have more than maximum allowed currency on hand. Remove all currency and checks from closed registers and leave registers open. Remove all valuable merchandise from display windows. Remove attractive merchandise (guns, knives, jewelry, furs, liquor) to safe location.	(Same as "1") *In addition,* reduce amount of currency in each operating checkstand to bare minimum needed to operate, and increase schedule for collecting from checkstands and depositing in safe. Temporarily discontinue such services as money orders, utility bill payments, etc., and put up notice at courtesy booth reading: "We regret that unsettled conditions in the neighborhood forces us to temporarily discontinue money order and utility bill services." Remove all company records, currency and checks, except for minimum needed to operate, to another location designated by Office, Control Point or DOM. Use 2 employees to make this transfer.	(Same as "2") *In addition,* when store is closed and customers have left, clear register and remove *all* checks and detail tapes to location designated by Office, Control Point or DOM. Use 2 employees to make this transfer. *All* remaining cash should be placed in safe. Leave registers open.

E. Procedure (*continued*)

STAGE

	(1)	(2)	(3)
COMMUNICATIONS	During meeting with employees, man-in-charge works out method of in-store warnings over public address system which will alert employees to a situation but not alarm customers. In addition, a plan will be developed to evacuate customers fom store quickly and without confusion. A means of communication from roof spotters to man-in-charge will be set up. When calling in alert to designated individuals, man-in-charge indicates phone number in store which will be kept clear as much as possible.	(Same as "1") *In addition*, if telephone service is disrupted, man-in-charge has an employee leave store unobtrusively (preferably a Negro employee if in a Negro neighborhood) and call Office or Control Point to explain situation. Be sure employee carries Manager's calling card with his identification and hours noted on back.	(Same as "2")
DELIVERIES	Company deliveries will continue on regular scheduled basis unless store is notified. If outbreak occurs, all deliveries to stores in outbreak area or on runs including stores in outbreak area will be confined to daylight hours.	All Company deliveries will be confined to daylight hours. Stores will be advised of any emergency schedule instituted by Office or Control Point. Stores should plan to order only items required for immediate need.	All Company deliveries will be continued as outlined under "2" while stores are still open for business. When conditions dictate or public authorities or Company orders closing of stores, all deliveries will cease.

V. Post Riot Operations

The Company will do all in its power to repair any damage to its facilities and return them to operation at the earliest time. The Manager is thus responsible for:

1. Establishing a system to notify employees of store opening.

2. Establishing advertising and public reactions policies concerning the effects of the riots.

3. Gather necessary records and documentation (including before and after photos of damage) for insurance and tax purposes.

4. Cooperate in the re-establishment of sound community relations and aiding the victims of the disorders.

5. When the entire situation returns to normal, each employee who has stood by during the emergency should receive an acknowledgment such as a personal letter of appreciation in recognition of his services at a most difficult time. The same should be done with all others who gave assistance during the disturbance.

A manual should be prepared. The manual should contain all pertinent data with regard to handling a riot situation and it should be the only place a person would have to look to obtain this information.

Employees

It is important that all employes be fully informed and that their active interest be enlisted in support of the emergency plan. One method of doing this is to call the employes together in a general meeting or series of meetings, depending upon the size of the plant and the number of shifts involved. Another method is to present the program to the employees' representatives or department heads who, in turn, will carry the plan back to their departments. Employees will be much more inclined to cooperate in carrying out their part of the program if they are allowed to participate in the working out of the plan and if they understand the problems involved. A training program should be provided all employees. This can be done by personal briefings, lectures, motion pictures, or the issuance of written material. To function effectively the individuals involved must be properly drilled and should believe in the success of the undertaking.

First, instruct employees to refrain from conversations which might set off a riot situation. Explain to them the necessity of courtesy and respect toward all persons. Any sign of belligerence might place your store on a target list during an outbreak. Ask employees to report any suspicious activities in or near your store. When they do, notify the police immediately. Tests of the plan are part of the employees training. They not only learn, but actual operating tests disclose weak spots in the organization plan.

Recent history has proven that immense loyalties and devotion to duty exist among employes. Some of them recognize their daily bread and butter depends upon a business being available so that their job security will be guaranteed. Others act out of respect for and gratitude to their employer. For these and various other reasons, employees have volunteered to make themselves available to operate a business under the most difficult circumstances. Recognizing this dedication and availability, management is in a better position to make decisions to remain open, closed, or manned in the immediate or surrounding areas of a riot situation.

RIOT CONTROL ORGANIZATION

As a detailed discussion of the disaster control organization is found elsewhere in this book, mention is made here only as a reminder of the importance of such a predesigned command group. Basically it will consist of the following:

1. The director of disaster control has full authority for the development of riot emergency plans and who will be in charge of operations when the plans are put into effect. In large establishments the person designated should be one of the key executives of the firm. In small firms the responsibility should be handled by the owner or manager.

2. Area disaster chiefs are the assistants to the director and

in charge of control at their particular facility or their area of a facility.

3. Disaster control committee. The chairman of the committee should be the director of disaster control. The area disaster chiefs should be members, and other persons could be included as needed. The committee should hold regular meetings and have experts in various fields of interest appear and brief the committee members. Reserves or alternate personnel should also be provided for in the event certain key individuals are absent.

   The director and his committee are charged with the responsibility of preparing all necessary plans prior to a riot and executing them during a riot.

4. Security, fire, first aid, and maintenance teams. The riot emergency operations plan is based on the assumpion that, in the event of a riot the firm and its employees will depend primarily on their own efforts to protect life and property. The bulk of responsibility for the proper functioning of the program will fall on the various disaster teams selected to function during emergencies.

The size and number of teams needed will vary widely according to the size and peculiarities of the facility involved. In a small facility a maintenance team may consist of one person, whereas in large establishments there may be separate teams, of many members each, of engineers, electricians and carpenters. Select men already in the type of work for which teams are to be formed. Do not assign persons to these teams unless they volunteer and really appear interested. In the facility riot manual should appear general instructions and a checklist for the various disaster teams.

Each team should review the instructions and engage in practice runs on the same premises in which they would operate in the event of an emergency.

Arrangements should be made for these team members to obtain information and instructions from various government and other groups, such as police and fire departments, the American Red Cross on first aid, the local office of civil defense, and certain fire association groups.

Special badges should be prepared for all people who work on this program. The badges should be attractive and large enough to be easily seen. The printing should identify the wearer with the company and the program.

LIAISON

Police

The first step that should be taken to plan for protection during a riot is to meet with police department officials who are experts in security and riot control. Have them make a thorough check of the facility. Discuss with them ways to tighten security and steps to take in case of trouble.

To assure immediate contact in an emergency, you should have a label on every telephone with police and fire department numbers. This will save precious time when your employees are excited.

Because telephone service can be disrupted during widespread civil disturbances, discover now whether there are possible alternate methods of quickly reaching the police and fire departments. One such possibility is the installation of a burglar alarm system which could bring immediate police assistance. If you already have this protection, determine whether the system should be updated with equipment which is more difficult to disconnect and with secret trip devices.

Do more than ask the police for advice and help. Reciprocate by doing everything within your power to aid the police. The police in these days of crisis, more than ever, need the dedicated support of other public officials, the business community and the public at large. This would seem to be an appropriate time to pledge once again our full moral backing to local law enforcement agencies.

Give more than mere moral support. Take an active interest in the police and their problems and the manner in which you can make their job easier and more effective. Do everything possible to improve their relations with all segments of the community. Keep them advised of any danger signals that may come to your attention. Remember that it is the police which stand as the first line against mob rule.

Fire Department

Along the same lines, meet with fire department officials, who are experts in fire prevention. Have them make a complete inspection of the facility. Get their recommendations on how to make your building less susceptible to fires. Ascertain what you can do to cooperate with them in any emergency.

PRE-RIOT PROTECTIVE MEASURES

New Construction

If the store or plant is not yet built or is to be remodeled, consideration should be given to certain measures that will give added protection such as the following:

1. The use of fewer windows which are placed high to hinder access through them.
2. The use of the wired glass having a fine stainless steel wire that is barely noticeable. A brick cannot be thrown through it. Indeed it is necessary to cut through it with wirecutters. This is a substantial deterrent as a mob usually is not so equipped, nor is it inclined to make a patient attack. Another new duoplate-type glass, laminated with plastic, has to be punched through a little at a time, for although the glass breaks, the plastic must be torn.
3. Use heavy locks and door frames.
4. Install automatic sprinklers both inside the facility and on the roof.

Display Windows and Outside Doors and Windows

The display windows located on street level are the store's most vulnerable area. If the special glass just mentioned is not used, consideration should also be given to the use of plywood panels to cover the display windows and outside doors on the ground floor. It would require the installation of special frames or other holding devices to support the panels securely.

Scissor Gates

Scissor gates have proved ineffective under riot conditions.

Some of these gates, which are pulled across store fronts, are supported by tracks at both top and bottom and some operate without any tracks whatever. However, all types are found to be ineffective, and the gates were readily pulled from the store fronts, either by a group of men working together or in stubborn cases by attaching one end of a rope to the gates and the other end to an automobile, which is then driven off.

Fire Equipment and Techniques

Modernize your fire-fighting equipment, and drill your employees on fire control techniques. Fire extinguishers should be hung in a conspicuous place next to a red or distinctive marker. Too often extinguishers are hidden from view by a stack or boxes or other obstacles. They should not be within easy reach of the public.

Extinguishers should be checked regularly to see that they are in good working order. Directions on their use appear on all extinguishers. Some facilities, in addition to having fire extinguishers placed at strategic points, maintain a fire control center where several extinguishers are kept along with buckets of sand and other fire fighting equipment.

Sprinklers could save many buildings. While gasoline would float and spread, the material with which it would come in contact would be rendered useless as fuel, and the buildings themselves could be saved.

All employees must know the exact location of all fire fighting equipment and how to use it. According to fire prevention officials, a dry chemical powder extinguisher is most effective against gasoline fires and other flaming liquids. For a regular surface fire a two and one-half stored water pressure extinguisher is normally recommended.

The danger of the semiexplosive molotov cocktail has been greatly exaggerated by motion pictures and television. It is ordinarily made up of a pint or less of gasoline in a bottle with a cloth wick. If the bottle is not broken after impact, the molotov cocktail can be snuffed out by a coat thrown over it or by a dry chemical powder extinguisher. In fighting a gasoline fire or other flaming liquids, it is important to remember to work around the edges of the blaze to prevent spreading. If the bottle breaks or explodes, setting several small fires, within moments each becomes a regular fire which can be controlled either by water or a dry chemical powder extinguisher.

Insurance

Review the present insurance program and make sure it provides protection against losses incurred during riots, looting and civil commotion. Also have enough to cover all possible loss.

Experience shows that the property stolen is so roughly handled that its marketability is destroyed even if it should be recovered. In addition the pattern of attack is such that there is little hope of salvage of the property which is not removed.

During a recent riot the following pattern developed in the looting and burning. Criminal types, operating alone or in groups of two to four, would break into and loot an establishment, skimming off the most salable and desirable merchandise. They were followed, in successive waves over the three-day looting period, by groups of opportunists, people who under ordinary circumstances would not steal but who, seeing the opportunity, would "help themselves." Eventually many of these repeatedly looted stores were burned to distract attention from the next area in which the pattern was repeated.

Records

In recent riots irreplaceable business records, including accounts receivable, inventory records and cash on hand were lost in the destructive havoc of the looting and burning. Fire safes, purchased to protect records, were ripped open by looters by methods described by police as "amateur." Most of them were what is called "chop jobs," this is where the safe is tipped over and the bottom hacked away. There is every reason to believe that the destruction of credit records was deliberate and systematic. Likewise there is every reason to expect similar destruction in any future riot.

One method of protecting vital papers, such as insurance policies, inventory records and tax data, is to keep them in a fireproof safe of sufficient strength to withstand the assault of amateurs. Another method is to plan for their removal. to a place of safety off the premises.

Even so, if all records are kept at one location the possibility of their complete destruction exists. It therefore is desirable to keep a duplicate set of vital records at another location. Such records might include inventories of goods and equipment, insurance policies, titles, accounts receivable and other necessary information.

Marking Merchandise

Merchandise and other property and equipment should be marked so that ownership can be proved in the event of the theft or loss of such property.

Communication

During a riot, communication is essential. To assure its existence it must be provided for in advance. A central communications room should be established, preferably in or immediately adjacent to the telephone switchboard, to give complete control over incoming and outgoing calls. This will provide the most expeditious method of controlling and coordinating the efforts of the various individuals and emergency squads throughout the facility and will avoid needless duplication of effort. If there is a loudspeaker system in the plant, it should be controlled from this central communications room.

Several systems or combinations of such systems may be used. A wired paging system utilizing voice in which the operator is contacted by telephone, already in use in many facilities, can be very useful. The use of walkie talkies, with one being located in the communications center, can be similarly valuable. The use of a regular radio transmitting and receiving system, with radio pagers being carried by key personnel, offers another combination.

The instruments carried by individuals may be one-way only; that is the person will receive the message but cannot talk back to the operator. With one-way receivers the telephone must be utilized to contact the operator. Two-way instruments may be used and, although much more expensive than the one-way receivers, in time of emergency they do a much better job.

A messenger system should also be organized for use in the event of electrical power failure. Several able-bodied young men should be assigned to this part of the operation equipped with bicycles, motor scooters, or other means of locomotion if the facility is quite large and covers an extensive area. These messengers can carry written messages, supplemented with verbal descriptians of what is occurring, and can make known any addition needs as they observe them.

Shelter Area

Shelter areas should be designated for riot emergency purposes. A shelter area is a place where customers and employees can be temporarily and safely accommodated when riot conditions make it undesirable to evacuate them from the facility. Such a shelter should be in a protected area, have toilet facilities, drinking water, seating accommodations, first aid supplies, communications faciilties and escape routes.

First Aid

A survey of the facility should be made to ascertain the adequacy of first aid stations, including supplies such as bandages, splints and stretchers. There should be one area, the location of which is well known, where first aid supplies are kept. During a riot it should be manned by a member of the first aid team, all of whose members have received instruction from the United States Public Health Service, the American Red Cross or some other qualified organization.

RIOT OPERATIONS

Security Measures

When a riot erupts there are certain measures that should immediately be taken:

1. Remove all guns, ammunition and knives from the selling floor and store them in a secure place.

2. Remove merchandise from window displays.
3. Close all liquor stores and departments.
4. Turn on electric lights on ground floor and parking lots so area is well lighted.
5. Place fire extinguishers for various types of fires near entrances and other strategic spots where fires are most likely to be set or where an incendiary device might be thrown into the facility. Make sure that extinguishers are kept filled.
6. Station uniformed guards on the ground floor within the store or plant and somewhat back from the street. Uniformed guards outside the store at such a time have a tendency to excite mobs.

Closing

As a precautionary step, law enforcement may or may not recommend closing your establishment. Management is not necessarily bound by such recommendations. It is important, nevertheless, to work closely with local authorities and to maintain liaison with them at all times in impending trouble spots.

If it is decided to close the business there should be an established plan. The person who has authority to make the decision must be designated. Duties should be assigned and employees trained. When announcement of the shutdown is made, the facility should immediately be placed on emergency operation. Stores in particular must plan for such an emergency as they must consider not only their employees, but their customers who must either be evacuated or protected.

Announcements should be prepared in advance to cover various emergencies, they should be worded and made so that they will not generate panic.

If necessary or desirable under the circumstances, customers could be reassured and asked to walk—not run—to the exits and to leave the building in an orderly manner. Should circumstances prevent the use of certain exits, these doors could be locked and customers should be told to leave by other exits.

The announcement may be made over the store's voice-paging system. In stores with only bell-paging systems, power megaphones could be used to advise of the early closing. Two employees using megaphones, and keeping the power low so as not to excite, might walk through the store, starting in the basement and working to the top floor, advising all present of the store closing.

Employees should be instructed, upon hearing the shutdown message to wind up their operations and depart from the store in an orderly manner, being careful not to say or do anything which would excite others. All funds and valuable items should be removed from cash registers, display and fixtures and taken to a more secure location. Register drawers should be left open.

Security Guard

Even though a decision to close the premises to the public is made, another decision will have to be made. Should security guards and store employees remain to protect the building? Most of the stores in Los Angeles which closed their doors and sent their employees home became the targets of looters. Over one hundred were destroyed, looted or set afire. Others which suddenly tried to employ guard agencies were denied manpower because of their unavailability and many other reasons. On the other hand, those stores which kept guards within the premises or had store employees who remained were not attacked.

Recent experience in other areas has also demonstrated the value of maintaining a security force to protect property. Indeed the absence of any protection is an invitation to the opportunistic type of thief who participates in a riot.

If employees are kept at the store, they should remain within the store throughout the night. To attempt to move through the riot area invites attack. At night the building and surrounding parking lots should be kept totally lighted. Remove all window displays so that persons inside can see out and police can see inside. These employees must be briefed on the various doors which might afford them an escape route in the event of dire emergency. They might be warned to expect crank telephone calls and to ignore such calls and the threats accompanying them. Their locked cars should be placed in strategic places in front or at the rear of the building for possible emergency use.

They must be familiarized with all fire extinguishers and their location. Do not leave fire extinguishers on the walls; centralize them where they will be immediately available. Make sure they are in fully charged conditions.

Those who remain should carry the telephone numbers of the police and fire departments on their persons and must know the locations of all telephones, including pay units, within the stores. They should patrol the inner facilities and be particularly alert to any noises on the roof. They must have quick access to the roof so that they can combat any fires set thereon. Most important, they must have direct communication with the security department and responsible management, so that at no time will they feel isolated during any hour of the day or night. This is a strong morale factor.

The police department should be kept informed of plans to maintain a holding force at the store. Some method of identifying this force must be established, particularly if the security men are to be armed so that they will not be mistaken for snipers. If possible, frequent patrols should be maintained to give moral support.

Contrary to popular belief, older guards may prove more effective in a riot than younger men. Some elderly Negro guards were used in the Watts area. Because of their age, they were able to and were successful in appealing to their people through logic and reasoning with the result that they protected the facilities to which they had been assigned. In the case of younger guards, their appeals, if any, were ignored by the rioters, and they were ordered from the premises of the business establishments which were then attacked.

During the Los Angeles riot, a loyal, dedicated employee who volunteered his services to act as an observer, mix with the crowd, and stand guard in civilian clothes, was able to minimize the damage to a store. He appealed to the looters and was successful in driving them from the store area. He was also in a position to fight the fire which a lone arsonist had set.

Armed Guards

There is disagreement on the question of whether store personnel should be armed and stationed in well-lighted areas in order to deter rioters. Some express concern that persons unfamiliar with riot situations, untrained in the use of firearms and excited by riot conditions might make errors in judgment which would aggravate the situation and result in needless deaths and loss of property. Some are of the view that weapons will incite the mob to violence.

Unquestionably, untrained inexperienced individuals should not handle weapons in such a situation. By the same token the premature and unjustified display of firearms can antagonize, but it is rather absurd to say that the sight of an armed guard protecting property and life from a wild looting mob will antagonize that mob. Such a sight will frustrate and frighten the mob, indeed may bring its members to their senses, for irrespective of the bravado of the mob, its members are basically cowards.

Experience has proven that the most effective deterrent to a rampaging mob is force. It is the only language that the mob understands and respects. Any attempt to appease will as in international relations be considered a sign of weakness and merely whet the appetite of the mob for more plunder.

It is thus advisable to use armed experienced guards to protect property and lives threatened by the mob. The guards should be armed with shotguns or preferably liquid gas dispensers such as "Chemical Mace," which are the most effective weapons in such situations. The armed guards should be inside. A man with a gun on a roof in an area where sniping is going on could easily be mistaken for a sniper and be shot by the police.

Withdrawal

Even in those instances where a security force remains, there may arise a time when a retreat is in order. After all, the lives of the guards are more precious than any property. The decision to abandon the premises must be left with the guard force itself. Only they are in a position to judge the danger and the need for flight. No restrictions can be placed on their discretion.

Acknowledgment

When the entire situation returns to normal, each employee

who has stood by during the emergency should receive an acknowledgment such as a personal letter of appreciation in recognition of his services at a most difficult time.

INDEX